Cave Biodiversity

Cave Biodiversity

Speciation and Diversity of Subterranean Fauna

EDITED BY

J. Judson Wynne

FOREWORD BY

Stuart Pimm

JOHNS HOPKINS UNIVERSITY PRESS   BALTIMORE

Printed in the United States of America on acid-free paper
9 8 7 6 5 4 3 2 1

Johns Hopkins University Press
2715 North Charles Street
Baltimore, Maryland 21218
www.press.jhu.edu

Library of Congress Cataloging-in-Publication Data

Names: Wynne, J. Judson, editor.
Title: Cave biodiversity : speciation and diversity of subterranean fauna / edited by J. Judson Wynne ; foreword by Stuart Pimm.
Description: Baltimore : Johns Hopkins University Press, 2022. | Includes bibliographical references and index.
Identifiers: LCCN 2021056398 | ISBN 9781421444574 (hardcover) | ISBN 9781421444581 (ebook)
Subjects: LCSH: Cave animals. | Cave ecology.
Classification: LCC QL117 .C38 2022 | DDC 591.75/84—dc23/eng/20211217
LC record available at https://lccn.loc.gov/2021056398

A catalog record for this book is available from the British Library.

Special discounts are available for bulk purchases of this book. For more information, please contact Special Sales at specialsales@jh.edu.

To my mother, Lynn, and my sister—
and my grandparents (in memoriam).
This would not have been possible without your love,
support, and much cajoling over the years.

Contents

Contributors

MARIA E. BICHUETTE
Laboratório de Estudos Subterrâneos
Departamento de Ecologia e Biologia Evolutiva
Universidade Federal de São Carlos
Rodovia Washington Luís km 235
PO Box 676
13565-905 São Carlos
São Paulo, Brazil

EVIN T. CARTER
Environmental Sciences Division
Oak Ridge National Laboratory
1 Bethel Valley Road
Oak Ridge, TN 37830 USA

PROSANTA CHAKRABARTY
Museum of Natural Science
Department of Biological Sciences
Louisiana State University
Baton Rouge, LA 70803 USA

KENNETH JAMES CHAPIN
Department of Ecology and Evolutionary Biology
University of Arizona
PO Box 210088
Tucson, AZ 85721 USA

DANTÉ B. FENOLIO
Center for Conservation & Research
San Antonio Zoological Society
3903 N. St. Mary's Street
San Antonio, TX 78212 USA

ANDREW G. GLUESENKAMP
Center for Conservation & Research
San Antonio Zoological Society
3903 N. St. Mary's Street
San Antonio, TX 78212 USA

JOZEF GREGO
Horná Mičiná 219
97401 Banská
Bystrica, Slovakia

FRANCIS G. HOWARTH
Hawai'i Biological Survey
Bishop Museum
1525 Bernice Street
Honolulu, HI 96817 USA

LEONARDO LATELLA
Museo di Storia Naturale of Verona
Lungadige Porta Vittoria 9
I-37129 Verona, Italy

MATTHEW L. NIEMILLER
Department of Biological Sciences
The University of Alabama in Huntsville
301 Sparkman Drive NW
Huntsville, AL 35899 USA

KAREN A. OBER
Department of Biology
College of the Holy Cross
1 College Street
Worcester, MA 01610 USA

T. KEITH PHILIPS
Department of Biology
Western Kentucky University
1906 College Heights Boulevard
Bowling Green, KY 42101 USA

JOHN G. PHILLIPS
Department of Biological Sciences
University of Idaho
875 Perimeter Drive MS 3051
Moscow, ID 83844 USA

DAPHNE SOARES
Federated Department of Biological Sciences
New Jersey Institute of Technology
323 Dr Martin Luther King Jr Boulevard
Newark, NJ 07102 USA

J. JUDSON WYNNE
Department of Biological Sciences
Center for Adaptive Western Landscapes
Northern Arizona University
Box 5640
Flagstaff, AZ 86011 USA

YAHUI ZHAO
Key Laboratory of Zoological Systematics and Evolution
Institute of Zoology
Chinese Academy of Sciences
No. 1-5 Beichen West Road
Chaoyang District, Beijing 100101 PR China

Foreword

Niaux Cave

"Please! Understand this is a difficult walk," our French guide admonishes. "It is one kilometer and will be wet underfoot. You must squeeze through tight passages. This is not for those who are claustrophobe. You cannot turn back." We nod in agreement, turn on our dim lights, and follow her into the darkness.

Outside, high on a mountainside in the French Pyrenees, it is a windy, cold winter's day. Inside the cave, the temperature is moderate, cool, humid, and constant—likely so for millennia. And 14 millennia ago, someone was at work deep within this cave. At Niaux are some of the most wonderful cave drawings—among the very few that one can see as originals. Horses, bison, ibex, and an exquisite weasel adorn the walls. All but the last are long gone from these mountains—and indeed beyond.

Of course, I marvel at the artwork. The likenesses capture the essence of the species, their magic, their power, and their energy. The weasel was done using a few strokes, a minimalist effort Picasso might envy. There are other lessons for me as a conservation scientist. Over the millennia, we have driven most of these species extinct. The drawings depict what Europe was like before our depredation.

"Why did the artists go so deep into the cave?" someone asks. There may not be a convincing answer. What is clear is that they did, penetrating as deep as one might go into this very alien world. They had technology sophisticated enough to keep their torches lit for long periods.

So, man had gone where none had gone before. Except that the handprints that adorn many cave walls show equal-sized first and ring fingers. That's more typical of women than men.

Caves—dark, remote, alien though they may be—are not beyond human impacts. We are just as capable of destroying what lives there as we are the flora and fauna of the rich agricultural lands long cleared of their native forests in the valley below.

Vulnerabilities

A consistent theme of this book is the high fraction of species that are at risk of extinction in caves. I am not a cave biologist. I'm meeting most of the authors of this book for the first time via the chapters they wrote here. They work in places I have never visited, with biological knowledge and exploration skills I do not possess. That they and I have connections, even common language, is alarming. There's too much common ground. What we know about the horses, bison, and ibex applies to the insects, snails, salamanders, and fishes.

For instance, in Chapter 5, Karen Ober and colleagues find that extinction threatens most of the trechine beetles in the United States. Perhaps 95% of the obligate cave species in the United States are in danger. In China, the *IUCN Red List of Threatened Species* deems a third of the cavefish species threatened with extinction. These are higher fractions than for the country's other vertebrates. Even hidden away, deep underground, there is no protection against human actions. The causes are familiar. Habitat loss and degradation, overexploitation, invasive species, and climate change are problems above and below the ground.

We physically destroy cave environments in different ways. We spend time in caves, explore them, exploit them as tourist destinations. In Chapter 1's overview of the various cave environments, Francis Howarth and Judson Wynne note that "humans have used caves for refuge, living quarters, harvesting resources, and burials since ancient times." The mouth of the cave at Niaux was—and is—a place to shelter from the cold.

We go further inside: cave exploration is a popular sport. It has conflicting outcomes. We find out much more about the caves' biodiversity but, in doing so, may harm species, even with our breath.

In Chapter 7, Yahui Zhao and colleagues provide a detailed assessment of conserving the rich diversity of China's cavefishes. A third of the world's cavefish species live there—most in single caves.

Here and elsewhere, the largest caves may be popular tourist destinations—no longer dark but bright with artificial illumination. There may be paved walkways. As China's middle class has rapidly expanded, tourism has exploded. Scenic parks, such as Jiuzhaigou in Sichuan, have gone from seeing tens of thousands of visitors 25 years ago to over 5 million visitors a year today. Spectacular caves see comparable changes. Ancient Alu Cave, a large underground karst cave system, now receives millions of tourists each year.

This cave contains at least three kilometers of paved concrete trails, artificial lighting, and speakers for music. In US national parks, Carlsbad Caverns in New Mexico and Mammoth Cave in Kentucky receive about half a million visitors each year.

Caves are particularly common in limestone areas. Rock mining threatens cave faunas in China (Chapter 7) and Italy (Leonardo Latella, Chapter 4). In the former, mining limestone for concrete is a significant component of the local economy.

The most pervasive concerns, however, are changes to the caves' water supplies. Limestone areas are porous, of course, and with that comes vulnerability to pollution from industry and agriculture.

In Chapter 3, which discusses subterranean mollusks, Jozef Grego concentrates on Europe's limestone ecosystems, especially those of southwestern Europe. A third of the world's freshwater mollusks occur here. Dam projects have flooded large parts of this area, and the caves found there. Many springs and habitats have been destroyed. Here, as elsewhere, there has been unrestrained enthusiasm to build yet more dams, though Grego sees hope in increasing public opposition.

Outside of these direct changes, human actions are changing land-use patterns—clearing forests, in particular, and altering watersheds' hydrologic patterns. Climate disruption continues unabated as we pump more carbon dioxide into the atmosphere and destroy the forests that can sequester it.

The final insult to hydrology is the withdrawal of groundwater. Zhao and colleagues consider this to be one of the greatest threats to Chinese cavefishes, as discussed in Chapter 7. They find that groundwater is the primary water source for human consumption. Water extraction for industry, agriculture, and residential needs is rapidly increasing. Similarly, in Chapter 6, Matthew Niemiller and colleagues find water withdrawal to pose a significant threat to cave-dwelling salamanders.

Remarkably, invasive species—one of the most severe threats to terrestrial biodiversity—pose a danger to subterranean species too. The most dramatic invasive species of my career is the brown tree snake, *Boiga irregularis*, an accidental introduction to the island of Guam. In short order, it eliminated almost all of Guam's terrestrial birds—the island's tree nesters were well within its reach. One memorable day, friends and I hiked to a remote cave. As we reached its entrance, we heard the calls of dozens of

Mariana swiftlets, *Aerodramus bartschi*, clicking away furiously to echolocate their way into the darkness. Even in their cave, they were not safe. Where it could, the snake would climb the walls and grab the birds as they flew by.

Not all invasive species are so obvious. In Chapter 1, Howarth and Wynne argue that the most serious invasion is white-nose syndrome, a fungal disease introduced to North America from Europe. They suggest the disease has resulted in the mortalities of more than 5 million bats.

Finally, our passion for exotic foods and pets is a general problem for conservation. As a species becomes rarer, it becomes more desirable. The price goes up, the incentives likewise, and as the species declines, a disastrous race to catch the last individuals ensues. People exploit cavefishes directly, particularly in rural areas. Zhao and colleagues in Chapter 7 write that "some cavefish, also known as 'oil fish,' are traditionally harvested as an important food resource in rural communities. Regionally, oil fish soup is considered a delicacy and has recently surged in popularity. . . . Market price can be as much as 50 times higher than other commercially available fish." Communities may also harvest cavefishes for the aquarium trade and traditional Chinese medicine.

A Passion for Beetles—and Many Other Things Too

There is an apocryphal story told about the British biologist J. B. S. Haldane. When asked by an Oxford cleric what he had deduced about the creator from studying his creation, Haldane replied, "an inordinate passion for beetles." (There are just so many species of beetles.)

I am not qualified to assess the details of the science in this book. But the authors' passions are clear enough. They describe limestone caves and lava tube caves, and within them, the different habitats—from the light at the entrances to the darkest depths of the caves (Chapter 1). Then, in Chapters 3 to 7, the authors show us different sets of species that live within the caves. And those species fascinate.

Of course, the species challenge us as scientists. How did they get there? How did they speciate? How did they lose adaptations to the light and develop ones for the dark? Why are so many species of cavefish here and not there? Such discussions are the grist for this book's mill. For biologists, these questions are endlessly fascinating. They are made so pointed here because of the singular fact of species living permanently in the dark.

Do not overlook a wider connection. Each chapter enthuses about the species it contains. Every species is a wonder; the sets of them shout the variety of species. Biodiversity is remarkable—so varied, so exceptional. In the apocryphal story, the answer might just as easily have been "a passion for Chinese blind cavefishes" or "European subterranean mollusks."

Back in the Niaux Cave, I am sure the artist had a sense of wonder. Yes, draw food being hunted. But why draw the species with such drama? And why did she draw a weasel? This book's key message is its celebration of biodiversity and, alas, of how our human actions threaten it, even in the deepest, darkest places.

Stuart Pimm
Doris Duke Professor of Conservation
Duke University

Preface

The subterranean realm represents one of the most sensitive and least studied biomes on earth. This realm often harbors unique animal communities adapted to life underground. How these organisms evolve into subterranean-adapted forms and the subsequent organization of animal communities in complete darkness have become central themes in cave biology. The subterranean environment, particularly habitat and substratum, influence both diversity and diversification of subterranean-dwelling taxa. In this book, these facets of biospeleology are investigated, and some of the factors driving diversity and speciation through a cross section of subterranean-adapted fauna are examined.

Working with a panoply of taxonomists, geneticists, and ecologists, I aspired to provide thorough analyses and overviews related to the biological diversity of subterranean ecosystems, as well as to provide descriptions of how animal populations evolve and develop tight evolutionary linkages with the subterranean environment. My goal is that this book will complement and enhance the growing body of comprehensive works on cave ecology, including the books *Encyclopedia of Caves* and *Cave Ecology* and the innumerable journal articles on this topic.

I realize edited volumes are often hampered by having "multiple voices"—as each chapter is penned by a different author or group of authors. I attempted to partially address this through consultations with contributing authors, as well as working closely with lead authors during the editing process.

This book consists of seven peer-reviewed chapters. The first two provide overviews on cave biological diversity and diversification/adaptation in the subterranean realm, while the remaining five chapters explore the diversity, distributions, and conservation needs of several subterranean taxonomic groups. Specifically, Chapter 1 discusses how diversity is governed by biotic and abiotic factors, and how biological diversity is often driven by complex interactions of those factors. This chapter also provides definitions for the evolutionary groups that inhabit caves (i.e., troglobionts, troglophiles, trogloxenes, etc.) and an in-depth explanation of the cave zonal environment. It is my hope the discussions therein will serve as a thread that weaves the other six chapters together. Moreover, because many cave species remain poorly studied or unknown, Chapter 1 should be particularly useful to scholars and decision-makers by providing them with some of the information to help identify caves a priori that may be of conservation concern and warrant biological inventory and assessment.

The second chapter provides a "state of the state" synopsis regarding how the subterranean environment influences evolutionary pathways toward troglomorphy. This chapter was written through the lens of next-generation molecular approaches and population genetics and aims to synthesize much of this work conducted on subterranean taxa over the past ~20 years. Through this review, the authors identify seven evolutionary models influencing how subterranean-dwelling populations evolve into troglobiontic species. They also highlight a few topics deemed particularly relevant to tackling unanswered questions on adaptation, evolution, and speciation of subterranean organisms, outline some of the limitations in addressing these questions, and provide some recommendations for future research.

Chapter 3 is a monograph developed by one of the leading global authorities on subterranean-dwelling Mollusca. Taking a global view, he illustrates our current knowledge of subterranean Mollusca taxonomy, their habitat requirements, and the distributional ranges of many species groups. As most hypogean taxa were described based solely on empty shell morphology, and frequently only from the type locality, researchers and conservation biologists are hindered in their ability to evaluate such species in terms of vulnerability to human disturbance and prioritizing management needs. Finally, the author proposes new ideas and research directions that should be endeavored to address many of the data deficiencies so that evidence-based conservation measures may be developed, proposed, and ultimately adopted.

The Chapter 4 chronicles the most highly troglobiontic subfamily of beetles in Italy, the Cholevinae. In this group, numerous Italian species are subterranean-dwelling, and many are troglobionts. Written by one of the leading Cholevinae taxonomists, this chapter both summarizes and examines the distributions and adaptive strategies of this fascinating group of carrion beetles. This chapter also reveals how this group, which comprises an important component of Italy's biological diversity, has largely been overlooked in conservation initiatives by the Italian government, the European Union, and the International Union for Conservation of Nature (IUCN). Of note, of the 128 known troglobiontic species in Italy, only one species currently receives some level of protection from human activities, and none have been assessed by the IUCN.

Chapter 5 examines the impressive diversity of eastern North American subterranean trechine beetles. Belonging to the family Carabidae, the Trechini tribe are the most speciose group of terrestrial troglobionts in North America. The authors explore both geographic and taxonomic distribution of these relict troglobiontic species focusing on how geology and hydrology influence biological diversity. Due to the combined restricted distributional ranges and anthropogenic threats, most eastern North American cave trechines are likely imperiled. Future research and management recommendations are provided to help amass the data necessary to best manage and protect this highly cave-specialized group of ground-dwelling beetles.

The sixth chapter represents a review on subterranean-dwelling salamanders. Specifically, the authors summarize our current knowledge of the diversity, diversification, biogeography, and natural history of one of the two vertebrate groups adapted to life in complete darkness. They also examine how climate, geology, hydrology, and dispersal capabilities have influenced speciation, and they offer future research directions on the ecology and evolution of this spectacular subterranean vertebrate group.

Addressing the second subterranean-adapted vertebrate group, the last chapter is a synthesis on the diversity and distribution of Chinese cavefishes. With 148 known species, the South China Karst supports the highest diversity of cavefishes in the world. The authors review Chinese cavefish diversity, discuss some of the distributional ranges, provide glimpses into cavefish community structure in China, and examine the current conservation status of Chinese cavefishes. As many of these species are considered "single cave" endemics and China continues to experience rapid economic growth, most species are likely imperiled. To date, nearly one-third are listed as "threatened" under the *IUCN Red List of Threatened Species*. The authors also outline future goals for management, educational outreach, and research.

While I found editing this book to be both a pleasure and privilege, the underpinning and urgent message that reverberated through these pages was both inescapable and disconcerting. Most subterranean ecosystems and the endemic species they support are unremittingly threatened by human activities. The extinction crisis looms large underground. If we do not enact effective conservation measures in many parts of the globe where sensitive species exist, many subterranean species will go extinct. Importantly, several of these species groups could be monitored as biological indicators of both ecosystem health and groundwater quality. Thus, by protecting these organisms, we are helping to safeguard a more healthful quality of life for human populations residing above these ecosystems.

I remain optimistic that humanity can indeed protect these bizarre subterranean animals, their communities, and their habitats through targeted educational outreach programs and effective community-based conservation programs. I am hopeful this tome will serve, in part, to inspire current and future biologists to charge to the front lines. Scientists are urgently needed to educate local communities concerning the importance of biological conservation, while also assisting decision-makers in developing effectual conservation strategies and initiatives. Through such an effort, we can stem the tide and protect what is left of our precious and wonderous natural world—and hopefully help save ourselves in the process.

Acknowledgments

The authors who contributed brought this book to life. I gratefully appreciate their patience and perseverance in working with me through the editorial and publication process. Additionally, the following scholars provided peer review for the chapters presented herein: Chris Beachy, Paulo Alexandre Vieira Borges, Carlos Juan Clar, Robert Davidson, Arnaud Faille, Andrzej Falniowski, Zoltán Fehér, Rodrigo Lopes Ferreira, Francis G. Howarth, Aron D. Katz, Valeria Lencioni, Brian Miller, Larry Page, Stewart Peck, Alexander Reschütz, Aldemaro Romero, Thomas G. Watters, and Kirk Zigler. Additionally, an anonymous reviewer offered insightful comments and emendations for the entire tome. In the aggregate, their contributions significantly elevated the impact of this work. Pat Kambesis contributed to an earlier version of Chapter 1.

I also wish to thank the various funding agencies and institutions, who are recognized in the individual chapters; through their support, the acquisition of much of the data and information required to develop this array of ecological compendiums was made possible.

1

Influence of the Physical Environment on Terrestrial Cave Diversity

Francis G. Howarth and J. Judson Wynne

Introduction

Subterranean habitats often support organisms morphologically, physiologically, and behaviorally adapted to the rigors of total darkness (Porter 2007; Romero 2009; Culver and Pipan 2013; Howarth and Moldovan 2018a). Subsequently, these systems are of great interest from an evolutionary and biogeographic perspective (Mammola 2019). Animals occurring in caves can be divided into six evolutionary categories based on their known or suspected association with caves (which are based upon Barr 1968; Peck 1970; Howarth 1982, 1983; Sket 2008). These are (1) *troglobiont*—terrestrial obligate cave dwellers that only complete their life cycle underground and exhibit morphological characteristics indicative of subterranean adaptation (i.e., troglomorphies); (2) *stygobiont*—the aquatic counterpart to troglobionts; (3) *troglophile*—species that occur facultatively within caves and complete their life cycles there but also occur in similar surface microhabitats (*stygophile* = aquatic counterpart); (4) *obligate troglophile*—taxa seemingly restricted to the subterranean environment but lack traits indicative of troglomorphy; (5) *trogloxene*—species that frequently use caves for shelter or food but return to the surface during their life cycle to forage or reproduce (*stygoxene* = aquatic counterpart); and (6) *accidental*—organisms that wander into the subterranean environment and typically are unable to establish viable populations. Plate 1.1 contains examples for five of six evolutionary groups. We emphasize that many species may not fit well into a particular category. Furthermore, cave biologists often make these designations on morphological characters with limited life history and occurrence data, especially when considering animals in tropical areas. Unfortunately, assigning a species to its correct group is difficult without detailed knowledge of its ecology and distribution. To address this dilemma, researchers rely on a troglobiontic interpretation whereby a species is considered troglobiontic if it displays morphological characters that appear to restrict it to subterranean habitats (Howarth and Moldovan 2018a).

The often-extreme phenotypic traits displayed by obligate subterranean-inhabiting animals have intrigued both biologists and the public since the first published accounts of blind cave animals over 450 years ago (Romero 2006b). These traits include reduction or loss of eyes, wings, and bodily color; thinning of the cuticle; elongation of appendages; and enhancement of sensory receptors (Howarth and Moldovan 2018a). Cave animals also adapt physiologically via enhanced water-balance mechanisms (Ahearn and Howarth 1982), enhanced tolerance to anoxia (Howarth and Stone 1990; Sarbu et al. 1996), broader diet, lower metabolism, lower fecundity, more precocious offspring, and potentially longer life span (Lunghi and Bilandžija 2022), as well as behaviorally by alterations in feeding, social behavior (including aggregation, responses to alarm substances, and antagonistic behavior), photic responses, loss of circadian rhythm, slower movement, and mating behavior (Romero 2006a; Howarth and Moldovan 2018a). Convergent evolution of these adaptive traits among unrelated taxa, in different cave types and in all biogeographic regions, indicates that subterranean environments exert selection pressure on animals colonizing underground habitats (Vandel 1965; Howarth and Moldovan 2018b).

Caves (sometimes called macrocaverns) are here defined as any natural underground void large enough for human entry. Some researchers (e.g., Moseley 2009; Mammola et al. 2020) include artificial cavities with caves, but because of the extreme disturbances in their creation and use by humans, their inclusion may confound our synthesis. On the other hand, since this chapter is focused on the factors affecting the diversity of cave communities, smaller subterranean voids (i.e., micro- and mesocaverns) that support cave life are included. Recent discoveries have expanded the diversity, biogeography, and geologic extent of subterranean habitats. Subterranean-adapted animals can now be expected to occur wherever suitable habitats exist, from beneath glaciers to tropical caves and in a myriad of geologic substrates (Deharveng and Bedos 2018). In aggregate, these discoveries provide us with a better understanding of the environmental factors influencing subterranean biodiversity and render caves valuable model systems in which to study evolutionary ecology (Mammola 2019).

Caves and cave-like voids can occur from near the surface—for example, the MSS, *milieu souterrain superficiel* (Juberthie et al. 1980; Juberthie 1983), and epikarst (Culver and Pipan 2014)—to extreme depths. The deepest known cave that contains macroinvertebrates is in limestone in the Arabika Massif, Georgia, which is about 2,200 m in depth (Sendra and Reboleira 2012). Lava tubes are even deeper; the thickest subaerial basaltic lava deposits occur on Mauna Loa in Hawai'i, which rises to 4,170 m above sea level. The depth of its submarine portion is even greater (Trusdell 2020). However, these deep lava tubes remain inaccessible and have not been surveyed. The deepest known multicellular organism is the nematode *Halicephalobus mephisto* Borgonie et al., 2011 discovered in cracks up to 3.6 km deep in South Africa, where they feed on bacteria (Borgonie et al. 2011). Bacterial communities are also known from exceptional depths elsewhere (Moser et al. 2005; Colman et al. 2017). Cave depths are usually reported as the measured depth of explored passage. Although depth can be determined using a combination of a high-resolution cave map and geospatial elevation models, the thickness of overburden is rarely calculated or reported.

The cave fauna in many karst areas remains poorly known or have not yet been surveyed, especially in the tropics (Deharveng and Bedos 2019). Compared to overlying surface habitats, the total subterranean fauna is extremely limited in numbers of taxa and functional groups (Gibert and Deharveng 2002). Curiously, many higher taxonomic groups found on the surface are absent in underlying caves, but this anomaly may be changing (see below). This disparity is due in large part to the filter created by reduced food resources, the harsh subterranean environment, and small, discrete habitat patches (Gibert and Deharveng 2002; Culver et al. 2006). Nevertheless, in many cave regions, the level of local endemism among troglobionts is exceptionally high (Gibert and Deharveng 2002; Mammola, Cardoso, et al. 2019; Deharveng and Bedos 2019).

Deharveng and Bedos (2018) provided an overview of the taxonomic diversity of troglobionts and troglophiles and included a rough estimate of the number of cave species per category. Among the Hexapoda (insects and their relatives) globally, they tallied more than 3,600 known troglobionts but acknowledged that, in many groups, the numbers are increasing rapidly as additional caves and regions are inventoried (Deharveng and Bedos 2018).

The cave biota of only a few regions have received the bulk of biological research (e.g., the Dinaric Karst of western Europe and the Cumberland Plateau of the southeastern United States; Culver et al. 2004). Even in these areas, the accumulation curves for the number of species did not reach an asymptote (Culver et al. 2004; Culver et al. 2006). For example, in well-studied regions like Slovenia and West Virginia, less than 20% of known caves have been investigated biologically, even in a cursory way (Culver et al. 2013). Additionally, the troglobiontic fauna of the Pilbara calcretes of Western Australia was estimated to be at least 1,500 species—of which only 680 have been formally described (Halse 2018).

Although the distribution and biodiversity of caves remain poorly known, we should not wait for completeness of sampling before attempting to ascertain the factors influencing diversity (Culver et al. 2004; Culver 2008). From what is known, it is possible to identify the main factors influencing the biological diversity of caves. This knowledge will help in planning more efficient biological studies in caves, as well as facilitate the development of data-driven conservation management programs.

Overall, subterranean diversity differs in scale. At the finest scale, diversity is characterized as the number of species inhabiting a single cave system, which includes the species occupying different environmental zones, habitats, and communities within the cave. At progressively larger scales, diversity theoretically includes enumeration of the total species, as well as the geographic and evolutionary factors affecting changes in species composition with increasing distance. This chapter focuses mainly on the influence of environmental factors affecting terrestrial biodiversity at the finer scale—that is, at the subterranean community level. Although there is overlap with the aquatic realm, the inclusion of the aquatic subterranean organisms is beyond the scope of this chapter, as such an effort would require a separate chapter.

Speleogenesis

Inhabitable caves and cave-like voids occur in many different geologic substrates and form by dissolution, erosion, thermal genesis, and biogenesis (White and Culver 2019). The distinction is somewhat artificial as all four processes affect the shape and size of the voids in most substrates and therefore influence cave communities. However, it is informative to examine caves by their host rock and principal processes governing their formation. We list the more common types of underground voids with notes on their speleogenesis and biology in Table 1.1 (on pages 40–43), and we describe here in greater detail the formation of the two most studied cave types: those formed by dissolution and volcanism.

Karst Caves

Karstic caves (formed via the dissolution of limestone, dolomite, and gypsum) are the best known and most often envisioned as conventional caves. Carbon dioxide dissolves in rainwater to form weak carbonic acid, which dissolves karstic substratum creating cavities. The overall formation of thick deposits of sedimentary limestone occurs over geologic time scales mostly through the layering and compaction of ancient marine life (e.g., shells, corals, and algae). The purity of these deposits and their geologic history affect the dissolution process and therefore the biological communities supported. Over geologic time scales, high temperature and pressure may metamorphose limestone to marble. Tectonic activity can fracture and expose limestone to weathering and exacerbate dissolution.

Since the carbon dioxide is principally from biological sources, caves occurring within karst may also be considered biogenic. Two distinct processes are recognized in the formation of karst caves (also known as solution caves): (1) hypergenic (aka epigenic), in which surface water affects solution (Fig. 1.1); and (2) hypogenic, in which sulfur-laden water (mainly hydrogen sulfide and sulfuric acid) upwells from deep aquifers and intersect soluble rock from below (Fig. 1.2; Polyak and Provencio 2000). Hypergenic solution may occur in two phases: (1) solution by percolating water within the vadose zone, and (2) solution by groundwater of different temperatures mixing, usually within the phreatic zone (Klimchouk 2009). Sulfidic acids are produced by deep communities of chemoautotrophic bacteria digesting sulfur and other minerals in the rock (Colman et al. 2017) and can dissolve limestone more aggressively than carbonic acid (Polyak and Provencio 2000). Heat, such as from a geothermal source, can affect the speed of dissolution (Bakalowicz et al. 1987).

Dissolution by hypergenesis begins by water flowing into cracks and crevices in the soluble rock. Voids that carry a greater volume of water enlarge by solution, while conduits carrying less remain small and may become plugged. Thus, larger conduits enlarge at the expense of smaller ones. However, variation in flow dynamics caused by increased rainfall or seasonal

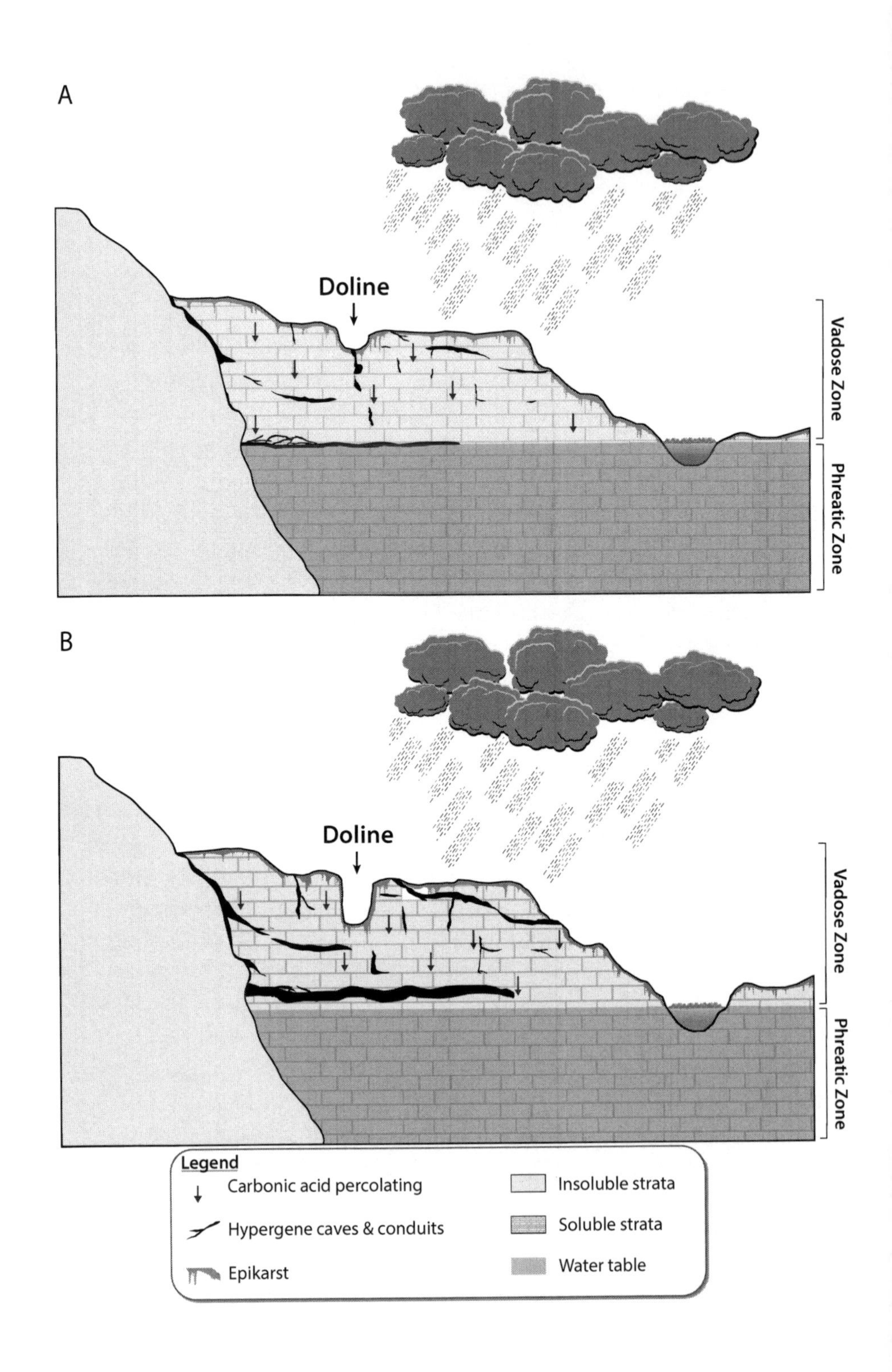
A
Doline
Vadose Zone
Phreatic Zone
B
Doline
Vadose Zone
Phreatic Zone
Legend
Carbonic acid percolating
Hypergene caves & conduits
Epikarst
Insoluble strata
Soluble strata
Water table

flooding can open previously blocked conduits (Polyak and Güven 2000). This dynamic process increases the variability in the size of voids and thus the void space available for cave life. Habitat availability may be further expanded by spalling and collapse of passages. In this way, solution caves typically maintain a vast anastomosing system of voids of varying sizes, which can support cave communities.

Lava Tubes

Lava tube caves (aka pyroducts) form as distributary channels during the eruption of basaltic lava flows. Basaltic lava has low viscosity and flows away from the eruptive vent. Two types of basaltic lava occur: ʻaʻā and pāhoehoe. Lava tubes can form in both types but are more characteristic of pāhoehoe flows. Pāhoehoe has a lower viscosity, flows like a river, and cools with a smooth ropey surface. Conversely, ʻaʻā is more viscous, flows like a military tank tread with a rough clinker surface crust that tumbles ahead of, and is buried by, the advancing flow. The difference between the two lava types depends on crystal formation on cooling, which is related to the effusion rate, ground slope, composition, temperature, and water content. That is, molten basalt is neither ʻaʻā nor pāhoehoe until it crystallizes (Kauahikaua et al. 2003).

Basaltic lava flows take the path of least resistance downslope from the vent, and as the flow advances, it cools at the edges, eventually confining the flowing lava to a channel. Initially, near the vent, erupting lava has a high effusion rate, and as it flows downhill over older, cooler substrates, it cools and forms ʻaʻā. Younger lava that covers the still hot ʻaʻā surface remains hot enough to become pāhoehoe as it cools. A pāhoehoe roof may form over the ʻaʻā channel, thereby insulating the lava in the tube and

Figure 1.1. (*opposite*) A simplified representation of the evolution of an hypergenic karst system. [A] Newly exposed soluble bedrock is infiltrated through cracks and fissures by acidified rain and surface water, and the subsequent solution in the vadose zone forms sinkholes, swallets, dolines, and other karst features, which divert surface water underground. Additional solution occurs mostly in the phreatic zone by mixing of water at different temperatures. [B] Over time, caves and voids enlarge by solution. As the base level and water table (phreatic zone) are lowered by erosion, more limestone is exposed to solution creating multilevel caverns. Modified from Audra and Palmer (2015).

A

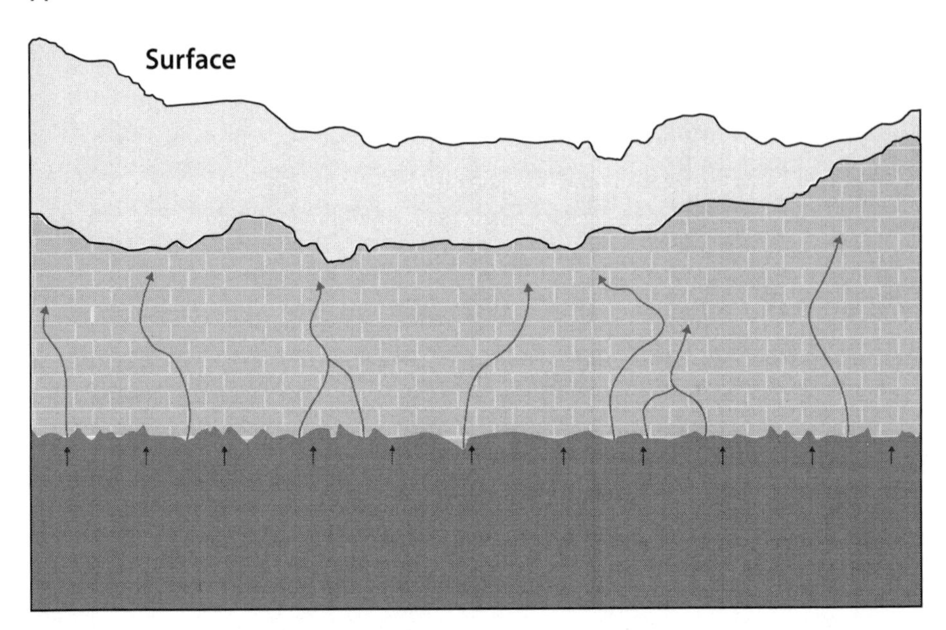
Surface

B

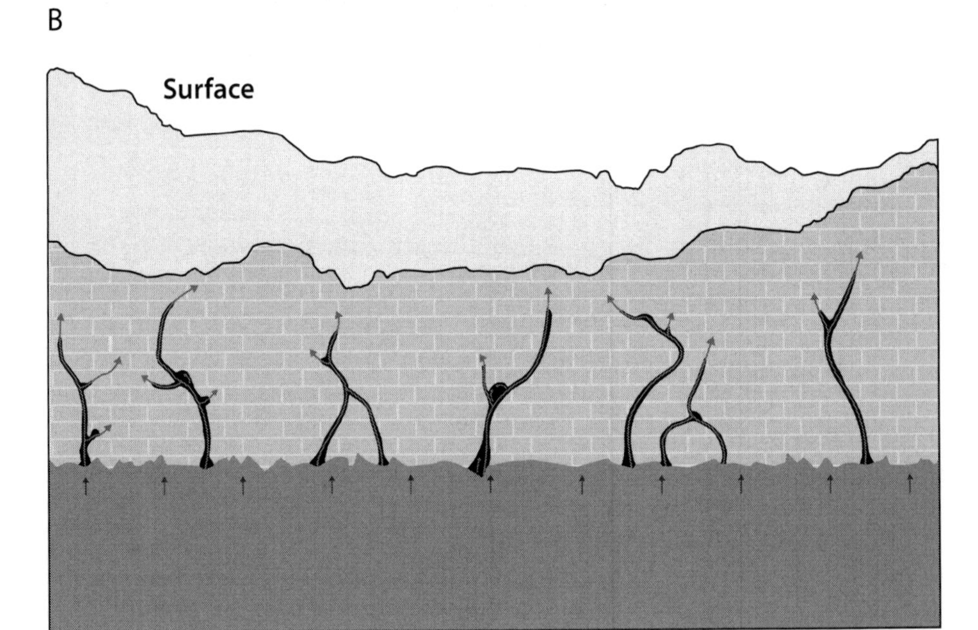
Surface

Legend
Hydrostatic upwelling of H_2S water
Hydrostatic upwelling through fractures
Hypogene conduits
Confining stata
Soluble strata
Aquifer

allowing it to travel further. Active lava tubes are remarkably efficient insulators, usually losing less than 1°C per kilometer (Kauahikaua et al. 2003). Away from the vent, the effusion rate is less, and pāhoehoe is favored especially on gentle slopes (Kauahikaua et al. 2009).

Three distinct mechanisms can create a permanent roof over lava channels: (1) a solid crust forms over the channel from each edge or downstream from an already formed roof; (2) spatter and overflows produce arched roofs over the channel; and (3) floating plates of crust become interlocked across the channel, anchoring the newly formed roof (Peterson et al. 1994). Once established, these roofs thicken by accretion of lava on the lower surface and by overflows covering the upper surface. Surges or reductions in flow volume can destroy a developing roof; thus, roofs generally survive only in steady, low to moderate volume flows that last long enough for the roof to form and stabilize—usually, this requires several months or longer. Skylights along the tube act as pressure valves, which allow overflows to thicken the roof downstream without destroying the established tube. Caves can be enlarged by inflation during changes in flow rate and by erosion of the substrate by flowing lava (Kauahikaua et al. 2003).

Near the flow front, fluid pāhoehoe often spreads out in thin, sheet-like flows. Such flows advance at the edge by extruding lava toes that crust over and then break again to create new toes. Thus, the flow edge advances like a giant amoeba. Overflows of lava build, flow on flow, layer on layer. These layers poorly fuse to older surfaces, so that there are numerous gaps preserved between each flow unit, making an extensive system of interconnected void spaces throughout young pāhoehoe flows (Fig. 1.3). Escaping gas and fluid pressure can inflate voids within the flow, often raising the ground level up to several meters in height upslope from the flow front. Older toes feeding the advancing edge can expand to become distributary tubes. Remnant sections of these distributary tubes are often numerous

Figure 1.2. (*opposite*) A generalized model depicting two stages of dissolution of karst by hypogenic upwelling of sulfidic water from a constrained aquifer. [A] Hydrostatic upwelling of sulfidic water enters fractures within karst. [B] Sulfidic water then enlarges the fissures into conduits via solution. These conduits may ultimately evolve into a complex maze network. Modified from Klimchouk (2009).

Figure 1.3. View of the 1974 pāhoehoe lava flow from Mauna Ulu eruption, Kīlauea, Hawai'i, where a road has been cut through. Voids pictured include buried 'a'ā clinker, lava tube, and poorly cemented flow units. Also note the barren surface of the flow. Courtesy of Francis G. Howarth.

within flows and occur as upper-level mazes in lava tubes and as shallow mazes near the surface.

When the eruption episode subsides, these conduits can remain filled with cooled lava or partially drain to form tubes. Active lava flows are extremely dynamic with the enclosed tubes frequently changing shape and volume depending on flow rate, slope, collapse, obstructions, temperature, and gas content. Resultant tubes reflect these dynamics (e.g., forming a single sinuous tube and an anastomosing maze in different sections of the cave). Additional voids and cave-like spaces within the lava can be created by tree molds, earthquakes, and cooling cracks. In this way, basaltic lava flows can cover large areas and create abundant underground habitats for cave animals.

Subterranean Habitats

Subterranean habitats include any naturally occurring underground voids that support life. Inhabitable caves and cave-like voids exist in many different shapes and sizes, from huge caverns hundreds of meters in diameter and kilometers in length to tiny spaces barely a millimeter in diameter. From a biological perspective, these voids can be grouped in three size classes: caves (aka macrocaverns), mesocaverns, and microcaverns (*sensu* Howarth 1983). Caves and caverns are any natural underground void large enough for human entry and generally accessible to humans. Mesocaverns are intermediate sized voids, too small for human exploration but large enough to contain sufficient nutrients to sustain populations of macroinvertebrates and some vertebrates. Microcaverns are the smallest voids that can support life. The boundary between caves and mesocaverns depends on the size and skill of the human explorer (i.e., >30 cm wide). Generally, accessible cave-sized passages tend to be relatively well-aerated.

The boundary between meso- and microcaverns is delimited by the physical properties of liquid water, its purity, and its interaction with the substrate. Mesocaverns are large enough to allow water to flow freely, whereas movement of water in microcaverns is largely governed by capillary action. Capillarity is governed by water quality and degree to which the parent rock attracts water. The distinction is significant since the restriction of flow in microcaverns limits passage of nutrients and organisms. Furthermore, unless flowing water is present, microcaverns would tend to plug and fill. Microcaverns support communities of microorganisms and meiofauna.

Cave Environmental Zones

Caves are often zonal, with three distinct zones defined largely by the presence of light: the entrance, twilight, and dark zones. The entrance zone extends from the limit of the mixing of surface and cave environments to the farthest point in the cave where light can support the growth of vascular plants. The twilight zone extends from the inner edge of the entrance zone to the edge of total darkness. Photosynthesis in the twilight zone is limited to a few ferns, mosses, algae, and microorganisms able to grow in the reduced light. This zone is usually more humid than the entrance zone but subject to drought. As the name implies, the dark zone is perpetually dark. This zone can be further divided into two to three distinct subzones: transition, deep, and stagnant air (Howarth and Stone 1990; Howarth 1993). The transition zone is a dynamic zone in total darkness beyond the twilight zone where the microclimate is conspicuously affected by short-term weather events on the surface (Howarth 1982). While temperature is relatively constant, relative humidity can vary widely. The deep cave zone remains relatively stable, characteristically with the atmosphere constantly near or above water saturation. Often, one can identify the boundary between the transition and deep zone by the presence of a film of water covering the walls, floor, and ceiling and a foggy atmosphere in the deep zone. The stagnant air zone (where present) is located beyond the deep cave zone where fresh air is exchanged with the surface only slowly. The relative humidity remains at or above 100%, whereas carbon dioxide and oxygen concentrations may fluctuate dramatically from decomposition of organic material (Howarth and Stone 1990; Howarth 1993). This zone is uncommon in larger caves, except in deep, dead-end, downward sloping passages where decomposing organic matter occurs; however, it may be characteristic of more isolated voids within the rock.

The presence and extent of each zone is governed by cave passage shape, size, distance from any entrances, local climate, and the input of moisture and nutrients (Howarth 1982). For example, the boundary between each zone is sometimes created by a constriction or change in shape of the cave passage, such as an n- or u-shaped passage. However, these boundaries can be dynamic and shift with the seasons and/or from the input of water or nutrients. Vagile animals track these boundary shifts to inhabit their selected environment (Howarth 1982). Each zone and habitat within caves can support unique biological communities. The extent and species diversity of each zone are highly variable and unique to each cave.

Biotic inventories of cave faunas often focus on obligate cave species (e.g., Culver and Pipan 2009)—as these species are typically endemic and often of management concern. A few studies have examined faunal changes from the entrance to the deep zone (e.g., Prous et al. 2004; Tobin et al. 2013; Wynne et al. 2018; Mazebedi and Hesselberg 2020). However, the definitions of the entrance and twilight zones are either not given or inconsistently applied, making comparisons on ecology and diversity within these zones among caves and across regions difficult.

Entrance Zone Habitats

The entrance habitat is an ecotone—that is, the transitional boundary between two distinct environments, in this case between the surface photic zone and the aphotic underground (Prous et al. 2004). Researchers often define the entrance boundary at the "dripline" (Prous et al. 2015). However, the extent of the entrance zone is highly variable. The entrance-cave ecotone can extend outwardly beyond the dripline to the limit of the mixing of surface and cave environments. Each entrance is unique and varies in its shape, size, and aspect: that is, from a tiny opening with no true entrance zone to a huge pit enclosing a diverse luxuriant forest (Fig. 1.4). When present, the terrestrial community of plants and animals harbors a subset of species from both the surface and underground, as well as species found only in entrance community (Northup and Welbourn 1997; Prous et al. 2004, 2015; Wynne 2013; Rabelo et al. 2021).

Species diversity is strongly governed by entrance size, shape, depth, and aspect. Large entrances, especially pits, often support higher species diversity than found in neighboring surface habitats because pits tend to accumulate organic and inorganic wind-borne debris and are sheltered with a lower evaporation rate, which provides more nutrients and moisture to the community (Rabelo et al. 2021). In addition, pits (e.g., on the Hawaiian Islands) can provide protection from disturbance, such as from larger

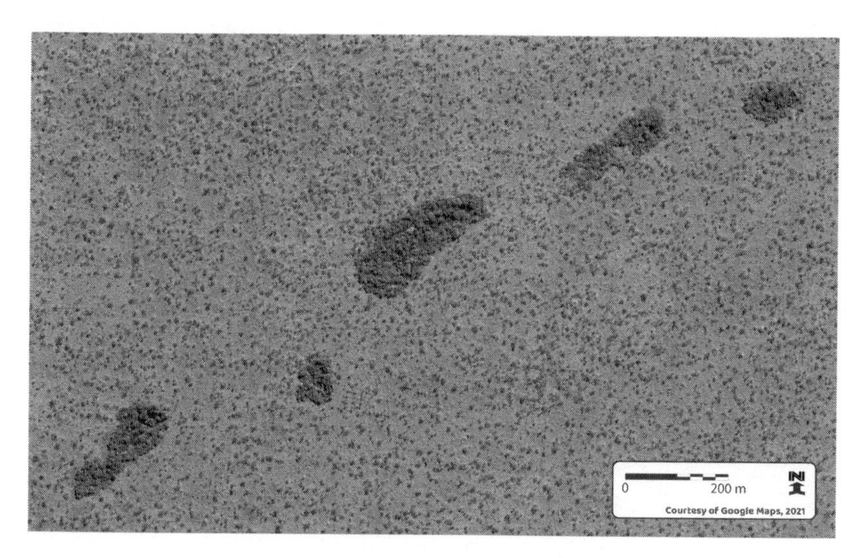

Figure 1.4. Portion of Undara Lava Flow, Mt. Surprise, Queensland, Australia, as seen from space. The ovate features are rainforest growing in pit entrances to the Undara Lava Tube System, which contrast sharply with the surrounding dry savanna habitat. Rainforest plants and animals either colonized from distant locations or are surviving remnants from an earlier wetter period. Image from Google Earth; Imagery © 2020 CNES/Airbus; Imagery © 2020 CNES/Airbus, Maxar Technologies; Map data © 2020 United States.

herbivores and wildfires (Howarth and Stone 2020). Animals occupying entrances include herbivores feeding on plants and their associated predators and parasites, and animals seeking shelter. A few colonial species, notably, social Hymenoptera, build nests in entrances, and many vertebrates use caves seasonally for nesting, dens, or refuge. In addition, the decomposer community on the ground can be highly diverse.

Cave entrances may also provide an insulating effect from both human activities and the climatic shifts on the surface. On Rapa Nui, Wynne et al. (2014) identified communities of endemic relict arthropod species restricted to relict vegetation communities occurring within cave entrances—these species are posited to be isolated within the entrance habitat due to a human/drought-induced ecological shift. Similarly, cave entrances have been identified in the United States as providing habitat for relict arthropod species due to climatic shifts associated with the glacial-interglacial period (Benedict 1979; Wynne and Shear 2016). Incidentally, in El Malpais National Monument in western New Mexico, caves with the highest arthropod diversity contained moss garden habitats within cave entrances and beneath skylights (Wynne 2013). Relict vascular plant communities have also been identified within cave entrances in southwestern China; they are believed to be restricted to caves due to intense human disturbance (Monro et al. 2018). This is discussed in slightly more detail in the flora section below.

Twilight Zone Habitats

Twilight zone communities are distinct from those in the entrance zone and generally not as diverse (Mammola 2019). Photosynthesis is reduced, and most energy input is allochthonous—arriving by transport via air currents, water, guano, frass, and animals. This zone is highly sensitive to local climate and changing seasons; importantly, this zone is often more humid during warm, moist weather conditions, but it dries out when the outside air is colder than the cave temperature (the "winter effect"; e.g., Howarth 1982; Tobin et al. 2013).

As one would expect, animal community composition reflects seasonal changes. During warm weather, many vertebrate species use the twilight zone for nesting, denning, and reproduction, as well as for refuge from heat. Numerous cave-roosting bat species select twilight zone habitats for establishing maternity colonies (e.g., Sherwin et al. 2000; Pennay 2008; Graening et al.

2017; Rosli et al. 2018), which can result in substantial guano deposition (e.g., Iskali and Zhang 2015). Resident animals, mostly arthropods and other invertebrates, feed on guano, frass, and carrion deposited by animals using the cave. A few animals graze on algae and biomineral oozes, and a few are predators. The availability of moisture and allochthonous nutrient input are strong regulators of the abundance and diversity. Temperature also strongly affects animal behavior. As the temperature cools, many resident species reduce or cease activity. In caves in temperate regions, many species roost or hibernate in the twilight zone to deep within caves.

Aphotic Zone Habitats

The dark, or aphotic, zone is highly complex, especially in long caves. Many distinct habitats may be present, each supporting a unique community. The constant darkness exerts selection pressures, and relatively few animals are able to colonize this zone. In many regions, bats (e.g., Gould 1988; Avila-Flores and Medellín 2004; Wynne and Pleytez 2005) and birds (i.e., cave swiftlets; Price et al. 2004) use caves for maternity roosts and nesting, respectively. Their colonies can be quite large, and their guano supports a highly diverse community of scavengers and predators (Ferreira et al. 2000; Gnaspini and Trajano 2000; Iskali and Zhang 2015; Ferreira 2019). This zone can be divided into two or three subzones: transition, deep, and/or stagnant air zones.

TRANSITION ZONE HABITATS. This zone is more conspicuous in tropical caves since the diurnal surface temperature often dips below the cave temperature each night. Cold air entering caves picks up moisture and dries the cave passage as it warms to cave temperature (the tropical winter effect; Howarth 1982). In caves in temperate regions, the transition zone often shifts with the seasons, being narrow to nonexistent in the summer and expanding deep into caves in winter. In temperate caves, drier surface air enters the cave, which can lead to seasonal desiccation (Barr 1968; Barr and Kuehne 1971). Many, if not most, cavernicoles track these environmental changes and migrate to their optimum microclimate (Barr 1968; Howarth 1982; Ferreira et al. 2018).

Among invertebrates, trogloxenous/troglophilous crickets and trogloxenous moths roost in this zone. In caves that do not host large colonies of these organisms, diversity is generally low compared to entrance and deep cave zone environments. The low diversity is due to the low availability of food resources and possibly to the environmental instability within this zone (Tobin et al. 2013). Most resident invertebrates are troglophiles. A few troglobionts may enter this zone as accidentals (such as those flushed out of mesocaverns by floods or collapse) or to exploit temporary food resources.

Many vertebrates roost in the transition zone. Bat colonies can become substantial, and their guano can support highly diverse communities of detritivores, especially in the tropics (Gnaspini and Trajano 2000; Moseley et al. 2012; Deharveng and Bedos 2019; Ferreira 2019).

DEEP AND STAGNANT AIR ZONE HABITATS. The deep and stagnant air zones comprise the principal habitats for troglobionts (e.g., Howarth and Stone 1990), although trogloxenes and troglophiles are also present. Separate habitats, each supporting a diverse community of cavernicoles in both zones, can often be characterized by the main food resources present. These include flood debris, plant root masses, fine organic matter deposited by percolating groundwater, biomineral oozes created by either or both chemoautotrophic and heterotrophic microorganisms, guano, and organic matter introduced by trogloxenes and accidentals. Some food resources are diffused and temporarily available as small randomly scattered patches (e.g., carcasses).

Overview of Cave Biodiversity

Microorganisms

Recent DNA sequencing has revealed diverse subterranean communities composed of Bacteria, Actinobacteria, Archaea, Fungi, and rarely some algae and Cyanobacteria (Barton and Northup 2007; Northup et al. 2011). The composition of these communities varies according to the biotic and abiotic environments, especially availability of moisture and nutrients, concentration of gasses in the atmosphere, pH, and minerals present in the substrate (Barton et al. 2014; de Paula et al. 2020). These communities can grow as biogenic oozes on the walls of caves and mesocaverns, especially where humidity remains at or above saturation. Chemoautotrophs that use mineral ions in the substrate to produce energy can contribute significant nutrients to support communities of larger organisms (Sarbu et al. 1996; Engel 2007).

Flora

Plants that use light for photosynthesis are restricted to the entrance and twilight zones of caves. The entrance zone, if sufficiently large, often supports lush vegetation—more diverse than the neighboring surface environment (Howarth and Stone 2020). Importantly, large dolines have been identified as refugial habitats for native tree species in a highly anthropogenically altered landscape in southwestern China (Su et al. 2017). Cave entrances can also provide relict habitats for vascular plants and bryophyte communities. Relict endemic fern and moss communities were identified on Rapa Nui (Ireland and Bellolio 2002; Wynne et al. 2014) and relict moss communities in western New Mexico (Lindsey 1951; Northup and Welbourn 1997; Wynne 2013). In the South China Karst, at least 31 vascular plant species were identified as restricted to cave entrances—due to governmental programs in the mid-1950s to mid-1970s that resulted in surface vegetation being cleared and removed (Monro et al. 2018). A few saprophagous plants have colonized deep subterranean habitats, notably whisk ferns, *Psilotum* species, which grow on biomineral oozes in the deep zone in the tropics (Howarth, unpublished data). In addition, plants on the surface over caves often send roots deep underground in search of water and nutrients, especially in the tropics, semiarid regions, and early successional habitats. Deeply penetrating plant roots (often reaching depths of more than 30 m) supply abundant food resources for subterranean communities (Hoch and Howarth 1989; Stone et al. 2004; Howarth et al. 2007; Souza-Silva et al. 2011; Wynne 2013).

Meiofauna

Tiny animals live on moist surfaces, microcaverns, interstices in fractured rock, and aquatic habitats underground. They are quite difficult to sample in caves, but a few studies indicate these animals are widespread and important components of subterranean communities. Crustacea is the best-known group and includes Bathynellacea and Copepoda (Sket 2004; Pipan and Culver 2013; Pipan et al. 2018).

Macroinvertebrates

Terrestrial invertebrates that are visible to the naked eye are often the most conspicuous animals in caves and the principal quest of most biological surveys (Wynne et al. 2019). They are found in most cave habitats, with each species occupying their habitat and niche. Most major arthropod groups (especially insects, arachnids, crustaceans, and myriapods) are represented by taxa able to colonize caves. In some regions, terrestrial mollusks, annelids, and planarians are also present. Terrestrial macroinvertebrates occur in caves as troglobionts, troglophiles, trogloxenes, and accidentals. Resident cavernicoles descended from representatives of the local surface fauna, and most ancestors were nocturnal and lived in damp, dark habitats, such as riparian, littoral, talus, and other moist cryptic surface habitats.

Vertebrates

Caves provide refuge for many terrestrial vertebrates. Larger caverns sometimes support substantial populations of roosting bats and birds (Gould 1988; Gnaspini and Trajano 2000; Sherwin et al. 2000; Avila-Flores and Medellín 2004; Price et al. 2004; Wynne and Pleytez 2005; Pennay 2008; Graening et al. 2017; Rosli et al. 2018). These colonies support diverse communities of coprophagous and saprophagous microorganisms and animals. Bats and birds are able to use deeper cave habitats, as they rely on echolocation to navigate in darkness. Numerous other vertebrates may use the deeper recesses of caves seasonally for dens or refuge (refer to Strong 2006; Wynne 2013; Wynne and Voyles 2014). Each species selects its microclimate and zone from the twilight to the deep zone.

Factors Influencing Terrestrial Cave Diversity

The number and abundance of species in caves are determined by both biotic and abiotic parameters (Lunghi and Manenti 2020). Biotic factors include living organisms (and their interactions) and organic matter. Abiotic parameters encompass the physical environment. These include climatic factors (e.g., light, water, humidity, temperature regime, atmospheric composition, and airflow), inorganic nutrients, types of substrates, and geologic history. Most research has focused on obligate cave fauna largely because of the keen interest of evolutionary biologists to understand the processes driving subterranean adaptation. In addition, unlike most troglophiles and trogloxenes, which are often widely distributed, obligate cave species and many species of troglophiles and trogloxenes typically are restricted to very narrow geographic ranges—either a specific geologic formation or single cave system.

The following section presents an overview of the main factors identified as influencing biological diversity in caves (Fig. 1.5). These factors do not

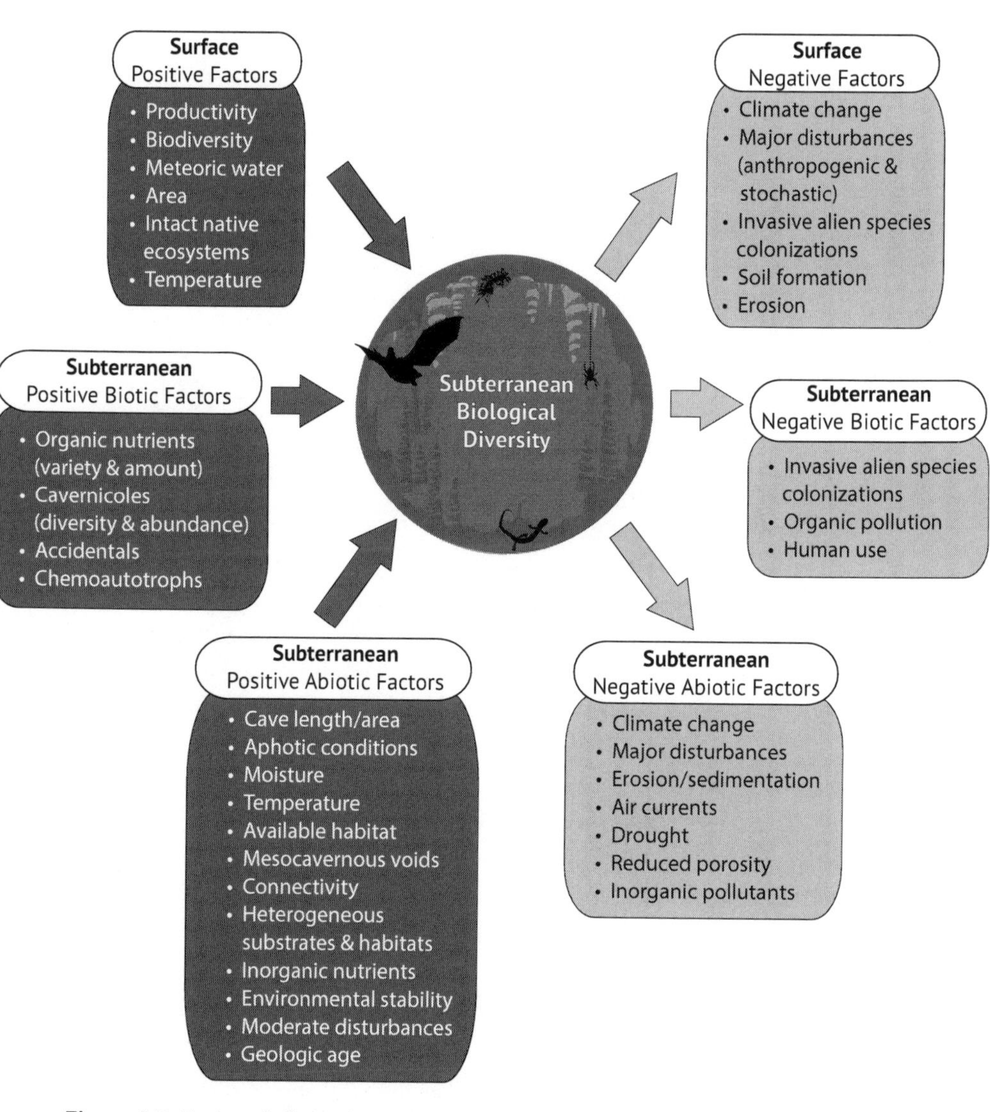

Figure 1.5. Factors influencing terrestrial biodiversity within caves. Dark gray boxes (*left*) include factors that enhance biodiversity, while light gray boxes (*right*) contain parameters that negatively affect diversity. Arrows point in direction of effect.

act independently but often interact in complex ways. We begin with a discussion on the effects of the surface environment on cave biodiversity and then move underground.

Surface Environments over Caves

Subterranean habitats are rigidly constrained by the surrounding rock, which can somewhat buffer the enclosed communities from weather and stochastic events on the surface. Ultimately, natural and anthropogenic changes on the surface (e.g., weather, climate, ecological succession, herbivory, invasive alien species, fire, and anthropogenic activities) affect the environment, as well as impact the input of nutrients, water, and pollutants (Borges et al. 2012; Simões et al. 2015; Christman et al. 2016; Zepon and Bichuette 2017; Castaño-Sánchez et al. 2020).

AREA. The areal extent of the land over contiguous cavernous terrains correlates, at least relatively, with the size of available subterranean habitats (Culver et al. 1973, 2006; Bregović and Zagmajster 2016; Christman et al. 2016). Larger surface areas would support more resources and potentially more opportunities for the development of underground habitats.

PRODUCTIVITY. Greater productivity of nutrients on the surface potentially would provide more nutrients entering cave environments (Culver et al. 2006; Zepon and Bichuette 2017; Christman et al. 2016). A proportion of organic nutrients produced on the surface would be transported underground by water and gravity. Greater rainfall during the growing season increases productivity and potentially transports more nutrients underground. Additionally, greater surface productivity would support larger and more diverse populations of troglophiles, trogloxenes, and accidentals, which would carry more nutrients underground. Mammola, Cardoso, et al. (2019) found that mean annual surface temperature (a proxy for productivity) was positively correlated with cave biological diversity in temperate caves. Similarly, Rabelo et al. (2021) found mean annual surface temperature was positively associated with total cave diversity. Using mean annual actual evapotranspiration as a surrogate for productivity, Culver et al. (2006) and Bregović and Zagmajster (2016) obtained differing results. Culver et al. (2006) found a positive correlation, while Bregović and Zagmajster (2016) reported only a weak association. Confounding the relationship between surface productivity and subterranean diversity is the eventual accumulation of soil over caves (Howarth 1996).

Often, the land surface overlying caves and other subterranean habitats appears barren or nearly so, with a thin soil layer and sparse vegetation (Fig. 1.6). Soil accumulation is slow because wind-borne debris and organic detritus eventually fall or wash underground. Where soil can accumulate on the surface, the amount of nutrients available to cave communities may be reduced. Thick or poorly drained soils act as traps, holding water and nutrients, thus preventing their movement deeper underground (Howarth 1996; Christman et al. 2016). Local rainfall is important to maintain moisture in caves (Culver et al. 2006; Tobin et al. 2013; Bregović and Zagmajster 2016). The amount of soil cover and its infiltration rate may explain, in part, the inconsistent results from studies correlating surface productivity with species diversity in underlying caves (e.g., Culver et al. 2006; Bregović and Zagmajster 2016).

Where soil does not accumulate on the surface over cavernous substrates, such as on talus, limestone terrains, and young basaltic lava flows, most of the organic nutrients produced on the surface fall or are washed underground. Rain falling on rocky surfaces quickly sinks into crevices or rapidly evaporates. Thus, these surface environments are often extremely xeric, and colonization by pioneer plants and animals is slow. For example, Figure 1.7 shows two views of the same area on the 1881 Mauna Loa lava flow at approximately 1,600 m above sea level. Figure 1.7A illustrates the surface environment looking downslope. The hill in the background is a remnant rainforest patch, which escaped being covered by lava. Even though the local climate is montane rainforest, only a few scattered shrubs and ferns less than a meter tall have colonized the flow after more than a century. Animals living in neighboring forests are largely unable to survive on the barren lava. However, to survive in the harsh surface environment, the small shrubs have produced exceptionally large and luxurious roots to obtain water and nutrients (Fig. 1.7B). This figure was photographed 8 m underground and located directly beneath the site shown in Figure 1.7A. The 1 m tall shrubs are each supported by a robust branching root mass that extends at least 12 m below the surface. Here, the "rainforest" is underground; animals able to colonize subterranean voids within this flow would be able to exploit vast new resources (Howarth 2019). In fact, seven obligate subterranean-adapted species have already colonized this cave. Young cavernous basalt is widespread

Figure 1.6. Sparse vegetation on surface habitats above caves and voids. [A] Aerial view of tropical tower karst, Chillagoe, North Queensland, Australia; [B] Talus slope, Crater Lake National Park, Oregon. Images courtesy of Francis G. Howarth.

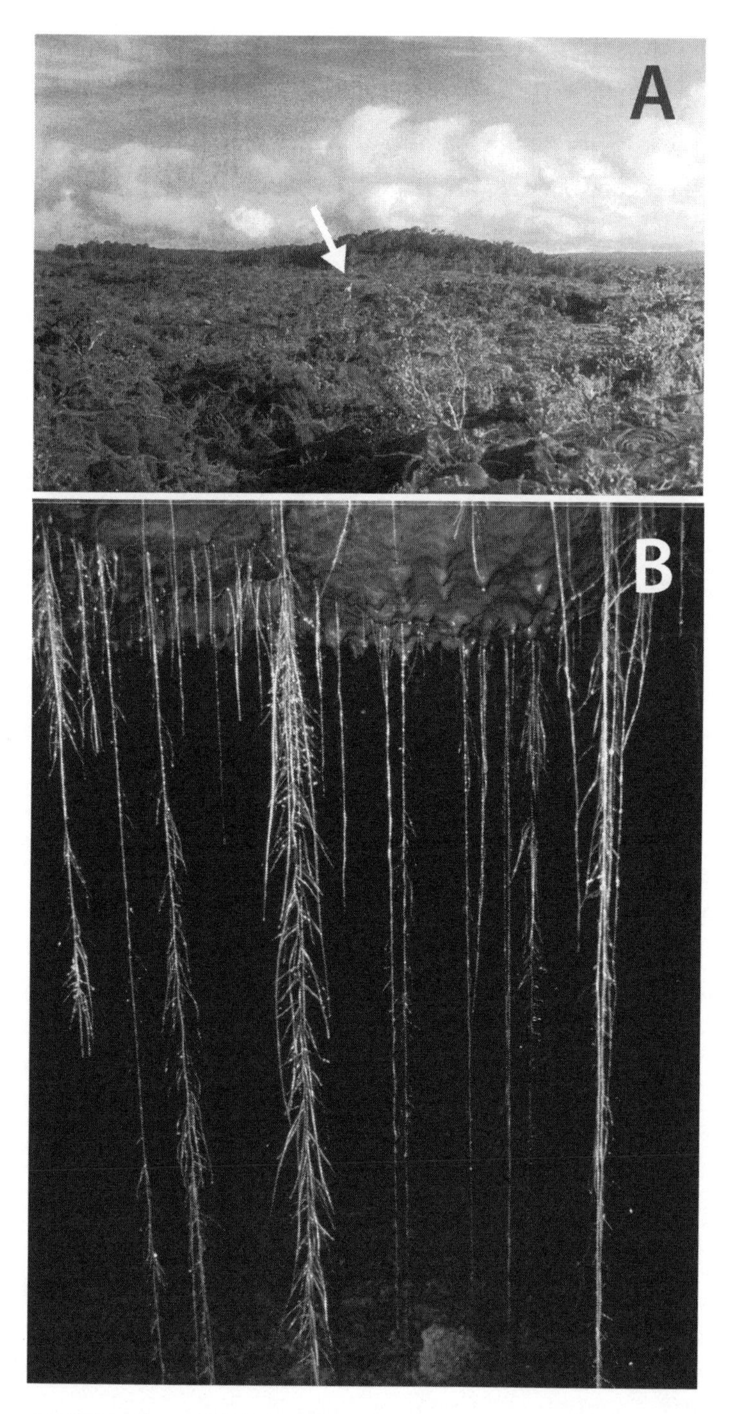

Figure 1.7. [A] Surface of 1881 lava flow from Mauna Loa directly over Emesine Cave passage. Person standing left of the center of photograph for scale (*location indicated with arrow*). [B] Same view as [A] but 8 m below ground level. The small shrubs, approximately 1 m tall, which grow on the nearly barren surface, require huge root systems, over 12 m long, to survive the harsh conditions. Images courtesy of Francis G. Howarth.

on Mauna Loa and Kīlauea volcanoes on the island of Hawai'i and provides subterranean habitats orders of magnitude larger than many overlying surface habitats, in both area and nutrient availability.

In a study of small limestone caves within a nature reserve in Brazil, Pellegrini et al. (2016) measured within-cave environmental parameters, including cave size, types of substrates and nutrients, environmental stability, and terrestrial cave arthropod community composition at three spatial scales—radii of 50 m, 100 m, and 250 m around each entrance. Upon analyzing these data with the land cover within each buffer area, they found that the best predictor of cave arthropod community composition was percent cover of limestone outcrop within each buffer zone. This corroborated the hypothesis that soil cover influences cave biodiversity. Furthermore, their results supported the scale-dependence hypothesis that predictions of cave community composition improved at larger spatial scales.

DISTURBANCE. Both natural and anthropogenic events on the surface can negatively affect subterranean diversity, both directly and indirectly (Zepon and Bichuette 2017; Deharveng and Bedos 2019). Natural disturbances include flooding, severe weather, wildfires, erosion, pedogenesis, and ecological succession. Anthropogenic impacts include destruction of native surface vegetation by land-use changes, urbanization, agriculture, human-caused wildfires, introduction of invasive alien species, mining, and climate change. Disruptions not only reduce productivity and diversity of surface communities, but they also affect caves by introduction of toxins, increased sedimentation of voids, and changes in nutrient and water availability (Borges et al. 2012; Wynne et al. 2014; Deharveng and Bedos 2018; Jaffé et al. 2018; Castaño-Sánchez et al. 2020).

Positive Biotic Factors in Underground Habitats

Organic Nutrients

As the amount, quality, and variety of available nutrients increases, subterranean habitats can sustain larger populations and greater diversity of organisms (Culver and Sket 2000; Culver et al. 2006; Zepon and Bichuette 2017; Pacheco, Souza-Silva, et al. 2020). Nutrients include allochthonous organic material transported into caves via percolating water and gravity, by animals as guano, frass, or carcasses, and by plant roots. Additionally, autochthonous nutrients are produced locally by chemoautotrophic microorganisms (e.g., Sarbu et al. 1996; Flot et al. 2010; Naaman 2011). Recognition of the importance of nutrients as a determinant of subterranean diversity has led to the strategy of concentrating efforts during biological surveys to areas where abundant nutrients occur (Culver 2008; Wynne et al. 2018). In a comprehensive study of microinvertebrates in a cave in southern Mexico, Palacios-Vargas et al. (2011) found that species abundance was greatest in the most nutrient-rich resource (guano), whereas the highest species diversity occurred in leaf-litter habitats (i.e., the most variable habitat due to the varied input of different litter types). Animals that enter caves accidentally are often overlooked as a source of food for resident species, but their contribution can be significant. Their presence explains, in part, the relatively high proportion of predators in the cave community (Howarth 1983; Resende and Bichuette 2016).

Although deep cave habitats appear barren, the apparent scarcity is relative. The driving force allowing successful colonization of underground habitats is the availability of suitable food resources (Howarth 1993; Simões et al. 2015; Zepon and Bichuette 2017). However, the scattered nature of resources within a dark, three-dimensional maze presents unique obstacles for animals attempting to exploit them. In theory, if caves are food-poor or food is difficult to locate, one would expect cavernicoles to show little specialization in feeding preferences and to be forced to feed on whatever was available (Gibert and Deharveng 2002). Smrž et al. (2015) tested this idea on a small group of cavernicoles and surprisingly found that most species displayed stronger feeding preferences (i.e., trophic niche specialization) than expected. Palacios-Vargas et al. (2011) also provided evidence of feeding preferences among cave invertebrates. Zepon and Bichuette (2017), studying caves in southeastern Brazil, concluded that the type of food resource influenced the composition of the community present and that leaf-litter accumulations supported the highest abundances and species diversity. Thus, cave ecosystems, like most surface communities, support both generalists and specialists.

Cavernicoles

The diversity and abundance of animals in caves may facilitate colonization by additional species because each species in a community creates novel niches that allow some of their associated species to become established (Mueller-Dombois and Howarth 1981). Successful newcomers may directly or indirectly cause extinctions among the associated established species. Conversely, if suitable resources are present, the colonizing species will integrate into the community, thus increasing diversity (Mueller-Dombois and Howarth 1981; Zepon and Bichuette 2017).

Negative Biotic Factors

Invasive Alien Species

Invasive alien invertebrates and vertebrates can negatively impact both ecosystem structure and function. However, the presence and impacts of alien species in caves is rarely reported or studied. A relatively small percentage of the invasive alien species in surface habitats has entered caves, which may result partly because sun-loving and drought-tolerant species are more likely to survive transport with humans. However, invasive species pose a serious threat to native cavernicoles, and the problem will likely worsen with intensifying global trade.

Currently, the most serious invasion is white-nose syndrome (WNS), a fungal disease caused by *Pseudogymnoascus destructans* (Blehert & Gargas) Minnis & D. L. Lindner that was introduced to North America from Europe (Leopardi et al. 2015). By 2012, the US Fish and Wildlife Service estimated that the disease had resulted in the mortalities of more than 5 million bats (USFWS 2012). Subsequently, cave-roosting bat populations have been reduced regionally, which we suggest has negatively impacted the ecosystems where they roost. This fungal pathogen has now been documented in at least 38 states in the United States and seven Canadian Provinces (WNSRT 2022).

In southern North America, the red imported fire ant (*Solenopsis invicta* Buren, 1972) is considered a serious threat to endangered cave invertebrates. Trogloxenic rhaphidophorine cave crickets are especially vulnerable to depredations by fire ants (Taylor et al. 2005). This cave-roosting cricket provides a major food resource for endangered troglobionts. Other invasive ants (e.g., *Paratrechina longicornis* (Latreille, 1802) and *Anoplolepis gracilipes* (Smith, F., 1857)) are widespread in the tropics and commonly enter caves for food and water. Similarly, Borges et al. (2012) listed the introduction of invasive species as an important threat to the cave fauna of the Azores.

Other invasive troglophilic invertebrates with wide distributions include the nemertine worm (*Argonemertes dendyi* (Dakin, 1915)), the pulmonate snail (*Oxychilus alliarius* (Miller, 1822)), the garden millepede (*Oxidus gracilis* (C. L. Koch, 1847)), a terrestrial isopod (*Porcellio scaber* Latreille, 1804), two spiders (*Nesticella mogera* (Yaginuma, 1972) and *Dysdera crocata* C. L. Koch, 1838), and the American cockroach (*Periplaneta americana* (Linnaeus, 1758); Howarth 1981; Reeves 1999, 2001; Wynne et al. 2014). Their impacts on cave ecology are suspected but not well documented (Howarth 1981; Howarth and Moore 1984).

Additionally, the invasive brown tree snake, *Boiga irregularis* (Merrem, 1802), has caused the extinction of the majority of forest birds on Guam, while the Vanikoro swiftlet, *Aerodramus vanikorensis bartschi* (Mearns, 1909), survives in only a few caves in which they nest on smooth walls where the snake cannot reach them (Wiles et al. 2003). Importantly, Howarth and Stone (2020) described the impacts of the invasive black rat, *Rattus rattus* (Linnaeus, 1758), on cave resources in Hawai‘i, including its role in the extirpation of colonies of endemic trogloxenic moths. The effects of the black rat are expected to impact caves similarly throughout its global distribution. Moreover, alien trees send their roots deep into cave habitats where they compete with native species, displace root-inhabiting obligate arthropod communities, and provide food for other alien species (refer to Howarth et al. 2007 for Hawaiian examples).

The use of living disease organisms as biopesticides to control agricultural pests poses serious risks to subterranean species. These agents are naturally occurring soil pathogens that are cultured and applied to crops. When introduced to new regions, they can become invasive, and they would survive well in damp cave environments if carried underground with runoff from agricultural fields (Howarth 2000).

Human Use

Humans have used caves for refuge, living quarters, harvesting resources, and burials since ancient times. The impact of human use ranges from relatively benign and temporary to severely affecting the biological diversity of native cave species. The urge to explore is an innate human behavior, and cave exploration is a popular sport. Exploring caves can result in two conflicting outcomes. The sport can potentially reduce biodiversity—for example, people might trample and break cave resources, disturb roosting animals, increase airflow by opening entrances, modify passages, and introduce foreign materials and toxins (Ferreira et al. 2020; Souza-Silva et al. 2020a). In addition, occasional acts of vandalism in caves can do lasting damage to cave resources and biodiversity. Conversely, cave exploration has provided the necessary baseline data on cave geography that have significantly aided cave research, including biodiversity studies.

Fortunately, impacts by human visitors within caves are generally restricted to accessible passages. Since the major habitat for cave animals occurs in the less accessible passages and mesocaverns, the long-term impacts on cave biodiversity resulting from direct human use may be limited. However, biodiversity studies can be significantly compromised by disturbance of research sites and study organisms, both inadvertently and by vandals, because biodiversity studies must be conducted in passages accessible to other persons visiting the cave.

Tourism development involves a similar conflict between benefit and harm. Commercial caves severely modify at least a portion of significant caves, which can damage or destroy local cave resources, alter cave climate, and introduce light and pollutants into the cave (Whitten 2009; Souza-Silva et al. 2011, 2020a; Borges et al. 2012; Pacheco, de Oliveira, et al. 2020). Concomitantly, when done responsibly, cave tourism can provide an invaluable service by educating the public on the wonders of caves, cave-roosting bats, and the need to protect cave species (Cigna 2016).

Caves used for waste disposal can lead to negative impacts of cave resources and biodiversity (Souza-Silva et al. 2020a). Subterranean voids are used both intentionally and unintentionally as cesspools for human waste. Pit entrances are used intentionally as convenient dump sites for trash disposal. Much of this material, including toxins, eventually disperses into the cave. Runoff from urban areas, roads, and agricultural land can sink into underground habitats. Increasing agricultural land cover within 50 m of caves reduced species richness and phylogenetic diversity (Jaffé et al. 2018). Raedts and Smart (2015) described the history of Horse Cave in Kentucky, which was a popular tourist cave supporting high biological diversity until animal waste from a dairy entered the cave. The effluents polluted the underground river, and the stench made the cave unenterable. Numerous other organic pollutants and pesticides enter caves via human activity (refer to Castaño-Sánchez et al. 2020 for a review on this topic).

Extraction of bird and bat guano, saltpeter, and other resources disturbs cave environments, introduces foreign material, and impacts cave diversity (Jaffé et al. 2018). Furthermore, if disturbance forces large colonies of birds or bats to abandon their roost, the guanobiont community will be extirpated along with any endemic species (Ferreira and Pellegrini 2019); thus, colony loss may be currently the most severe threat to tropical subterranean biodiversity (Deharveng and Bedos 2019). Also, in the tropics, whole tower karsts have been removed to ground level for cement production, thereby causing the extinction of troglobionts (Whitten 2009; Deharveng and Bedos 2019). Also, the flooding of caves by the construction of dams and reservoirs has extirpated endemic cave species (D. J. Robinson 1978; Briggs 1981; Leithauser and Holsinger 1985; Ferreira and Pellegrini 2019).

Positive Abiotic Factors

Inorganic Nutrients

In a broad sense, water is the most important inorganic nutrient for living organisms. Liquid water is required for cellular activity in living organisms. The amount of liquid water and water vapor is generally positively correlated with biological diversity from the entrance to the deep zone, with each species occupying a moisture level suitable for its survival (Tobin et al. 2013; Simões et al. 2015; Jaffé et al. 2018). Troglobionts are restricted to areas where the potential evaporation rate is negligible (i.e., relative humidity is near or even above saturation; Howarth 1982; Howarth and Moldovan 2018b). Liquid water (often aided by pH concentration) also dissolves essential minerals, making them available to cave organisms. Mineral composition, availability, and abundance affect the distribution and diversity of microorganisms and plant roots, thereby influencing the distribution, abundance, and diversity of subterranean communities (Barton and Northup 2007; de Paula et al. 2020).

Available Habitat

As the size of suitable habitat increases in area or volume (i.e., the species-area relationship), it should support larger populations and more species (Culver and Sket 2000; Culver et al. 2006; Simões et al. 2015; Christman et al. 2016; Zepon and Bichuette 2017; Jaffé et al. 2018; Mammola, Cardoso, et al. 2019). Also, a larger habitat provides greater protection from extinction since there is a greater chance of the presence of refuges during disturbances (Zepon and Bichuette 2017), and larger populations are typically more resilient to stochasticity. The larger habitat would have more resources and niches, all other factors being equal. The latter hypothesis is difficult to study in caves because most of the habitat is inaccessible and unknown. However, Ferreira and Pellegrini (2019) documented a direct correlation of the species-area relationship from a small cave in Brazil before and after it was partially inundated by a dam. Christman and Culver (2001), Culver et al. (2006), Bregović and Zagmajster (2016), and Christman et al. (2016) used surrogate values (i.e., number of caves per county and areal extent of limestone) to test the relationship with some success. However, Christman and Culver (2001) noted that the number of caves per county may not be a good proxy for area, as caves vary greatly in size, and that the number of caves likely measures a combination of habitat size, diversity, and, especially, sampling intensity. Borges et al. (2012) found a positive correlation between troglobiontic diversity and length of lava tubes in the Azores. Similarly, Simões et al. (2015) and Rabelo et al. (2021) demonstrated that linear extent of limestone caves is positively correlated with invertebrate diversity and believed this relationship is due to the larger number, greater stability, and increased heterogeneity of habitats farthest from an entrance. In a comparison of caves of different lithologies, Souza-Silva et al. (2011) reported a positive relationship between diversity and cave length across the four cave types studied; ferruginous caves exhibited the steepest slope followed by carbonate and then siliciclastic caves, while this relationship was not significant for magmatic caves. This relationship was also observed by Schneider and Culver (2004), Graening et al. (2006), and Graening et al. (2012), who found positive relationships between arthropod species richness and cave length in several areas in the temperate southeastern United States. Additionally, in tropical caves of southern Mexico, Brunet and Medellín (2001) described a positive correlation between bat species richness and area of cave roosts—they further suggested variability of relative humidity and the presence of conical depressions (i.e., vugs) in cave ceilings were driving this correlation.

Extent of Mesocaverns

Related to size is the presence of accessible subterranean voids of variable sizes (Zepon and Bichuette 2017). A critical requirement for habitat suitability in caves is the degree of interconnectedness of these voids. The more extensive an interconnected system of voids is for nutrient and water transport, the larger and more diverse the communities are that it can support (Howarth 1983; Souza-Silva et al. 2011, 2020b; Jiménez-Valverde et al. 2017). Mesocaverns provide surface area for troglobiontic invertebrates orders of magnitude larger than caverns do (e.g., compare the available surface area within a large passage to that in a similar-sized void filled with a jumble of breakdown). However, solitary vugs, gas vesicles, crevices, and other perpetually isolated voids are likely to be sterile (Sendra et al. 2014).

Young basaltic lava flows have abundant interconnected voids of various sizes, and the caves therein can support more species than older lava tubes (Howarth 1996). The diversity of troglobionts in calcretes has been correlated to the degree of fracturing and the presence of interconnected voids (Halse 2018), while arthropod diversity in iron ore caves in Brazil was correlated with lithology and void structure (Souza-Silva et al. 2011; Jaffé et al. 2018). Ferreira et al. (2018) further stressed the importance of a vast system of interconnected voids in iron ore caves of Brazil, which supported an exceptionally high level of subterranean diversity. Likewise, the invertebrate fauna of talus caves was positively correlated with the size, extent, and interconnectedness of the voids (Souza-Silva et al. 2020b).

Habitats in limestone formations are intermediate in their extent and interconnectedness of voids since dissolution enlarges larger passages at the expense of smaller ones and because small voids tend to fill with soil and debris. However, limestone deposits are frequently layered and fractured. These and other imperfections dissolve at different rates creating voids of various sizes. In addition, as caverns enlarge, portions collapse, increasing the surface area within the void. Thus, mesocavernous voids are important habitats in limestone caves (Mammola, Cardoso, et al. 2019).

The extent of mesocaverns in sandstone and many piping caves depends on characteristics of the parent rock since the loosened clastics are flushed out of the cave. Fine-grained sandstone would likely have fewer mesocaverns, whereas parent rock composed of cemented boulders or soft and resistant strata in alternate layers would contain abundant mesocaverns.

Habitat Heterogeneity

Associated with both the species-area relationship and connectedness of voids, greater habitat variability would be expected to support more species because the species could segregate into communities that occupy different microhabitats (Pacheco, Souza-Silva, et al. 2020; Zepon and Bichuette 2017). Bregović and Zagmajster (2016) examined elevational range as a proxy for habitat variability of caves in southeast Europe; they found a statistically significant positive correlation with biodiversity of troglobiontic beetle diversity, which they interpreted as an increased ability of the species to migrate in response to climate change. Jaffé et al. (2018) also reported a correlation between diversity and elevation. In a study of large talus caves between 827 and 1,790 m elevation in Brazil, Souza-Silva et al. (2020b) found that higher elevation caves supported more species. They attributed the increase in diversity to the more stressful and unstable surface environment at higher elevations, which resulted in a larger number of animals seeking refuge in caves. However, further studies are needed to more robustly quantify how elevation contributes to habitat variability.

Although deep cave environments appear relatively homogeneous and contain few disparate terrestrial habitats (Bregović and Zagmajster 2016), heterogeneous substrates, different-sized voids, and patches of dissimilar food resources can support unique communities (Palacios-Vargas et al. 2011; Simões et al. 2015; Zepon and Bichuette 2017; Howarth and Moldovan 2018b). In addition, early colonizers of temporary food patches often shape the eventual composition of the community (Lavoie 1981). The dynamic nature of community development and succession allows more species to coexist in a given habitat. For example, Zepon and Bichuette (2017) grouped limestone substrates into 11 categories for sampling and analysis. Rocky substrates and substrates with visible food sources supported greater diversity, while homogeneous substrates (i.e., soil) supported the lowest diversity. They also reported that community structure on the walls and ceiling differed from the cave floor community (Zepon and Bichuette 2017).

Habitat Stability

Habitat stability over time allows species to survive and improves adaptations to the cave environment (Culver et al. 2006; Bregović and Zagmajster 2016; Ferreira et al. 2018). Long-term stability permits additional species to colonize, as well as species to diversify in situ, thus increasing biological diversity (Wessel et al. 2013). Paradoxically, regularly occurring disturbances that have low to intermediate severity (such as lava flows, seasonal floods, and seasonal roosting by trogloxenes) can increase diversity by restructuring habitats, resetting succession, and preventing overpopulation of dominate species (Wessel et al. 2013; Simões et al. 2015; Zepon and Bichuette 2017).

Temperature

Higher average surface temperatures, which generally correspond to higher cave temperatures, are positively correlated with higher cave diversity—at least locally (Culver et al. 2006; Mammola, Cardoso, et al. 2019). However, this relationship may not hold when comparing cave faunas in different regions and climates. Because the temperature within the deep cave zone is believed to be constant, troglobionts are assumed to be adapted to a narrow range of temperatures (Mammola, Piano, et al. 2019); yet, there are exceptions. On Hawai'i Island, the cave wolf spider (*Lycosa howarthi* Gertsch, 1973) occurs in high elevation lava tubes from 15°C to 20°C in low elevation caves, as well as in a geothermally heated cave at over 35°C. Additionally, temperatures in individual passages in a single cave may differ by 5°C or more depending on the location in relation to other passages. For example, upper-level dead-end passages can be heat traps remaining hotter than the main passage. Elevated carbon dioxide can also trap heat. In caves with different temperature regimes, potential evaporation rate is a better predictor of the presence of troglobionts than temperature (Howarth 1982).

Geologic Age

Colonization time of the ancestors of troglobionts is almost always older than the caves currently occupied because cave formation processes are dynamic, but some iron ore caves (Ferreira et al. 2018) and lava tubes (Hoch and Howarth 1999) may be exceptions. Younger caves become available as older passages erode or become unavailable. Erosion and downcutting of soluble strata expose more substrate for dissolution. Thus, most solution caves generally deepen with age with older remnant cave passages remaining above younger active caves (Culver et al. 2006). In contrast, as volcanism progresses, younger lava tubes are created over older tubes, so that younger active caves are overlaid on older tubes (Howarth 1996). Species accumulation in a community is dynamic, with additional species continually attempting to colonize new habitats, especially if suitable resources are present. Thus, over time if there are sufficient resources and available habitat, species diversity will increase. For example, the more than 100-year-old 1881 Mauna Loa lava flow, which contains Emesine and Kaumana Caves, harbors about one-half the number of expected troglobionts compared to the 550-year-old 'Ailā'au lava flow, which includes Thurston, Kazumura, and Ainahou lava tubes (Howarth, unpublished data). Additionally, in their analysis of the biogeography of terrestrial troglobionts, Culver et al. (2006) demonstrated that troglobiont diversity was highest in comparatively old regions in terms of length of time that cave habitats were continuously available. Correspondingly, Ferreira et al. (2018) attributed the high diversity in iron ore caves in Brazil to their great age.

Speleogenesis

Differing modes of speleogenesis in diverse parent rocks directly or indirectly affect faunal diversity by influencing the moisture regime, extent of mesocaverns, habitat size, heterogeneity, stability, and age. Thus, one would expect that the composition of the parent rock and mode of formation would be a strong predictor of diversity, but sampling bias obscures the evidence. By far, limestone caves in temperate regions have received the most attention. As other caves and regions are sampled, our understanding of the relationship between diversity and different types of caves will evolve. Souza-Silva et al. (2011) compared caves of four different lithologies (carbonate, ferruginous, magmatic, and siliciclastic) in the Atlantic Forest of Brazil and found significant differences in diversity. These differences were attributed to the extent of mesocaverns, productivity, nutrient input, habitat area, and heterogeneity, which were in part controlled by lithology.

The mode of formation of solution caves can also influence diversity. Sendra et al. (2014) and Jiménez-Valverde et al. (2017) reported that hypogene limestone caves with no connection to surface waters did not harbor obligate subterranean-adapted species even when suitable habitat was available. They attributed the limited fauna to long-term isolation from surface habitats following cave formation, which restricted colonization. However, diverse communities of chemoautotrophic microorganisms would be present during formation to produce sulfuric acid (Engel 2007; Sendra et al. 2014). Conversely, Sarbu et al. (1996) and Engel (2007) reported high invertebrate diversity in sulfidic caves that were connected to meteoric water for a sufficient time. Engel (2007) attributed species richness to the productivity of a plentiful, chemosynthetically produced food source, possibly geologic age, and the hydrologic and geochemical stability of the aquifer.

Negative Abiotic Factors

Darkness

The most conspicuous environmental parameter in the deep subterranean environments is, of course, total darkness. Normal cues used by surface animals to locate food, find mates, escape predation, and disperse are typically absent or not useful in a dark, three-dimensional underground maze. Thus, behavior must play a considerable role in adapting to caves. Darkness limits the number of species in a region that potentially could successfully colonize or use subterranean habitats (Fernandes et al. 2016). However, when the absence of light is constant, its influence on diversity may be limited to its interaction with other factors. An exception, which alters community structure in the deep cave zone, is the introduction of artificial lighting for tourism (Cigna 2016).

Major Disturbances

Severe disturbances that negatively impact the cave environment and/or are largely irreversible could reduce biological diversity by reducing the carrying capacity of the habitat. Such disturbances may also depress survival and reproductive success of native species. Anthropogenic impacts include the creation of new entrances, tourism development, conversion or loss of surface habitats, introduction of nonnative species, floods, and mining (Trajano 2000; Whitten 2009; Borges et al. 2012; Simões et al. 2015; Ferreira et al. 2018; Jaffé et al. 2018; Ferreira and Pellegrini 2019; Zhao et al., Chap. 7). Natural disturbances include unusual geologic events, such as extreme volcanism, earthquakes, floods, and wildfires.

Erosion

The natural process of weathering and removal of rock and clastics can both create more cave habitat and decrease available habitat. Erosion of the substrate negatively affects cave biodiversity by increasing sedimentation, which reduces porosity (i.e., the interconnectivity of voids). Shallow mesocavernous voids are rare in the wet tropics, compared to temperate regions, since erosion and soil formation is much more rapid in the moist, warm climate (Howarth 1996; Deharveng and Bedos 2018). In the tropics, shallow habitats are well developed in substrates where they are continuously recreated (e.g., talus slopes, canaliculi in iron ore landscapes, and lava flows).

In Hawai‘i, these shallow habitats fill within a few millennia in mesic to wet climates (Howarth 1996).

Erosion also causes collapse and spalling within caves, which can open new entrances and change passage shape, thereby increasing air currents and reducing humidity. However, Simões et al. (2015) found that more non-troglobiontic species occupied caves with larger entrances than occurred in caves with small entrances. They attributed the relationship to larger entrances, which allow more nutrients and animals to enter the caves.

Drought

The importance of water and humidity in positively influencing diversity is well recognized and described above under positive abiotic factors (Culver et al. 2006; Simões et al. 2015; Bregović and Zagmajster 2016; Howarth and Moldovan 2018b). It follows, then, that desiccation of cave habitats due to reduced input or storage capacity of water and/or increased airflow would negatively impact biological diversity (Howarth 1983; Culver et al. 2006; Tobin et al. 2013; Bregović and Zagmajster 2016).

Climate Change

The continuing warming of the earth's climate will negatively affect cave diversity (Chevaldonné and Lejeune 2003; Mammola, Piano, et al. 2019). Temperature within deep cave zones is relatively constant and near the mean annual surface temperature above the cave (e.g., Wynne et al. 2008; McClure et al. 2020). Heat exchange between the surface and caves is largely influenced by incoming water and airflow (i.e., by convection). Temperature change in shallow caves would be more rapid (Mammola, Piano, et al. 2019), while temperatures of caves with greater depths and thick rock overburden are expected to change more slowly from conduction due to a lag effect (Titus et al. 2010). Changes in precipitation patterns and increase in storm frequency and severity will change the water regime in caves, as well as increase erosion. As temperatures rise, the potential evaporation rate will increase, making caves more prone to desiccation (Howarth 1982). Furthermore, the impacts of climate change on the surface (such as increased wildfire severity and frequency, and ecological succession) will directly or indirectly affect cave diversity (Zepon and Bichuette 2017).

Inorganic Pollutants

Chemical pollutants enter caves in runoff from roadways and urban and agricultural areas, as well as by being transported by humans and animals. Inorganic pesticides and toxic metals can be especially damaging to cave diversity (Castaño-Sánchez et al. 2020; Zhao et al., Chap. 7).

Constraints on Understanding Cave Biological Diversity

Many serious gaps remain in our knowledge of subterranean biology. These lacunae hinder our understanding of subterranean diversity, ecology, and evolution. For example, the Racovitzan impediment (Ficetola et al. 2019), which posits that the biota in many, if not most, caves and cave regions remains to be surveyed, still haunts cave biological diversity studies and community-level analyses. Most limestone caves have not been adequately surveyed, and even in relatively well-studied regions, new undescribed species continue to be discovered (Culver et al. 2013). For example, during surveys in the Edwards Plateau, Texas, USA, previously known to harbor three endangered troglobionts, Krejca and Weckerly (2008) determined that between 10 and 22 visits would be required to confirm, with 95% confidence, that a target species was present in a given cave. The impediment is especially evident in non-limestone caves, which have been largely ignored until recently.

The Racovitzan impediment should be expanded to include unstudied taxa; many cave taxa remain poorly studied or unknown because of the difficulty of collecting adequate specimens for study, lack of taxonomic authorities, or bias from assumptions that individuals detected in caves were accidental occurrences (Vas and Kutasi 2016; Howarth et al. 2020; Lunghi et al. 2020). For example, several major taxonomic arthropod groups have been added to the known obligate cave fauna. These include true crickets, Gryllidae in Orthoptera (Gurney and Rentz 1978; Desutter-Grandcolas 1993); earwigs in Dermaptera (Brindle 1980); true bugs, Cixiidae, Reduviidae, and Mesoveliidae in Hemiptera (Gagné and Howarth 1975; Hoch and Howarth 1989); moths, Erebidae and Tineidae in Lepidoptera (G. S. Robinson 1980; Howarth et al. 2020); and wasps, Embolemidae in Hymenoptera (Olmi et al. 2014). Although some of the initial records are decades old, these groups are largely considered exceptional occurrences and outside the remit of inventories. What other taxonomic groups are we missing?

Another problem hindering biodiversity studies is this question: "What is a species?" In addition to the confounding philosophical differences between "splitters" and "lumpers" in taxonomy that either conflate or reduce the number of taxa recognized, there is also the issue of species boundaries in closely related diversifying groups (Miller 2007). The Hawaiian cave fauna provides some examples of the latter phenomenon. Several taxonomic groups in Hawaiian caves consist of a kaleidoscope of different morphologies and behaviors, each of which might represent separate species. Recent molecular, morphological, and behavioral studies have clarified the status of some groups, but precise species boundaries remain obscure for many taxa (Wynne et al., Chap. 2). For example, cave moths in the genus *Schrankia* on the islands of Hawai'i and Maui include a bewildering diversity of forms, from truly subterranean-adapted individuals found deep in lava tubes that are pale, blind, and flightless, with reduced wings and eyes, to fully volant epigean individuals living in cave entrances and twilight zones. Molecular analysis revealed that all forms belong to a single polymorphic species that freely hybridizes (Medeiros et al. 2009). Thus, *Schrankia howarthi* Davis and Medeiros, 2009 represents a case of gene flow between cave and surface populations during subterranean adaptation, and therefore could be interpreted as adaptive shift in progress (Juan et al. 2010; Howarth 2019). In contrast, molecular, morphological, and behavioral studies have shown that *Caconemobius* crickets (Otte 1994; Croom et al. 2008) and *Oliarus* planthoppers (Hoch and Howarth 1999; Wessel et al. 2013) on the island of Hawai'i are considered separate species, with some caves harboring two or more sympatric species occupying different niches. Other groups, including cambalid millipedes, linyphiid sheet-web spiders, phorid flies, and emesine thread-legged bugs on Hawai'i Island, appear to represent separate, distinct cave populations, but these distinctions have not been quantified. The recommendation arising from this variation is that conservation programs should stress perpetuating populations so that the evidence regarding the true diversity of a taxon is not lost but preserved for study. Otherwise, much of the evidence of evolution of new species will be lost, and the view that each species is a distinct entity in nature will be erroneously reinforced (Miller 2007; Mammola et al. 2020).

A critically related issue is the importance of well-maintained accessible voucher collections that document each of these distinct populations

(Raven and Miller 2020; Wynne et al. 2021). Voucher specimens anchor the history of cave diversity studies, as well as permit comparisons of community structural changes over time.

Implications for Conservation and Management

The next few decades may be the last chance to conduct all taxa biological inventories of subterranean communities to determine the pre-Anthropocene biological diversity (Dirzo et al. 2014; Raven and Miller 2020). High priority should be given to inventory subterranean habitats in poorly known regions to address knowledge gaps (Wynne et al. 2021). All taxa and microhabitats should be sampled (Wynne et al. 2018; Pacheco, Souza-Silva, et al. 2020); however, for efficiency, initial inventories may need to focus on the most promising substrates and microhabitats—in particular, the selection of sites chosen based on the influential factors described in this chapter. Biological surveys are critical for understanding cave diversity, developing effective protective management programs, and answering fundamental research questions in biospeleology (Wynne et al. 2018; Mammola et al. 2020; Pacheco, Souza-Silva, et al. 2020). Assisting in this effort is the availability of powerful databases and geographic information systems that allow researchers to search and analyze complex databases (Culver et al. 2006; Miller 2007; Mammola 2019; Mammola et al. 2020).

In addition, biospeleologists should launch programs to track the presence and status of known cavernicoles, monitor habitat changes within sensitive caves, and document species losses. Cave researchers should also document alien species in caves and, when possible, describe their potential impacts. By monitoring and documenting both habitat changes and species invasions, extirpations, and extinctions, cave biologists will accrue the data to help determine the causes of decline for individual species. Knowing specific causes can be used to improve conservation programs to protect surviving species and communities.

Conservation programs should also place a premium on keeping common species common—most importantly, to maintain their function in the community and prevent substantial changes to the ecosystem (Deharveng and Bedos 2019). Secondarily, common species add aesthetic and scientific value to the occupied caves, as well as provide study species for education and field and laboratory research. All of which affords greater justification for continued protection for caves and their resources.

Table 1.1. Principal categories of caves with notes on speleogenesis and biology

Cave category	Formational processes	Biodiversity references
Caves formed by dissolution		
Subsurface limestone caves / Vadose caves	Slow solution of calcium carbonate ($CaCO_3$) by carbonic acid, aided by humic and fulvic acids, and dissolved in meteoric (or hypergenic) water sinking underground through fissures. Passages are often canyon shaped (Audra and Palmer 2015; Fig. 1.1).	Many; see text
Phreatic caves	Solution by mixing of groundwater at or below water table. Passages often cylindrical tubes.	Many; see text
Hypogene limestone caves	Solution by sulfuric acid upwelling from deep aquifers (Klimchouk 2009; Sendra et al. 2014; Colman et al. 2017).	Many; see text, except where caves do not connect with hypergenic water (Engel 2007; Sendra et al. 2014)
Aeolianite caves	Solution of lithified coralline sand dunes (calcarenite) by processes analogous to subsurface limestone caves. Often formed in mixohaline zone along coasts (Mylroie 2008; Mylroie and Mylroie 2011).	Anchialine fauna (Mylroie and Mylroie 2011)
Calcretes	Complex multiple cycles of dissolution and precipitation of limestone in arid regions over long time spans. Terrestrial habitats mainly shallow to deep mesocaverns (Halse 2018).	Highly speciose (Halse 2018)
Evaporite karst (aka gypsum caves)	Gypsum ($CaSO_4$ 2 H_2O) deposits most commonly occur where shallow seas have evaporated. This mineral is more soluble than limestone (Johnson 1996; Klimchouk 2019; White and Culver 2019).	Mazebedi and Hesselberg 2020
Salt karst / Halite caves	Large deposits of salt (NaCl) occur usually as evaporites. Since salt readily dissolves in water, caves formed are usually short-lived (Frumkin 2013).	Very few cavernicoles—caves are short-lived and the salt concentration is too high (Juberthie 2000)
Dolomite caves	Dolomite ($CaMg(CO_3)_2$) forms by replacement of some calcium by magnesium. It is less soluble than limestone, but over time large caves can form (Martinez and White 1999).	

Table 1.1. (continued)

Cave category	Formational processes	Biodiversity references
Quartzite sandstone caves	Silica (SiO_2) dissolves very slowly in groundwater; over time caves form by solution at crystal junctions, which weakens the mineral making it subject to erosion often by piping q.v. (Wray 2003; Barton et al. 2014).	Sharratt et al. 2000; Barton et al. 2014; Souza-Silva et al. 2020a
Other solution caves	Over geologic time scales, solution caves can form in many different types of rock, even in bedrock assumed to be resistant to solution, such as granites and gneiss (Osborne et al. 2013; Holler 2019).	Mostly trogloxenes, a few troglobionts (Souza-Silva et al. 2020b)
Caves formed mainly by erosion		
Littoral caves	Erosion of rock at point of weakness (fractures, faults, or bedding planes) by wave action (Bunnell 1988).	Mostly trogloxenes
Tectonic-crevice caves and slip-block caves	Mass movement of bedrock or regolith; subsidence of regolith into subsurface voids; slip block—crevasses and voids created by slippage along cliff faces with deep crack parallel to cliff face, which is often enlarged by subsequent erosion (Holler 2019).	Bats; roosting colonies and associated biota
Talus (aka solifluction caves)	Talus: voids among rockfall or fractured rock masses by solifluction and freeze-thaw cycles (i.e., accumulations of coarse clastic deposits with interstitial cavities; Holler 2019; Fig. 1.6).	Troglobionts, troglophiles, etc. (Olmi et al. 2014; Jiménez-Valverde et al. 2015; Souza-Silva et al. 2020b)
Piping caves (aka suffosion caves)	Mechanical removal of clays or clastics from weakly consolidated sediment from beneath a more resistant layer by movement of groundwater (Holler 2019).	Juberthie (2000)
Aeolian caves (aka tafone)	Shallow caves hollowed out of steeply sloping or vertical rockfaces initially by expansion of salt or ice crystals in crevices (haloclasty) that break the rock surface. Fragments are removed by wind (Roqué et al. 2011).	Trogloxene roosts

(*continued*)

Table 1.1. (continued)

Cave category	Formational processes	Biodiversity references
Caves formed by biological processes (biospeleogenesis)		
Iron ore caves	Exceptionally old, and their formation is complex (Ferreira et al. 2018). Iron ore deposits with Fe^{3+} are largely insoluble, but bacterial fermentation can reduce the iron to Fe^{2+}, which is more soluble (Parker et al. 2018; Auler et al. 2019). Other ferruginous caves form by slow dissolution of carbonate and silica intrusions (Ferreira et al. 2018).	Ferreira et al. 2018; Jaffé et al. 2018; Parker et al. 2018; Auler et al. 2019
Coral caves	Coral growth creates overhanging shelves and shelter caves (Holler 2019).	Refuge habitats for marine life
Tufa caves	Tufa is deposited from mineral-rich water emerging from springs and draining limestone-rich deposits. Deposition is enhanced by photosynthesizing organisms, which remove CO_2 from the water thus causing the $Ca(CO)_2$ to precipitate (Pedley et al. 2009).	Given shallow nature, only trogloxenes and troglophiles expected
Caves and voids formed by volcanism		
Lava tubes (aka pyroducts)	Roofing over molten basaltic lava flow channels. See text.	Many; see text
Voids between flow units / Poorly cemented and inflated flow units	Pāhoehoe basalt often advances at the flow front by amoeba-like toes, which poorly cement to the substrate. Escaping gas, sagging, and shrinkage on cooling can enlarge the void and create cavities ranging from less than one to a few meters in height and width. These occur in great numbers within pahoehoe lava flows (Fig. 1.3).	Subset of lava tubes; q.v.
Pressure ridge caves (aka tumuli)	Cooling surface of pāhoehoe flows buckle from the movement of lava underneath and form ridges several meters to tens of meters in height and triangular in cross section. They often partially drain creating small caves (Kauahikaua et al. 2003).	Subset of lava tubes; q.v.
Spatter cones, splatter ridges, hornitos	Vertical, conical structures formed by lava ejected through an opening in the crust of lava flow or upwelling of lava through the roof of a lava tube (Kauahikaua et al. 2003).	Subset of lava tubes; q.v.

Table 1.1. (continued)

Cave category	Formational processes	Biodiversity references
Cooling cracks and lava fractures	Molten basalt shrinks and breaks on cooling, creating cracks and voids.	Subset of lava tubes; q.v.
Gas vesicles	Escaping gas may form isolated bubbles as the lava cools; however, in pāhoehoe, these bubbles may coalesce and form an open-cell sponge (Kauahikaua et al. 2003).	A subset of lava tubes when interconnected; q.v.
ʻAʻā clinker	Buried ʻaʻā clinker contains numerous voids among the blocks of lava (Fig. 1.3).	Subset of lava tubes; q.v.
Vents and great cracks	Eruptive vents often drain creating deep complex pits. Deep earth cracks form along and parallel to rift zones on shield volcanoes. Both vents and deep cracks provide deep cave-like habitats.	Subset of lava tubes; q.v.
Tree molds	Trees buried in flow burn away leaving hollow mold.	Subset of lava tubes; q.v.
Cooling cracks	As lava flows cool and shrink, cooling cracks permeate through the flow increasing the interconnectedness of voids.	Subset of lava tubes; q.v.
Caves and voids formed within ice		
Glacier caves	Formed by surface meltwater sinking into the glacier through crevasses, moulins, and fissures. They are enlarged by geothermal melting, as well as by pressure and friction at the contact between the ice and bedrock (Giggenbach 1976; Smart 2006; Puccini and Mecchia 2013; Kováč 2018; Gulley and Fountain 2019).	Highly diverse communities of microorganisms; a few aquatic species; terrestrial species poorly sampled (Howarth 2021)

Furthermore, we recommend the adoption of standardized definitions of the biotic zones within caves and have proposed a system based on measurable biological and environmental parameters. Within most caves, the boundary between each zone is conspicuous to the experienced researcher (Howarth 1993). However, thorough descriptions of the zones sampled are rarely given in the published literature. As mentioned earlier, this is especially true for the entrance and twilight zones—the extent and boundaries of which are either not given or inconsistently applied, which makes comparisons difficult (e.g., Tobin et al. 2013; Wynne et al. 2019; Mazebedi and Hesselberg 2020).

Conservation management programs need to shift their priorities from individual caves to subterranean habitats. An abundance of data confirms that accessible cave passages are not the main habitat supporting subterranean biodiversity. Therefore, conservation programs and protected areas need to include sufficient areas around cave footprints as buffer areas. The size and location of the buffer area should be based on geologic and biotic factors that indicate the presence of subterranean voids, and on sufficient area of surface habitat to support functional cave ecosystems (Mammola et al. 2020; Rabelo et al. 2021). Conservation of cave clusters should be prioritized because geographic distance is often the main factor determining connectivity between troglobiontic communities (Jaffé et al. 2018).

ACKNOWLEDGMENTS

Pat Kambesis contributed to an early version of this chapter and made instructive suggestions concerning geology and figure composition. Paulo Alexandre Vieira Borges and Rodrigo Lopes Ferreira provided invaluable comments leading to the improvement of the work. The spider silhouette in Figure 1.6 provided courtesy of http://phylopic.org/.

REFERENCES

Ahearn, Gregory A., and Francis G. Howarth. 1982. "Physiology of Cave Arthropods in Hawaii." *Journal of Experimental Zoology* 222:227–238.

Audra, Philippe, and Arthur N. Palmer. 2015. "Research Frontiers in Speleogenesis. Dominant Processes, Hydrogeological Conditions and Resulting Cave Patterns." *Acta Carsologica* 44:315–348.

Auler, Augusto S., and Francesco Sauro. 2019. "Quartzite and Quartz Sandstone Caves of South America." In *Encyclopedia of Caves*, 3rd ed., edited by William B. White, David C. Culver, and Tanja Pipan, 850–860. Amsterdam: Elsevier, Academic Press.

Auler, Augusto S., Ceth W. Parker, Hazel A. Barton, and Gustavo A. Soares. 2019. "Iron Formation Caves: Genesis and Ecology." In *Encyclopedia of Caves*, 3rd ed., edited by William B. White, David C. Culver, and Tanja Pipan, 559–566. Amsterdam: Elsevier, Academic Press.

Avila-Flores, Rafael, and Rodrigo A. Medellín. 2004. "Ecological, Taxonomic, and Physiological Correlates of Cave Use by Mexican Bats." *Journal of Mammalogy* 85:675–687.

Bakalowicz, Michel J., Ford D. Clifford, T. E. Miller, Arthur N. Palmer, and Margaret V. Palmer. 1987. "Thermal Genesis of Dissolution Caves in the Black Hills, South Dakota." *Geological Society of America Bulletin* 99:729–738.

Barr, Thomas C., Jr. 1968. "Cave Ecology and the Evolution of Troglobites." In *Evolutionary Biology*, edited by Theodosius Dobzhansky, Max K. Hecht, and William C. Steere, 35–102. Boston: Springer.

Barr, Thomas C., Jr., and Robert A. Kuehne. 1971. "Ecological Studies in the Mammoth Cave System of Kentucky: II. The Ecosystem." *Annales de Spéléologie* 26:47–96.

Barton, Hazel A., and Diana Northup. 2007. "Geomicrobiology in Cave Environments: Past, Current and Future Perspectives." *Journal of Cave and Karst Studies* 69:163–178.

Barton, Hazel A., Juan G. Giarrizzo, Paula Suarez, Charles E. Robertson, Mark J. Broering, Eric D. Banks, et al. 2014. "Microbial Diversity in a Venezuelan Orthoquartzite Cave Is dominated by the Chloroflexi (Class Ktedonobacterales) and Thaumarchaeota Group I. 1c." *Frontiers in Microbiology* 5:615.

Benedict, Ellen M. 1979. "A New Species of *Apochthonius* Chamberlin from Oregon (Pseudoscorpionida, Chthoniidae)." *Journal of Arachnology* 7:79–83.

Blehert, David S., Alan C. Hicks, Melissa Behr, Carol U. Meteyer, Brenda M. Berlowski-Zier, Elizabeth L. Buckles, et al. 2009. "Bat White-nose Syndrome: An Emerging Fungal Pathogen?" *Science* 323:227.

Borges, Paulo A. V., Pedro Cardoso, Isabel R. Amorim, Fernando Pereira, Joao Paulo Constância, João C. Nunes, et al. 2012. "Volcanic Caves: Priorities for Conserving the Azorean Endemic Troglobiont Species." *International Journal of Speleology* 41:101–112.

Borgonie, Gaetan, Antonio García-Moyano, Derek Litthauer, Wim Bert, Armand Bester, Esta van Heerden, et al. 2011. "Nematoda from the Terrestrial Deep Subsurface of South Africa." *Nature* 744:79–82.

Bregović, Petra, and Maja Zagmajster. 2016. "Understanding Hotspots within a Global Hotspot—Identifying the Drivers of Regional Species Richness Patterns in Terrestrial Subterranean Habitats." *Journal of Insect Conservation and Diversity* 9:268–281.

Briggs, Thomas S., and Darrell Ubick. 1981. "Studies on Cave Harvestmen of the Central Sierra Nevada with Descriptions of New Species of *Banksula*." *Proceedings of the California Academy of Sciences* 42:315–322.

Brindle, Alan. 1980. "The Cavernicolous Fauna of Hawaiian Lava Tubes. 12. A New Species of Blind Troglobitic Earwig (Dermaptera: Carcinophoridae), with a Revision of the Related Surface-Living Earwigs of the Hawaiian Islands." *Pacific Insects* 21:261–274.

Brunet, Anja K., and Rodrigo A. Medellín. 2001. "The Species-Area Relationship in Bat Assemblages of Tropical Caves." *Journal of Mammalogy* 82:1114–1122.

Bunnell, David E. 1988. *Sea Caves of Santa Cruz Island*. Santa Barbara: McNally & Loftin.

Castaño-Sánchez, Andrea, Grant C. Hose, and Ana Sofia P. S. Reboleira. 2020. "Ecotoxicological Effects of Anthropogenic Stressors in Subterranean Organisms: A Review." *Chemosphere* 244:125422.

Chevaldonné, P., and C. Lejeusne. 2003. "Regional Warming-Induced Species Shift in North-West Mediterranean Marine Caves." *Ecology Letters* 6:371–379.

Christman, Mary C., and David C. Culver. 2001. "The Relationship between Cave Biodiversity and Available Habitat." *Journal of Biogeography* 28:367–380.

Christman, Mary C., Daniel H. Doctor, Matthew L. Niemiller, David J. Weary, John A. Young, Kirk S. Zigler, et al. 2016. "Predicting the Occurrence of Cave-Inhabiting Fauna Based on Features of the Earth Surface Environment." *PLOS ONE* 11:e0160408.

Cigna, Arrigo A. 2016. "Tourism and Show Caves." *Zeitschrift für Geomorphologie* 60:217–233.

Colman, Daniel R., Saroj Poudel, Blake Stamps, Eric S. Boyd, and John R. Spear. 2017. "The Deep, Hot Biosphere: A Retrospection." *Proceedings of the National Academy of Sciences* 114:6895–6903.

Croom, Henrietta B., Shelley James, Rodrigo Velasquez Gonzalez, Fred Stone, Francis G. Howarth, and Anna Karsin. 2008. "P-43. Hawaii's Nemobiine Crickets: Unique Components of Caves and High-Energy Marine Ecosystems." Poster presentation at the Hawaii Conservation Conference. July 29–31, 2008. Hawai'i Convention Center, Honolulu, HI.

Culver, David C. 2008. "The Struggle to Measure Subterranean Biodiversity." In *Frontiers of Karst Research Karst Waters Institute Special Publication* 13, edited by Jonathan B. Martin and William B. White, 54–59. Leesburg, Virginia.

Culver, David C., and Tanja Pipan. 2009. *The Biology of Caves and Other Subterranean Habitats*. Oxford: Oxford University Press.

Culver, David C., and Tanja Pipan. 2013. "Subterranean Ecosystems." In *Encyclopedia of Biodiversity*, 2nd ed., edited by Simon A. Levin, 49–62. Waltham, MA: Academic Press.

Culver, David C., and Tanja Pipan. 2014. *Shallow Subterranean Habitats: Ecology, Evolution, and Conservation*. Oxford: Oxford University Press.

Culver, David C., and Boris Sket. 2000. "Hotspots of Subterranean Biodiversity in Caves and Wells." *Journal of Cave and Karst Studies* 62:11–17.

Culver, David C., Mary C. Christman, Boris Sket, and Peter Trontelj. 2004. "Sampling Adequacy in an Extreme Environment: Species Richness Patterns in Slovenian Caves." *Biodiversity and Conservation* 13:1209–1229.

Culver, David C., Louis Deharveng, Anne Bedos, Julian J. Lewis, Molly Madden, James R. Reddell, et al. 2006. "The Mid-Latitude Biodiversity Ridge in Terrestrial Cave Fauna." *Ecography* 29:120–128.

Culver, David [C.], John R. Holsinger, and Roger Baroody. 1973. "Toward a Predictive Cave Biogeography: The Greenbrier Valley as a Case Study." *Evolution* 27:689–695.

Culver, David C., Peter Trontelj, Maja Zagmajster, and Tanja Pipan. 2013. "Paving the Way for Standardized and Comparable Subterranean Biodiversity Studies." *Subterranean Biology* 10:43–50.

Deharveng, Louis, and Anne Bedos. 2018. "Diversity of Terrestrial Invertebrates in Subterranean Habitats." In *Cave Ecology*, edited by Oana Teodora Moldovan, Ľubomír Kováč, and Stuart Halse, 107–172. Switzerland: Springer, Nature.

Deharveng, Louis, and Anne Bedos. 2019. "Biodiversity in the Tropics." In *Encyclopedia of Caves*, 3rd ed., edited by William B. White, David C. Culver, and Tanja Pipan, 146–162. Amsterdam: Elsevier, Academic Press.

de Paula, Caio César Pires, Maria Elina Bichuette, and Mirna Helena Regali Seleghim. 2020. "Nutrient Availability in Tropical Caves Influences the Dynamics of Microbial Biomass." *MicrobiologyOpen* 2020:e1044.

Desutter-Grandcolas, Laure. 1993. "The Cricket Fauna of Chiapanecan Caves (Mexico): Systematics, Phylogeny and the Evolution of Troglobitic Life (Orthoptera, Grylloidea, Phalangopsidae, Luzarinae)." *International Journal of Speleology* 22:1–82.

Dirzo, Rodolfo, Hillary S. Young, Mauro Galetti, Gerardo Ceballos, Nick J. B. Isaac, and Ben Collen. 2014. "Defaunation in the Anthropocene." *Science* 345:401–406.

Engel, Annette S. 2007. "Observations on the Biodiversity of Sulfidic Karst Habitats." *Journal of Cave and Karst Studies* 69:187–206.

Fernandes, Camile Sorbo, Marco Antonio Batalha, and Maria Elina Bichuette. 2016. "Does the Cave Environment Reduce Functional Diversity?" *PLOS ONE* 11:e0151958.

Ferreira, Rodrigo Lopes. 2019. "Guano Communities." In *Encyclopedia of Caves*, 3rd ed., edited by William B. White, David C. Culver, and Tanja Pipan, 474–484. Amsterdam: Elsevier, Academic Press.

Ferreira, Rodrigo Lopes, and Thais Giovannini Pellegrini. 2019. "Species-Area Model Predicting Diversity Loss in an Artificially Flooded Cave in Brazil." *International Journal of Speleology* 48:155–165.

Ferreira, Rodrigo Lopes, Gonzalo Giribet, Gerhard Du Preez, Oresti Ventouras, Charlene Janion, and Marconi Souza Silva. 2020. "The Wynberg Cave System, the Most Important Site for Cave Fauna in South Africa at Risk." *Subterranean Biology* 36:73–81.

Ferreira, Rodrigo Lopes, Rogério Parentoni Martins, and Douglas Yanega. 2000. "Ecology of Bat Guano Arthropod Communities in a Brazilian Dry Cave." *Ecotropica* 6:105–116.

Ferreira, Rodrigo Lopes, Marcus Paulo Alves de Oliveira, and Marconi Souza Silva. 2018. "Subterranean Biodiversity in Ferruginous Landscapes." In *Cave Ecology*, edited by Oana Teodora Moldovan, Ľubomír Kováč, and Stuart Halse, 415–434. Switzerland: Springer, Nature.

Ficetola, Gentile Francesco, Claudia Canedoli, and Fabio Stoch. 2019. "The Racovitzan Impediment and the Hidden Biodiversity of Unexplored Environments." *Conservation Biology* 33:214–216.

Flot, Jean-François, Gert Wörheide, and Sharmishtha Dattagupta. 2010. "Unsuspected Diversity of *Niphargus* Amphipods in the Chemoautotrophic Cave Ecosystem of Frasassi, Central Italy." *BMC Evolutionary Biology* 10:171.

Frumkin, Amos. 2013. "Salt Karst." In *Treatise in Geomorphology*, vol. 6, *Karst Geomorphology*, edited by Amos Frumkin, 407–424. San Diego: Elsevier, Academic Press.

Gagné, Wayne C., and Francis G. Howarth. 1975. "The Cavernicolous Fauna of Hawaiian Lava Tubes, 7. Emesinae or Thread-Legged Bugs (Heteroptera: Reduviidae)." *Pacific Insects* 16:416–426.

Gibert, Janine, and Louis Deharveng. 2002. "Subterranean Ecosystems: A Truncated Functional Biodiversity." *BioScience* 52:473–481.

Giggenbach, Werner F. 1976. "Geothermal Ice Caves on Mt Erebus, Ross Island, Antarctica." *New Zealand Journal of Geology and Geophysics* 19:365372.

Gnaspini, Pedro, and Eleonora Trajano. 2000. "Guano Communities in Tropical Caves." In *Ecosystems of the World: Subterranean Ecosystems*, edited by Horst Wilkens, David C. Culver, and William F. Humphreys, 251–268. Amsterdam: Elsevier.

Gould, Edwin. 1988. "Wing-Clapping Sounds of *Eonycteris spelaea* (Pteropodidae) in Malaysia." *Journal of Mammalogy* 69:378–379.

Graening, G. O., Danté B. Fenolio, and Michael E. Slay. 2012. *Cave Life of Oklahoma and Arkansas*. Norman: University of Oklahoma Press.

Graening, G. O., Michael J. Harvey, William L. Puckette, Richard C. Stark, D. Blake Sasse, Steve L. Hensley, et al. 2017. "Conservation Status of the Endangered Ozark Big-Eared Bat (*Corynorhinus townsendii ingens*)—a 34-Year Assessment." *Oklahoma Biological Survey* 11:1–16.

Graening, G. O., Michael E. Slay, and Chuck Bitting. 2006. "Cave Fauna of the Buffalo National River." *Journal of Cave and Karst Studies* 68:153–163.

Gulley, Jason D., and Andrew G. Fountain. 2019. "Glacier Caves." In *Encyclopedia of Caves*, 3rd ed., edited by William B. White, David C. Culver, and Tanja Pipan, 468–473. Amsterdam: Elsevier, Academic Press.

Gurney, Ashley B., and David C. Rentz. 1978. "The Cavernicolous Fauna of Hawaiian Lava Tubes, 10. Crickets (Orthoptera, Gryllidae)." *Pacific Insects* 18:85–103.

Halse, Stuart A. 2018. "Research in Calcretes and Other Deep Subterranean Habitats outside Caves." In *Cave Ecology*, edited by Oana Teodora Moldovan, Ľubomír Kováč, and Stuart Halse, 415–434. Switzerland: Springer, Nature.

Hoch, Hannelore, and Francis G. Howarth. 1989. "Six New Cavernicolous Cixiid Planthoppers in the Genus *Solonaima* from Australia (Homoptera: Fulgoroidea)." *Systematic Entomology* 14:377–402.

Hoch, Hannelore, and Francis G. Howarth. 1999. "Multiple Cave Invasions by Species of the Planthopper Genus *Oliarus* in Hawaii (Homoptera: Fulgoroidea: Cixiidae)." *Zoological Journal of the Linnean Society* 127:453–475.

Holler, Cato. 2019. "Pseudokarst." In *Encyclopedia of Caves*, 3rd ed., edited by William B. White, David C. Culver, and Tanja Pipan, 836–849. Amsterdam: Elsevier, Academic Press.

Howarth, Francis G. 1981. "Community Structure and Niche Differentiation in Hawaiian Lava Tubes." In *Island Ecosystems: Biological Organization in Selected Hawaiian Communities, US/IBP Synthesis Series*, vol. 15, edited by Dieter Mueller-Dombois, Kent W. Bridges, and Hampton L. Carson, 318–336. Pennsylvania: Hutchinson Ross Publishing Co.

Howarth, Francis G. 1982. "Bioclimatic and Geologic Factors Governing the Evolution and Distribution of Hawaiian Cave Insects." *Entomologia Generalis* 8:17–26.

Howarth, Francis G. 1983. "Ecology of Cave Arthropods." *Annual Review of Entomology* 28:365–389.

Howarth, Francis G. 1993. "High-Stress Subterranean Habitats and Evolutionary Change in Cave-Inhabiting Arthropods." *American Naturalist* 142:S65–S77.

Howarth, Francis G., 1996. "A Comparison of Volcanic and Karstic Cave Communities." *Proceedings of 7th International Symposium on Volcanospeleology*, edited by Pedro Oromi, 63–68.

Howarth, Francis G. 2019. "Adaptive Shifts." In *Encyclopedia of Caves*, 3rd ed., edited by William B. White, David C. Culver, and Tanja Pipan, 47–55. Amsterdam: Elsevier, Academic Press.

Howarth, Francis G. 2000. "Non-target Effects of Biological Control Agents." In *Measures of Success in Biological Control*, edited by Geoff M. Gurr and Steve D. Wratten, 369–403. Dordrecht: Kluwer Academic Publishing.

Howarth, Francis G. 2021. "Glacier Caves: A Globally Threatened Subterranean Biome." *Journal of Cave and Karst Studies* 83. https://doi.org/10.4311/2019LSC0132.

Howarth, Francis G., and Oana T. Moldovan. 2018a. "The Ecological Classification of Cave Animals and Their Adaptations." In *Cave Ecology*, edited by Oana Teodora Moldovan, Ľubomír Kováč, and Stuart Halse, 41–67. Switzerland: Springer, Nature.

Howarth, Francis G., and Oana T. Moldovan. 2018b. "Where Cave Animals Live." In *Cave Ecology*, edited by Oana Teodora Moldovan, Ľubomír Kováč, and Stuart Halse, 23–37. Switzerland: Springer, Nature.

Howarth, Francis G., and Janet Moore. 1984. “The Land Nemertine *Argonemertes dendyi* (Dakin) in Hawaii (Nemertinea: Hoplonemertinea: Prosorhochmidae).” *Pacific Science* 37:141–144.

Howarth, Francis G., and Fred D. Stone. 1990. “Elevated Carbon Dioxide Levels in Bayliss Cave, Australia: Implications for the Evolution of Obligate Cave Species.” *Pacific Science* 44:207–218.

Howarth, Francis G., and Fred D. Stone. 2020. “Impacts of Invasive Rats on Hawaiian Cave Resources.” *International Journal of Speleology* 49:35–42.

Howarth, Francis G., Shelly A. James, David J. Preston, and Clyde T. Imada. 2007. “Identification of Roots in Lava Tube Caves Using Molecular Techniques: Implications for Conservation of Cave Arthropod Faunas.” *Journal of Insect Conservation* 11:251–261.

Howarth, Francis G., Matthew J. Medeiros, and Fred D. Stone. 2020. “Hawaiian Lava Tube Cave Associated Lepidoptera from the Collections of Francis G. Howarth and Fred D. Stone.” *Bishop Museum Occasional Papers* 129:37–54.

Ireland, Robert R., and Gilda Bellolio. 2002. “The Mosses of Easter Island.” *Tropical Bryology* 21:11–20.

Iskali, Goniela, and Yixin Zhang. 2015. “Guano Subsidy and the Invertebrate Community in Bracken Cave: The World’s Largest Colony of Bats.” *Journal of Cave & Karst Studies* 77:28–36.

Jaffé, Rodolfo, Xavier Prous, Allan Calux, Markus Gastauer, Gilberto Nicacio, Robson Zampaulo, et al. 2018. “Conserving Relics from Ancient Underground Worlds: Assessing the Influence of Cave and Landscape Features on Obligate Iron Cave Dwellers from the Eastern Amazon.” *PeerJ* 6:e4531.

Jiménez-Valverde, Alberto, José D. Gilgado, Alberto Sendra, Gonzalo Pérez-Suárez, Juan J. Herrero-Borgoñón, and Vicente M. Ortuño. 2015. “Exceptional Invertebrate Diversity in a Scree Slope in Eastern Spain.” *Journal Insect Conservation* 19:713–728.

Jiménez-Valverde, Alberto, Alberto Sendra, Policarp Garay, and Ana Sofia P. S. Reboleira. 2017. “Energy and Speleogenesis: Key Determinants of Terrestrial Species Richness in Caves.” *Ecology and Evolution* 7:10207–10215.

Johnson, Kenneth S. 1996. “Gypsum Karst in the United States.” *International Journal of Speleology* 25:167–193.

Juan, Carlos, Michelle T. Guzik, Damià Jaume, and Steven J. B. Cooper. 2010. “Evolution in Caves: Darwin’s ‘Wrecks of Ancient Life’ in the Molecular Era.” *Molecular Ecology* 19:3865–3880.

Juberthie, Christian. 1983. “Le Milieu Souterrain: Étendue et Composition.” *Mémoires de Biospéologiques* 10:17–65.

Juberthie, Christian. 2000. “The Diversity of the Karstic and Pseudokarstic Hypogean Habitats in the World.” In *Ecosystems of the World: Subterranean Ecosystems*, edited by Horst Wilkens, David C. Culver, and William F. Humphreys, 17–39. Amsterdam: Elsevier.

Juberthie, Christian, Bernard Delay, and Michael Bouillon. 1980. Extension du Milieu Souterrain Superficiel en Zone Non-Calcaire: Description d’un Nouveau Milieu et de son Peuplement par les Coleopteres Troglobies. *Mémoires de Biospéologie* 7:19–52.

Kauahikaua, Jim, Francis G. Howarth, and Kenneth Hon. 2009. “Lava Tubes.” In *Encyclopedia of Islands*, edited by Rosemary Gillespie and David A. Clague, 544–549. Berkeley: University of California Press.

Kauahikaua, Jim, David R. Sherrod, Katharine V. Cashman, and Christina Heliker. 2003. "Hawaiian Lava-Flow Dynamics during the Pu'u 'Ō'ō-Kūpaianaha Eruption: A Tale of Two Decades." *U.S. Geological Survey Professional Paper* 1676:63–87.

Klimchouk, Alexander B. 2009. "Principal Features of Hypogene Speleogenesis." In *Hypogene Speleogenesis and Karst Hydrogeology of Artesian Basins,* edited by Alexander B. Klimchouk and Derek Ford, 7–15. Simferopol: Ukrainian Institute of Speleology and Karstology.

Klimchouk, Alexander B. 2019. "Gypsum Caves." In *Encyclopedia of Caves*, 3rd ed., edited by William B. White, David C. Culver, and Tanja Pipan, 485–495. Amsterdam: Elsevier, Academic Press.

Kováč, Ľubomír. 2018. "Ice Caves." In *Cave Ecology*, edited by Oana Teodora Moldovan, Ľubomír Kováč, and Stuart Halse, 331–349. Switzerland: Springer, Nature.

Krejca, Jean K., and Butch Weckerly. 2008. "Detection Probabilities of Karst Invertebrates." In *Proceedings of the Eighteenth National Cave and Karst Management Symposium, St. Louis, Missouri, USA, 8–12 October 2007*, edited by William R. Elliott, 283–289. Austin: Texas Parks and Wildlife Department.

Lavoie, Kathleen. H. 1981. "Invertebrate Interactions with Microbes during the Successional Decomposition of Dung." In *Proceedings 8th International Congress Speleology. Bowling Green, Kentucky*, edited by B. F. Beck, 265–266.

Leithauser, Arthur T., and John R. Holsinger. 1985. *Ecological Analysis of the Kentucky Cave Shrimp, Palaemonias ganteri* Hay, *at Mammoth Cave National Park (Phase V).* Unpublished final report submitted to US Department of Interior, National Park Service, Contract Number CX-5000-1-1037.

Leopardi, Stefania, Damer Blake, and Sébastien J. Puechmaille. 2015. "White-nose Syndrome Fungus Introduced from Europe to North America." *Current Biology* 25:217–219.

Lindsey, Alton A. 1951. "Vegetation and Habitats in a Southwestern Volcanic Area." *Ecological Monographs* 21:227–253.

Lunghi, Enrico, and Helena Bilandžija. 2022. "Longevity in Cave Animals." *Frontiers in Ecology and Evolution* 10:874123.

Lunghi, Enrico, and Raoul Manenti. 2020. "Cave Communities: From the Surface Border to the Deep Darkness." *Diversity* 12:167–171.

Lunghi, Enrico, Gentile Francesco Ficetola, Yahui Zhao, and Raoul Manenti. 2020. "Are the Neglected Tipuloidea Crane Flies (Diptera) an Important Component for Subterranean Environments?" *Diversity* 12:333.

Mammola, Stefano. 2019. "Finding Answers in the Dark: Caves as Models in Ecology Fifty Years after Poulson and White." *Ecography* 42:1331–1351.

Mammola, Stefano, Isabel R. Amorim, Maria E. Bichuette, Paulo A. V. Borges, Naowarat Cheeptham, Steven J. B. Cooper, et al. 2020. "Fundamental Research Questions in Subterranean Biology." *Biological Reviews* 95:1855–1872.

Mammola, Stefano, Pedro Cardoso, Dorottya Angyal, Gergely Balázs, Theo Blick, Hervé Brustel, et al. 2019. "Local- versus Broad-Scale Environmental Drivers of Continental β-diversity Patterns in Subterranean Spider Communities across Europe." *Proceedings of the Royal Society B: Biological Sciences* 286:1–9.

Mammola, Stefano, Elena Piano, Pedro Cardoso, Philippe Vernon, David Domínguez-Villar, David C. Culver, et al. 2019. "Climate Change Going Deep: The Effects of Global Climatic Alterations on Cave Ecosystems." *The Anthropocene Review* 6:98–116.

Martinez, Myrna I., and William B. White. 1999. "A Laboratory Investigation of the Relative Dissolution Rates of the Lirio Limestone and the Isla de Mona Dolomite and Implications for Cave and Karst Development on Isla de Mona." *Journal of Cave and Karst Studies* 61:7–12.

Mazebedi, Richard, and Thomas Hesselberg. 2020. "A Preliminary Survey of the Abundance, Diversity and Distribution of Terrestrial Macroinvertebrates of Gcwihaba Cave, Northwest Botswana." *Subterranean Biology* 35:49–63.

McClure, Meredith L., Daniel Crowley, Catherine G. Haase, Liam P. McGuire, Nathan W. Fuller, David T. S. Hayman, et al. 2020. "Linking Surface and Subterranean Climate: Implications for the Study of Hibernating Bats and Other Cave Dwellers." *Ecosphere* 11:e03274.

Medeiros, Matthew J., Don Davis, Francis G. Howarth, and Rosemary Gillespie. 2009. "Evolution of Cave Living in Hawaiian *Schrankia* (Lepidoptera: Noctuidae) with Description of a Remarkable New Cave Species." *Zoological Journal of the Linnean Society* 156:114–139.

Miller, Scott E. 2007. "DNA Barcoding and the Renaissance of Taxonomy." *Proceedings National Academy of Science* 104:4775–4776.

Monro, Alexandre K., Nadia Bystriakova, Longfei Fu, Fang Wen, and Yigang Wei. 2018. "Discovery of a Diverse Cave Flora in China." *PLOS ONE* 13:e0190801.

Moseley, Max. 2009. "Size Matters: Scalar Phenomena and a Proposal for an Ecological Definition of 'Cave.'" *Cave and Karst Science* 35:89–94.

Moseley, Max, Teckwyn Lim, and Lim T. Tshen. 2012. "Fauna Reported from Batu Caves, Selangor, Malaysia: Annotated Checklist and Bibliography." *Cave and Karst Science* 39:77–92.

Moser, Duane P., Thomas M. Gihring, Fred J. Brockman, James K. Fredrickson, David L. Balkwill, Michael E. Dollhopf, et al. 2005. "*Desulfotomaculum* and *Methanobacterium* spp. Dominate a 4- to 5-Kilometer-Deep Fault." *Applied and Environmental Microbiology* 71:8773–8783.

Mueller-Dombois, Dieter, and Francis G. Howarth. 1981. "Niche and Life-Form Integration in Island Communities." In *Island Ecosystems: Biological Organization in Selected Hawaiian Communities, US/IBP Synthesis Series,* edited by Dieter Mueller-Dombois, Kent W. Bridges, and Hampton L. Carson, 337–364. Pennsylvania: Hutchinson Ross Publishing Company.

Mylroie, John E. 2008. "Late Quaternary Sea-Level Position: Evidence from Bahamian Carbonate Deposition and Dissolution Cycles." *Quaternary International* 183:61–75.

Mylroie, John E., and Mylroie Joan R. 2011. "Void Development on Carbonate Coasts: Creation of Anchialine Habitats." *Hydrobiologia* 677:15–32.

Naaman, I. 2011. "The Karst System and the Ecology of Ayalon Cave, Israel." PhD diss., Hebrew University of Jerusalem.

Northup, Diana E., and W. Calvin Welbourn. 1997. "Life in the Twilight Zone—Lava Tube Ecology, Natural History of El Malpais National Monument." *New Mexico Bureau of Mines and Mineral Resources Bulletin* 156:69–82.

Northup, D. E., L. A. Melim, M. N. Spilde, J. J. M. Hathaway, M. G. Garcia, M. Moya, et al. 2011. "Lava Cave Microbial Communities within Mats and Secondary Mineral Deposits: Implications for Life Detection on Other Planets." *Astrobiology* 11:1–18.

Olmi, Massimo, Toshiharu Mita, and Adalgisa Guglielmino. 2014. "Revision of the Embolemidae of Japan (Hymenoptera: Chrysidoidea), with Description of a New Genus and Two New Species." *Zootaxa* 3793:423–440.

Osborne, L. Armstrong R., Wasantha S. Weliange, Pathmakumara Jayasingha, A. S. Dandeniya, A. K. Prageeth P. Algiriya, and Ross E. Pogson. 2013. "Caves and Karst-Like Features in Proterozoic Gneiss and Cambrian Granite, Southern and Central Sri Lanka: An Introduction." *Acta Carsologica* 42:25–48.

Otte, Dan. 1994. *The Crickets of Hawaii: Origin, Systematics and Evolution.* Philadelphia: Orthopterists' Society.

Pacheco, Gabrielle Soares Muniz, Marcus Paulo Alves de Oliveira, Enio Cano, Marconi Souza-Silva, and Rodrigo Lopes Ferreira. 2020. "Tourism Effects on the Subterranean Fauna in a Central American Cave." *Insect Conservation and Diversity* 1–13. https://doi.org/10.1111/icad.12451.

Pacheco, Gabrielle Soares Muniz, Marconi Souza-Silva, Enio Cano, and Rodrigo Lopes Ferreira. 2020. "The Role of Microhabitats in Structuring Cave Invertebrate Communities in Guatemala." *International Journal of Speleology* 49:161–169.

Palacios-Vargas, José G., Gabriela Castaño-Meneses, and Daniel A. Estrada. 2011. "Diversity and Dynamics of Microarthropods from Different Biotopes of Las Sardinas Cave (Mexico)." *Subterranean Biology* 9:113–126.

Parker, Ceth W., Augusto S. Auler, Michael D. Barton, Ira D. Sasowsky, John M. Senko, and Hazel A. Barton. 2018. "Fe(III) Reducing Microorganisms from Iron Ore Caves Demonstrate Fermentative Fe(III) Reduction and Promote Cave Formation." *Geomicrobiology Journal* 35:311–322.

Peck, Stewart B. 1970. "The Terrestrial Arthropod Fauna of Florida Caves." *Florida Entomologist* 53:203–207.

Pedley, Martyn, Mike Rogerson, and Richard Middleton. 2009. "Freshwater Calcite Precipitates from In Vitro Mesocosm Flume Experiments: A Case for Biomediation of Tufas." *Sedimentology* 56:511–527.

Pellegrini, Thais Giovannini, Lilian Patrícia Sales, Polyanne Aguiar, and Rodrigo Lopes Ferreira. 2016. "Linking Spatial Scale Dependence of Land-Use Descriptors and Invertebrate Cave Community Composition." *Subterranean Biology* 18:17–38.

Pennay, Michael. 2008. "A Maternity Roost of the Large-Eared Pied Bat *Chalinolobus dwyeri* (Ryan) (Microchiroptera: Vespertilionidae) in Central New South Wales Australia." *Australian Zoologist* 34:564–569.

Peterson, Donald W., and Donald A. Swanson. 1974. "Observed Formation of Lava Tubes during 1970–71 at Kilauea Volcano, Hawaii." *Studies in Speleology* 2:209–222.

Peterson, Donald W., Robin T. Holcomb, Robert I. Tilling, and Robert L. Christiansen. 1994. "Development of Lava Tubes in the Light of Observations at Mauna Ulu, Kīlauea Volcano, Hawaii." *Bulletin of Volcanology* 56:343–360.

Pipan, Tanja, and David C. Culver. 2013. "Forty Years of Epikarst: What Biology Have We Learned?" *International Journal of Speleology* 42:215–223.

Pipan Tanja, David C. Culver, and Papi F. Kozel. 2018. "Partitioning Diversity in Subterranean Invertebrates: The Epikarst Fauna of Slovenia." *PLOS ONE* 13:e0195991.

Polyak, Victor J., and Necip Güven. 2000. "Clays in Caves of the Guadalupe Mountains, New Mexico." *Journal of Cave and Karst Studies* 62:120–126.

Polyak, Victor J., and Paula Provencio. 2000. "By-Product Materials Related to H2S-H2SO4-Influenced Speleogenesis of Carlsbad, Lechuguilla, and Other Caves of the Guadalupe Mountains, New Mexico." *Journal of Cave and Karst Studies* 63:23–32.

Porter, Megan L. 2007. "Subterranean Biogeography: What Have We Learned from Molecular Techniques?" *Journal of Cave and Karst Studies* 69:179–186.

Price, J. Jordan, Kevin P. Johnson, and Dale H. Clayton. 2004. "The Evolution of Echolocation in Swiftlets." *Journal of Avian Biology* 35:135–143.

Prous, Xavier, Rodrigo Lopes Ferreira, and Claudia M. Jacobi. 2015. "The Entrance as a Complex Ecotone in a Neotropical Cave." *International Journal of Speleology* 44:177–189.

Prous, Xavier, Rodrigo Lopes Ferreira, and Rogério Parentoni Martins. 2004. "Ecotone Delimitation: Epigean–Hypogean Transition in Cave Ecosystems." *Austral Ecology* 29:374–382.

Puccini, Leonardo, and Marco Mecchia. 2013. "Englacial Caves of Glaciar Perito Moreno and Glaciar Ameghino, Patagonia (Argentina)." In *Proceedings of the 16th International Congress Speleology, Czech Republic, Brno, July 21–28, 2013*, vol. 3, edited by Michal Filippi and Pavel Bosák, 292–297.

Rabelo, Lucas Mendes, Marconi Souza-Silva, and Rodrigo Lopes Ferreira. 2021. "Epigean and Hypogean Drivers of Neotropical Subterranean Communities." *Journal of Biogeography* 1–14. https:/doi.org/10.1111/jbi.14031.

Raedts, Caren, and Chris Smart. 2015. "Tracking of Karst Contamination Using Digital Mapping Technologies: Hidden River Cave Kentucky." *Proceedings of the 14th Multidisciplinary Conference on Sinkholes and the Engineering and Environmental Impacts of Karst*, edited by Daniel H. Doctor, Lewis Land, and J. Brad Stephenson, 327–336. National Cave and Karst Research Institute (NCKRI) Symposium 5.

Raven, Peter, and Scott Miller. 2020. "Here Today, Gone Tomorrow: Editorial." *Science* 370:149.

Reeves, Will K. 1999. "Exotic Species in North American Caves." In *Proceedings of the 1999 National Cave and Karst Management Symposium*, 164–166. Chattanooga: Southeastern Cave Conservancy.

Reeves, Will K. 2001. "Invertebrate and Slime Mold Cavernicoles of Santee Cave, South Carolina, USA." *Proceedings of the Academy of Natural Sciences of Philadelphia* 151:81–85.

Resende, L. P. A., and Maria Elina Bichuette. 2016. "Sharing the Space: Coexistence among Terrestrial Predators in Neotropical Caves." *Journal of Natural History* 50:2107–2128.

Robinson, D. J. 1978. "The Glenlyon Region, Some Facets of Its History." *Memoirs of Queensland Museum* 19:5–16.

Robinson, Gaden S. 1980. "Cave-Dwelling Tineid Moths: A Taxonomic Review of the World Species (Lepidoptera: Tineidae)." *Transactions British Cave Research Association* 7:83–120.

Romero, Aldemaro. 2006a. "Adaptation: Behavioural." In *Encyclopedia of Caves and Karst Science*, edited by John Gunn, 4–7. London: Routledge.

Romero, Aldemaro. 2006b. "Biospeleologists." In *Encyclopedia of Caves and Karst Science*, edited by John Gunn, 313–318. London: Routledge.

Romero, Aldemaro. 2009. *Cave Biology: Life in Darkness*. Cambridge: Cambridge University Press.

Roqué, Carles, Rogelio Linares, Roberto Rodríguez, and Mario Zarroca. 2011. "Granite Caves in the North-East of the Iberian Peninsula: Artificial Hypogea versus Tafone." *Zeitschrift für Geomorphologie* 55:341–364.

Rosli, Qhairil Shyamri, F. A. A. Kahn, Muhd Amsyari Morni, J. William-Dee, Roberta Chaya Tawie Tingga, and A. R. Mohd-Ridwan. 2018. "Roosting Behaviour and Site Mapping of Cave Dwelling Bats in Wind Cave Nature Reserve, Bau, Sarawak, Malaysian Borneo." *Malaysian Applied Biology Journal* 47:231–238.

Sarbu, Serban M., Thomas C. Kane, and Brian K. Kinkle. 1996. "A Chemoautotrophically Based Cave Ecosystem." *Science* 272:1953–1955.

Schneider, Katie, and David C. Culver. 2004. "Estimating Subterranean Species Richness Using Intensive Sampling and Rarefaction Curves in a High Density Cave Region in West Virginia." *Journal of Cave and Karst Studies* 66:39–45.

Sendra, Alberto, and Ana Sofia P. S. Reboleira. 2012. "The World's Deepest Subterranean Community—Krubera-Voronja Cave (Western Caucasus)." *International Journal of Speleology* 41:221–230.

Sendra, Alberto, Policarp Garay, Vicente M. Ortuño, José D. Gilgado, Santiago Teruel, and Ana Sofia P. S. Reboleira. 2014. "Hypogenic versus Epigenic Subterranean Ecosystem: Lessons from Eastern Iberian Peninsula." *International Journal of Speleology* 43:253–264.

Sharratt, Norma J., Mike D. Picker, and Michael J. Samways. 2000. "The Invertebrate Fauna of the Sandstone Caves of the Cape Peninsula (South Africa): Patterns of Endemism and Conservation Priorities." *Biodiversity and Conservation* 9:107–143.

Sherwin, Richard E., Dave Stricklan, and Duke S. Rogers. 2000. "Roosting Affinities of Townsend's Big-Eared Bat (*Corynorhinus townsendii*) in Northern Utah." *Journal of Mammalogy* 81:939–947.

Simões, Matheus Henrique, Souza-Silva Marconi, and Rodrigo Lopes Ferreira. 2015. "Cave Physical Attributes Influencing the Structure of Terrestrial Invertebrate Communities in Neotropics." *Subterranean Biology* 16:103–121.

Sket, Boris. 2004. "The Cave Hygropetric—A Little Known Habitat and Its Inhabitants." *Archiv für Hydrobiologie* 160:413–425.

Sket, Boris. 2008. "Can We Agree on an Ecological Classification of Subterranean Animals?" *Journal of Natural History* 42:1549–1563.

Smart, Chris. 2006. "Glacier Caves and Glacier Pseudokarst." In *Encyclopedia of Caves and Karst Science*, edited by John Gunn, 385–387. London: Routledge.

Smrž, Jaroslav, Ľubomír Kováč, Jaromír Mikeš, Vladimír Šustr, Alena Lukešová, Karel Tajovsky, et al. 2015. "Food Sources of Selected Terrestrial Cave Arthropods." *Subterranean Biology* 16:37–46.

Souza-Silva, Marconi, Luiz Felipe Moretti Iniesta, and Rodrigo Lopes Ferreira. 2020a. "Invertebrates Diversity in Mountain Neotropical Quartzite Caves: Which Factors Can Influence the Composition, Richness, and Distribution of the Cave Communities?" *Subterranean Biology* 33:23–43.

Souza-Silva, Marconi, Luiz Felipe Moretti Iniesta, and Rodrigo Lopes Ferreira. 2020b. "Cave Lithology Effect on Subterranean Biodiversity: A Case Study in Quartzite and Granitoid Caves." *Acta Oecologica* 108:1–10.

Souza-Silva, Marconi, Rogério Parentoni Martins, and Rodrigo Lopes Ferreira. 2011. "Cave Lithology Determining the Structure of the Invertebrate Communities in the Brazilian Atlantic Rain Forest." *Biodiversity and Conservation* 20:1713–1729.

Stone, Fred D., Francis G. Howarth, Hannelore Hoch, and Manfred Asche. 2004. "Root Communities in Lava Tubes." In *Encyclopedia of Caves,* edited by David C. Culver and William B. White, 477–484. Burlington: Elsevier, Academic Press.

Strong, Thomas S. 2006. "Vertebrate Species Use of Cave Resources in the Carlsbad Caverns Region of the Chihuahuan Desert." In *People, Places, and Parks: Proceedings of the 2005 George Wright Society Conference on Parks, Protected Areas, and Cultural Sites*, edited by David Harmon, 270–276. Hancock, MI: The George Wright Society.

Su, Yuqiao, Qiming Tang, Fuyan Mo, and Yuegui Xue. 2017. "Karst Tiankengs as Refugia for Indigenous Tree Flora amidst a Degraded Landscape in Southwestern China." *Scientific Reports* 7:1–10.

Taylor, Steven J., Jean K. Krejca, and Michael L. Denight. 2005. "Foraging Range and Habitat Use of *Ceuthophilus secretus* (Orthoptera: Rhaphidophoridae), a Key Trogloxene in Central Texas Cave Communities." *The American Midland Naturalist* 154:97–114.

Titus, Timothy N., J. Judson Wynne, Daniel Ruby, and Nathalie A. Cabrol. 2010. "The Atacama Desert Cave Shredder: A Case for Conduction Thermodynamics." *41st Annual Lunar and Planetary Sciences Conference*, Abstract #1096. Houston, TX: Lunar and Planetary Institute.

Tobin, Benjamin W., Benjamin T. Hutchins, and Benjamin F. Schwartz. 2013. "Spatial and Temporal Changes in Invertebrate Assemblage Structure from the Entrance to Deep-Cave Zone of a Temperate Marble Cave." *International Journal of Speleology* 42:203–214.

Trajano, Eleonora. 2000. "Cave Faunas in the Atlantic Tropical Rain Forest: Composition, Ecology, and Conservation." *Biotropica* 32:882–893.

Trusdell, Frank A. 2020. "Mauna Loa—History, Hazards, and Risk of Living with the World's Largest Volcano." Accessed July 20, 2020. https://www.usgs.gov/volcanoes/mauna-loa.

[USFWS] US Fish and Wildlife Service. 2012. "News Release: North American Bat Death Toll Exceeds 5.5 Million from White-nose Syndrome." Accessed June 2, 2022. https://www.fws.gov/story/2012-01/north-american-bat-death-toll-exceeds-55-million-white-nose-syndrome.

Vandel, Albert. 1965. *The Biology of Cavernicolous Animals*, translated by B. E. Freeman. Oxford: Pergamon Press.

Vas, Zoltán, and Csaba Kutasi. 2016. "Hymenoptera from Caves of Bakony Mountains, Hungary—an Overlooked Taxon in Hypogean Research." *Subterranean Biology* 19:31–39.

Wessel, Andreas, Hannelore Hoch, Manfred Asche, Thomas von Rintelen, Björn Stelbrink, Volker Heck, et al. 2013. "Founder Effects Initiated Rapid Species Radiation in Hawaiian Cave Planthoppers." *Proceedings National Academy of Sciences* 110:9391–9396.

White, William B., and David C. Culver. 2019. "Cave, Definition of," In *Encyclopedia of Caves*, 3rd ed., edited by William B. White, David C. Culver, and Tanja Pipan, 255–259. Amsterdam: Elsevier, Academic Press.

Whitten, Tony, 2009. “Applying Ecology for Cave Management in China and Neighbouring Countries.” *Journal of Applied Ecology* 46:520–523.

[WNSRT] White-nose Syndrome Response Team. 2022. Accessed May 27, 2022. https://www.whitenosesyndrome.org.

Wiles, Gary J., Jonathan Bart, Robert E. Beck Jr., and Celestino F. Aguon. 2003. “Impacts of the Brown Tree Snake: Patterns of Decline and Species Persistence in Guam’s Avifauna.” *Conservation Biology* 17:1350–1360.

Wray, Robert A. 2003. “Quartzite Dissolution: Karst or Pseudokarst?” Accessed July 20, 2020. https://www.speleogenesis.info/journal/publication.php?id=4494.

Wynne, J. Judson. 2013. “Inventory, Conservation and Management of Lava Tube Caves at El Malpais National Monument, New Mexico.” *Park Science* 30:45–55.

Wynne, J. Judson, and William Pleytez. 2005. “Sensitive Ecological Areas and Species Inventory of Actun Chapat Cave, Vaca Plateau, Belize.” *Journal of Cave and Karst Studies* 67:148–157.

Wynne, J. Judson, and William A. Shear. 2016. “A New Millipede (*Austrotyla awishashola*, n. sp., Diplopoda, Chordeumatida, Conotylidae) from New Mexico, USA, and the Importance of Cave Moss Gardens as Refugial Habitats.” *Zootaxa* 4084:285–292.

Wynne, J. Judson, and Kyle D. Voyles. 2014. “Cave-Dwelling Arthropods and Vertebrates of North Rim Grand Canyon, with Notes on Ecology and Management.” *Western North American Naturalist* 74:1–17.

Wynne, J. Judson, Ernest C. Bernard, Francis G. Howarth, Stefan Sommer, Felipe N. Soto-Adames, Stefano Taiti, et al. 2014. “Disturbance Relicts in a Rapidly Changing World: The Rapa Nui (Easter Island) Factor.” *BioScience* 64:711–718.

Wynne, J. Judson, Francis G. Howarth, Stefano Mammola, Rodrigo Lopes Ferreira, Pedro Cardoso, Tiziana Di Lorenzo, et al. 2021. “A Conservation Roadmap for the Subterranean Biome.” *Conservation Letters* e12834, https://doi.org/10.1111/conl.12834.

Wynne, J. Judson, Francis G. Howarth, Stefan Sommer, and Brett G. Dickson. 2019. “Fifty Years of Cave Arthropod Sampling: Techniques and Best Practices.” *International Journal of Speleology* 48:33–48.

Wynne, J. Judson, Stefan Sommer, Francis G. Howarth, Brett G. Dickson, and Kyle D. Voyles. 2018. “Capturing Arthropod Diversity in Complex Cave Systems.” *Diversity and Distributions* 24:1478–1491.

Wynne, J. Judson, Timothy N. Titus, Charles A. Drost, Rickard S. Toomey III, and Knutt Peterson. 2008. “Annual Thermal Amplitudes and Thermal Detection of Southwestern U.S. Caves: Additional Insights for Remote Sensing of Caves on Earth and Mars,” Abstract #2459, *39th Annual Lunar and Planetary Science Conference*. Houston, TX: Lunar and Planetary Institute.

Zepon, Tamires, and Maria Elina Bichuette. 2017. “Influence of Substrate on the Richness and Composition of Neotropical Cave Fauna.” *Anais da Academia Brasileira de Ciências* 89:1615–1628.

2

Evolutionary Models Influencing Subterranean Speciation

J. Judson Wynne, Matthew L. Niemiller, and Kenneth James Chapin

Introduction

Here we discuss how the environment, in particular habitat and geologic structure, influence evolutionary processes leading toward subterranean specialization. The most important variable driving the evolution of troglobiontic taxa is the presence of deep cave zone conditions. This environmental zone is characterized as a completely dark region with relatively stable temperature, low to no airflow, and a near water-saturated atmosphere with a negligible evaporation rate (Howarth 1980, 1982; Howarth and Wynne, Chap. 1). If these conditions are met and adequate energy and nutrients are available, the habitat becomes suitable to be colonized by a population that may ultimately become troglobiontic.

Animals often colonize the subterranean environment equipped with physiological and/or behavioral preadaptive (Christiansen 2012) or exaptive traits (i.e., exaptation *sensu* Gould and Vrba 1982), which provide them with an advantage toward becoming evolutionarily tied to the hypogean realm. Numerous researchers agree some degree of exaptive traits are required for successful colonization of subterranean habitats (e.g., Barr 1968; Culver 1982; Ashmole 1993; Holsinger 2000; Christiansen 2012). For example, six of nine Amblyopsidae fish species from the eastern United States are troglobiontic—two additional species are facultative cave inhabitants with some degree of troglomorphy, and one species is epigean (Niemiller and Poulson 2010; Niemiller, McCandless, et al. 2013). Although the epigean relative is incapable of subterranean life, it has small eyes, sensory receptors which permit feeding and orientation in an aphotic environment, and is nocturnal (Poulson 1963). Thus, Christiansen (2012) postulated that it may have been these exaptive traits that facilitated the success of ancestral amblyopsid colonizing the hypogean environment.

When a subterranean population is evolving toward a troglobiontic form, geology, topography, hydrology, and other abiotic factors (refer to Howarth and Wynne, Chap. 1), as well as the vagility of the colonizing organism, dictate whether the species range will become highly restricted to a single cave (i.e., a short-range endemic) or be distributed more broadly across a given geologic formation (i.e., a regional endemic). Understanding the potential of a given species range, as well as its dispersal potential, is predicated upon geologic substrate (rock being highly fractured or impermeable) and the presence or absence of topographic barriers (e.g., geologic formations, basins, and water courses) that may impede or facilitate dispersal.

As most troglobiontic taxa are thought to be incapable of establishing populations in epigean habitats, these organisms have been considered relicts of past ecosystems—now completely isolated from the surface environment (e.g., Assmann et al. 2010; Ćurčić et al. 2011). As with most phenomena in biology, the story is far more complex. Prior to modern genetic techniques, evincing clear evolutionary relationships between closely related epigean and hypogean species or closely related subterranean species has been challenging. Traditionally, systematists relied solely on Linnaean taxonomy to examine these relationships. Many of their interpretations ultimately became hypotheses tested with genetic analysis.

Like all closely related species, cryptic species will inevitably arise if isolation and/or selection pressures are sufficient (Witt et al. 2006; Zakšek et al. 2009; Zhang and Li 2014). Thus, an ancestral epigean species can evolve into multiple genetically distinct but morphologically indistinguishable clades. For example, the southern cavefish, *Typhlichthys subterraneus* Girard, 1859, from the eastern United States was considered the most widely distributed cavefish species in the world (Proudlove 2006; Niemiller and Poulson 2010). However, genetic analysis revealed there were several morphologically cryptic but genetically distinct lineages in this species complex (Niemiller et al. 2012).

As the use of molecular approaches have become widely accessible over the last two decades, researchers are increasingly examining molecular differences between morphologically similar populations. Many examples discussed here elucidated how species complexes often require molecular methodologies to unravel the complexities of distribution, dispersal, and divergence. Through such analyses, researchers may ascertain whether a given group represents an individual species, multiple species, or different clades.

Overall, the drivers that give rise to the evolution of troglobiontic taxa include vagility of the hypogean organism (or ancestral epigean organism), the availability of mesocaverns for dispersal, the subsequent presence of barriers to dispersal, isolation by distance, climatic shifts, natural selection, genetic drift, and mutation. Taking these environmental drivers and genetic mechanisms into account, we conducted a scan of the literature. We identified seven evolutionary models influencing how organisms diverge and adapt to the subterranean environment. These include the climatic relict hypothesis, adaptive shift hypothesis, divergence with gene flow, subterranean sympatric divergence, subterranean allopatric divergence, allopatric speciation with multiple subterranean colonization events, and subterranean dispersal with sympatric speciation. While this chapter is not intended to serve as an exhaustive review on this subject, it provides evidence for each of these evolutionary models using examples from recent studies that employed molecular techniques. Finally, we offer ideas for future research to best examine both cryptic species and species complexes using the latest molecular techniques.

The Environment Selects

In addition to troglobiontic taxa, other organisms that exhibit no characteristics of subterranean adaptation establish viable populations underground. However, their ability to successfully colonize subterranean environments is often hinged upon their possession of exaptive physiological and behavioral traits (Christiansen 2012). In the past, cave-dwelling animals were classified according to their perceived ability to live in surface or subterranean environments. The gradation of troglobiontic morphological evolutionary traits (or lack thereof) is principally applied to classify cave-dwelling animals. Because ecologists rarely had the data required to develop robust classification systems based upon a combination of physiological, behavioral, and morphological traits, systems based upon morphological traits alone have been employed. These classification systems ranged from the Schiner-Racovitza classification (which was the first proposed and included three groups; Racovitza 1907) to more elaborate systems proposed by

Sket (2008) and Trajano and Carvalho (2017). Refer to Howarth and Moldovan (2018) and Howarth and Wynne (Chap. 1) for thorough discussions on these classification systems.

For this chapter, we only discuss the following categories for subterranean-dwelling taxa, which are based upon Barr (1968), Howarth (1982, 1983), and Sket (2008). These are (1) *epigean species*—a surface-dwelling organism; (2) *troglobiont*—terrestrial obligate cave dwellers that only complete their life cycle underground and exhibit morphological characteristics indicative of subterranean adaptation (i.e., troglomorphies); (3) *stygobiont*—the aquatic counterpart to troglobionts; and (4) *troglophile*—species that occur facultatively within caves and complete their life cycles there but also occur in similar surface microhabitats. For a complete discussion of the six categories of subterranean-dwelling fauna (with examples), please refer to Howarth and Wynne (Chap. 1).

The ability for organisms to colonize underground and evolve into highly specialized forms is related to, at least to some degree, the absence of light and the availability and extent of suitable habitat (Culver and Pipan 2015). However, most cave ecosystems are not closed, nor do they function independently from the surface. Climatic shifts occurring on the surface, along with subsequent changes in precipitation patterns and temperature fluctuations, influence the cave environment (e.g., Titus et al. 2011; Mammola et al. 2019). The most important environmental variables driving habitat suitability for both cave communities and troglobiontic organisms is temperature and humidity. Subsequently, troglobiontic organisms are generally considered stenothermic (capable of persisting within a narrow and specific temperature threshold; e.g., Rizzo et al. 2015; Mammola et al. 2019; Pallarés et al. 2019; Rendoš et al. 2020) and stenohygric (requiring a near water-saturated atmosphere; e.g., Barr and Kuehne 1971; Ahearn and Howarth 1982; Tobin et al. 2013). Additionally, Deharveng (2004) reported the distribution of troglobiontic taxa in the tropics was negatively correlated with increasing temperatures and decreasing latitude, while Peck (1980) postulated the low number of troglobiontic taxa in the arid southwestern United States was likely attributed to the lack of deep zone conditions and adequate nutrients.

Moreover, subterranean-restricted organisms can migrate across multiple strata as environmental conditions are favorable. Movements of terrestrial species include between the epikarst / *milieu souterrain superficiel*

(MSS; Juberthie et al. 1980) and the deeper hypogean environment (Gers 1998; Pipan and Culver 2012; Culver and Pipan 2014; Kozel et al. 2019), near-surface soil and epikarst/MSS (Nitzu et al. 2014), surface soil habitats and the cave environment (Ducarme and Lebrun 2004), macro- and mesocaverns (Howarth 1983; Howarth and Wynne, Chap. 1), and from deep zone (where most troglobiontic taxa typically occur) to other zonal environments (Howarth and Stone 1990; Kozel et al. 2019). For aquatic organisms, movements have been documented between spring heads and the hypogenic environment (Hahn 2002; Dole-Olivier et al. 2009; Hichem et al. 2019).

Concerning movements between mesocaverns and open cave habitats (i.e., macrocaverns), Wynne (unpublished data) identified a highly troglobiontic mite *Linopodes* sp. (family Eupodidae) in the typically highly xeric earth cracks at Wupatki National Monument, Arizona. This species was detected in 2013 during an intense summer monsoon. At that time, both cave walls and ceiling within the deep zone were saturated with water. Unfortunately, these conditions have not been observed in subsequent monsoons. This morphospecies was not encountered during previous work on the monument (refer to Welbourn 1976), nor has it been observed since. Wynne believed these animals migrated into the open cave habitat from the interstices in the rock only because environmental conditions were favorable; furthermore, it seems these suitable habitat conditions occur infrequently.

In addition to appropriate microclimatic conditions, subterranean-restricted life (chemoautotroph-centered systems being the exception; Sarbu et al. 1996; Dattagupta et al. 2009) require the allochthonous conveyance of energy and nutrients from the surface. Allochthonous nutrient input includes flood detritus (Barr 1967; Silva et al. 2011; Pacioglu et al. 2019), bird and bat guano (Fenolio et al. 2006; Ferreira et al. 2007; Ladle et al. 2012), cricket frass (Hobbs 1992; Benoit et al. 2004; Peck and Wynne 2013), root mats in aquatic habitats (Jasinska et al. 1996), root masses in terrestrial settings (Stone et al. 2012), and the more nebulous dissolved and colloidal organic material via percolation of rainwater through the host rock (Pipan and Brancelj 2001). Flood detritus transported into caves may occur episodically with seasonal flooding or continually via sinking streams and rivers. Cave-roosting bats and birds spend much of their lives in caves, where they provide nutrients into the system as guano (i.e., feces) and carcasses (Ferreira and Martins 1999; Gnaspini and Trajano 2000). Guano/frass deposits in some cave systems represent a significant nutrient source and, as such, are consumed by a myriad of organisms (Fenolio et al. 2006; Ferreira et al. 2007; Peck and Wynne 2013). Finally, dissolved organic material has been positively correlated with macroinvertebrate species richness and abundance in both epikarst (Simon et al. 2007; Pipan et al. 2020) and interstitial habitats (Dumnicka et al. 2020). Perhaps the most impressive example is an arthropod community, which occurs at −2,140 m in Krubera-Voronja Cave, Western Caucasus; with at least 12 species in this community, allochthonous nutrients from the surface were identified as the center of the food web (Sendra and Reboleira 2012).

The principal ingress points for organisms colonizing the subterranean environment are fissures, cracks, crevices (i.e., meso- and macrocaverns), and talus slopes, as well as the edaphic environment. The aggregation of these smaller subterranean void spaces represents the primary habitat for many subterranean-adapted species (Howarth, 1983, 1993; Culver and Pipan 2015). What varies among subterranean habitats is the level of genetic isolation that populations experience from their epigean counterparts and adjacent subpopulations, and the drivers of speciation.

In the simplest scenario, individuals from an epigean population colonize the subterranean realm and ultimately become genetically isolated from the epigean population (i.e., allopatric speciation). This can occur over relatively short to geologic time spans as the subterranean population evolves into a troglobiontic phenotype (e.g., eye reduction to complete eye loss, depigmentation, and appendage elongation). We expect species to evolve similar subterranean-adapted traits (via both convergent and parallel evolutionary processes), and empirical work has provided several examples (e.g., Wilkens 2001; Wilkens and Strecker 2003; Bilandžija et al. 2012; W. Liu et al. 2017; Ober et al., Chap. 5).

With troglobiontic taxa, genetic relatedness or divergence into multiple species/clades across the landscape is determined by the ability of individuals within a related population or between populations to move freely, or relatively unimpeded, underground. This occurs via active or passive dispersal through the mesocaverns (Fig. 2.1; refer to Howarth 1983). Active dispersal involves subterranean taxa with moderate to high vagility to use the network of meso- and macrocaverns (i.e., cracks and crevices of various dimensions) as dispersal corridors, whereas passive dispersal occurs when organisms are dispersed via flood waters, moving through these cracks and crevices, that transport both individuals and eggs. For example, Katz et al. (2018) intimated

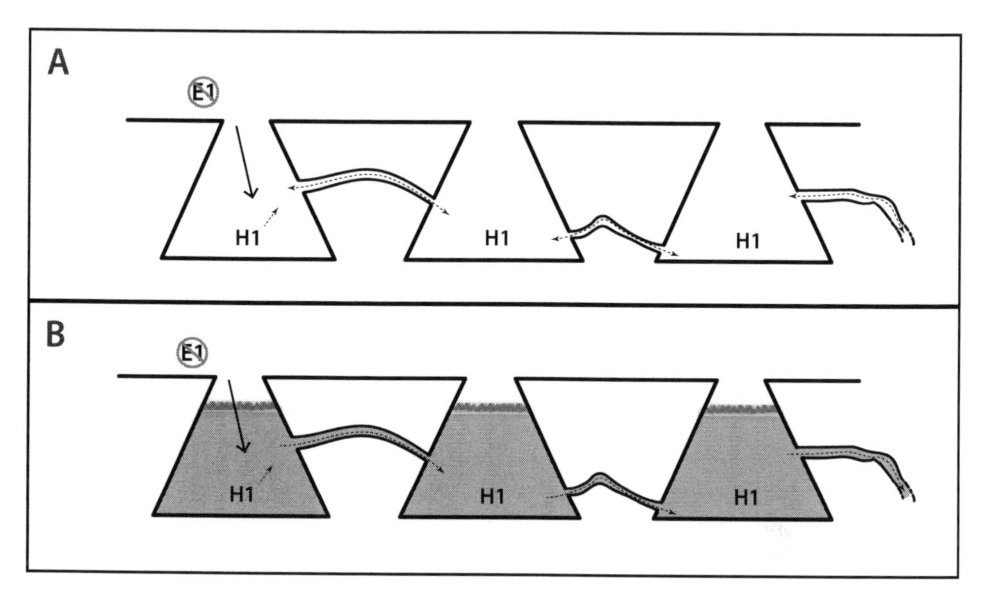

Figure 2.1. Modes of migration for subterranean fauna through the mesocaverns (i.e., the labyrinth of subterranean cracks, crevices, and pore space). E1 (epigean species #1) may or may not go extinct (*enclosed in a circle with a backslash*) on surface; H1 (hypogean or subterranean-adapted species #1) disperses through mesocaverns. [A] Active dispersal whereby organisms with moderate to high vagility can actively migrate. [B] Passive dispersal whereby flood waters facilitate dispersal of organisms with low to no vagility.

those shared haplotypes across multiple populations of a troglobiontic Collembola species (*Pygmarrhopalites* sp.) from the Mississippi River Valley were indicative of passive dispersal via seasonal flooding and groundwater. This was based upon a growing body of evidence that Collembola establish abundant and diverse communities in deep hypogean habitats (e.g., Deharveng et al. 2008; Shaw et al. 2011; Palacios-Vargas et al. 2018).

Another means of passive dispersal is phoresy (i.e., a nonparasitic process whereby one organism disperses by attaching to the body of another organism). Pseudoscorpions (Beier 1948; Binns 1982; Poinar et al. 1998; Francke and Villegas-Guzmán 2006; Del-Claro et al. 2009) and mites (Mitchell 1970; Houck and O'Connor 1991; S. Liu et al. 2016; Schäffer and Koblmüller 2020) are the most well-documented phoretic dispersers in epigean ecosystems. Both groups are represented by an impressive diversity of troglobiontic taxa globally.

Overall, phoresy remains an understudied dispersal mechanism in epigean environments (Bartlow and Agosta 2020). Concerning hypogean ecosystems, numerous notes have been published discussing this phenomenon between subterranean organisms (e.g., Ebermann and Palacios-Vargas 1989; Ćurčić et al. 2008; Souza et al. 2018); however, there have been no studies examining and/or quantifying the importance of phoresy for dispersal within subterranean environments.

Conversely, there are barriers that arise over geologic time, which can restrict dispersal and thus subdivide and isolate a subterranean population. When this occurs, genetic divergence results. While there may be others, Barr and Holsinger (1985) identified at least two types of dispersal barriers (fluvial and stratigraphic), which can arise within the distributional range of a given troglobiontic organism (Fig. 2.2). As expected, these barriers can affect subterranean-adapted organisms differently. For fluvial incumbrances, gene exchange between newly established subpopulations is most likely to be impeded or restricted for most troglobionts. For the Mississippi River Valley troglobiontic springtail, *Pygmarrhopalites* sp., Katz et al. (2018) evinced that molecular divergence times and patterns of genetic structure across these populations were driven by the downcutting of the Mississippi and Kaskaskia Rivers.

For aquatic species, early conventional wisdom asserted that stygobionts were likely unhindered by either fluvial and stratigraphic barriers (Barr and Holsinger 1985), as they can disperse via the hyporheic zone. This is predicated upon the notion that, if water-filled dispersal conduits exist, these animals are likely to disperse, and gene flow will continue. Recent genetic studies have produced mixed results. In a study of northern cavefish, *Amblyopsis spelaea* DeKay, 1842, the modern Ohio River was hypothesized to be a dispersal barrier; Niemiller, McCandless, et al. (2013) provided genetic evidence to support that the populations on either side of the river had diverged—with the timing of divergence corresponding to the formation of the modern Ohio River approximately 800,000 years ago. For an anchihaline subterranean-adapted brittle star (family Ophionereididae) from Quintana Roo, Mexico, Bribiesca-Contreras et al. (2019) sequenced cytochrome oxidase I nucleotide (COI) and found the species appears endemic to only one cave—despite a large number of surveys aimed to detect this species. They further asserted its presumed restricted range is attributed to a reduced developmental stage (i.e., embryos are lecithotrophic). Evidence of

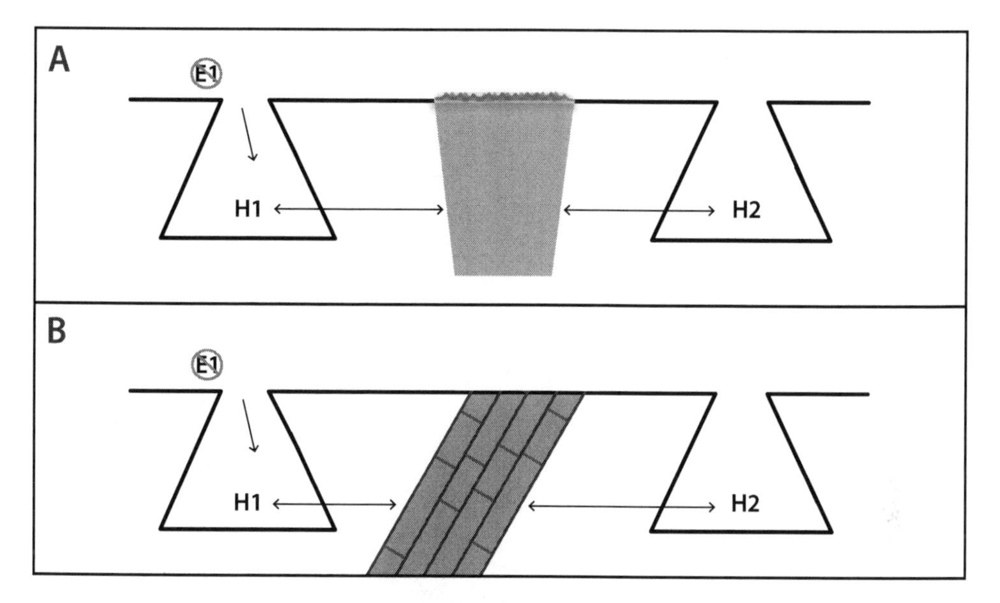

Figure 2.2. E1 (epigean species #1) may or may not go extinct (*enclosed in a circle with a backslash*) on surface; H1 (hypogean or subterranean-adapted species #1) dispersed through mesocaverns, a barrier arose and speciation resulted (which led to a divergence event resulting in a second hypogean species, H2). [A] Fluvial barriers include substantial watercourses (rivers and streams) that result in downcutting/incising or valley formation, and/or a riverbed and an associated high water table that subdivides a population and restricts gene flow. [B] Stratigraphic barriers occur due to uplift associated with faulting; an impermeable strata restricts immigration and emigration of individuals of a population. In both cases, when a dispersal barrier arises, subterranean-adapted organisms within a population can probably disperse as far as the dispersal barrier via the mesocaverns. Double-headed arrows indicate species movement from macrocavern to barrier and barrier to macrocavern via the mesocaverns.

genetic connectivity of another anchihaline stygobiontic species, the atyid shrimp, *Typhlatya mitchelli* H. Hobbs III & H. H. Hobbs Jr., 1976, was reported from the northern Yucatan Peninsula (Hunter et al. 2008); analyses of a three-gene concatenated dataset, consisting of COI, cytochrome *b* gene, and 16S, revealed identical haplotypes at multiple localities at distances up to 235 km. Additionally, gene flow of a troglobiontic gastropod species (*Neritilia cavernicola* Kano & Kase, 2004) was confirmed up to 200 km across open ocean (Kano and Kase 2004).

Stratigraphic barriers may restrict gene exchange for some troglobionts, have little effect on other troglobionts, and have no effect on stygobionts. Barr and Holsinger (1985) proposed that these barriers are composed of "noncaverniferous" strata (e.g., shales, sandstones, or other insoluble substrates) occurring between caverniferous (karst) strata. The interposition of noncaverniferous and caverniferous strata can create hurdles restricting the dispersal of some troglobionts.

Examining phylogenetic data from a *Nesticus* spider (family Nesticidae) complex in the southern Appalachian Mountains in the United States, Hedin (1997) identified stratigraphic barriers at two spatial scales. Firstly, divergence between breeding populations was high, indicating that some populations were confined to a specific cave or talus slope. At the landscape scale, at least one sibling species pair was identified as each being restricted to adjacent drainage basins, which provided support for basins serving as isolating topographic features on the landscape—albeit this was not universally observed across the complex. However, these conclusions may be limited by the technology available at the time of study.

For volcanic stratum, troglobionts may have a dispersal window when colonizing newly formed lava tubes and associated pore space. Once the mesocaverns are filled with surface soils migrating downward and settling into these pore spaces, terrestrial troglobiontic populations can become isolated (refer to Hoch and Howarth 1999).

Evolutionary Models Driving Subterranean Speciation

The same evolutionary mechanisms (drift, mutation, selection, and migration) and modes of speciation (e.g., allopatry, parapatry, and sympatry) that govern diversification of epigean organisms also drive evolutionary processes in subterranean fauna (refer to Urry et al. 2016). Nested within these mechanisms and modes are several evolutionary models we identified from the literature. These processes have been identified as influencing how subterranean organisms evolve toward a troglobiontic phenotype. We emphasize these models are not mutually exclusive; two or more models may contribute to the divergence of a given subterranean-adapted species and/or population.

These models can be largely organized in two important ways: (1) the location of divergence, and (2) the degree of gene flow/connectivity between populations. Successful subterranean colonization first involves divergence between epigean and hypogean populations at the surface-subterranean ecotone (Phase 1 speciation, *sensu* Holsinger 2000). Divergence associated with the transition from surface to subterranean existence has received the most attention historically and includes popular models, such as the climatic relict hypothesis and adaptive shift hypothesis. Speciation may also occur after a lineage has become established within the subterranean environment (Phase 2 speciation, *sensu* Holsinger 2000). These models can also be categorized based upon the degree of overlap and level of gene flow between populations, ranging from pure allopatric speciation (in which populations are geographically isolated and there is no gene flow between incipient species from the beginning of the divergence process) to sympatric speciation (where there is complete range overlap and panmixia of the ancestral population at the time of divergence; Fitzpatrick et al. 2008). These two extremes fall at either end of the spectrum of connectivity and gene flow, with parapatric speciation and divergence with gene flow models falling in between.

Through an examination of 35 phylogenetic studies on subterranean-dwelling taxa (Table 2.1), we identified at least seven evolutionary models that can lead to troglomorphy (Fig. 2.3). Of these, 15 studies were focused upon terrestrial organisms, while 20 analyzed phylogenetic relationships of aquatic species clades/complexes (Fig. 2.4). Regarding the aquatic species examined, 10 studies were based on cavefishes, including four studies on the population genetics of *Astyanax mexicanus* (De Filippi, 1853). While previous researchers (e.g., Barr and Holsinger 1985; Holsinger 2000; Juan et al. 2010) also provided thorough reviews on this topic, these efforts are now somewhat dated. Thus, reexamining this question through the lens of contemporary genetic techniques was warranted.

Climatic Relicts and Adaptive Shifts

Early on, cave biologists formulated two hypotheses to elucidate how organisms became restricted to the subterranean environment and evolved into troglobiontic forms. The first, the climatic relict hypothesis (CRH), postulated that a likely common epigean species colonized hypogean habitats and established a cave-dwelling population. This population then evolved into a troglobiontic species once the surface population went extinct due to glacial-interglacial climatic oscillations. As surface environmental conditions became inhospitable, the epigean population went extinct, while the

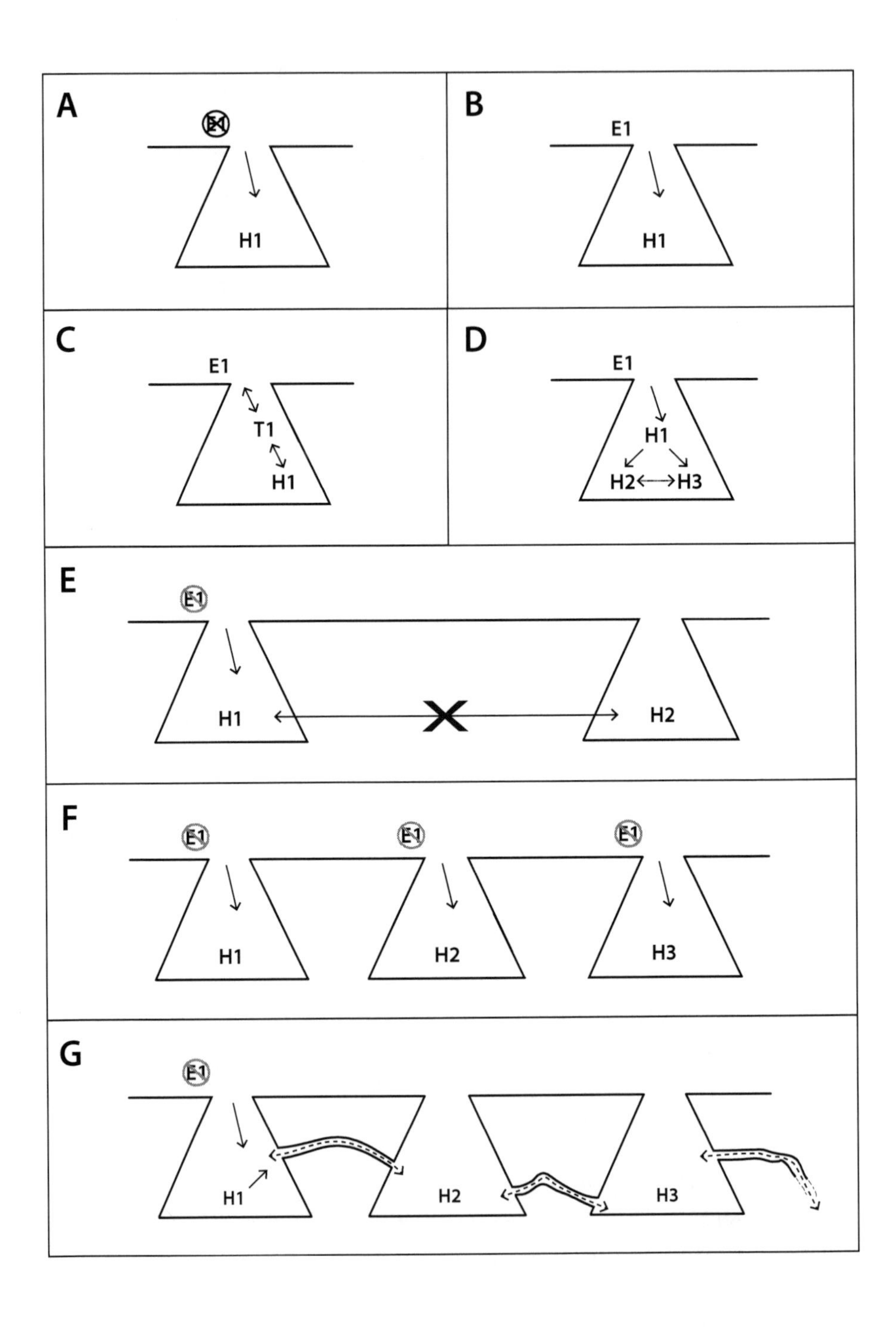
A
E1
H1
B
E1
H1
C
E1
T1
H1
D
E1
H1
H2
H3
E
E1
H1
H2
F
E1
E1
E1
H1
H2
H3
G
E1
H1
H2
H3

relict subterranean-dwelling population persisted and evolved into a troglobiontic form. CRH is most concerned with changing climatic conditions as the primary driver of allopatric speciation and has been used to explain the evolution of troglobiontic organisms, principally in temperate biomes.

Based upon the earlier work of Holdhaus (1933), Jeannel (1943), and Barr (1968), some researchers may have somewhat misappropriated CRH as a process to be applied only within the context of climatic oscillations of the Pleistocene. In fact, Jeannel (1943) merely suggested that it was unlikely that a trechine beetle group became troglobiontic before the Pliocene. Thus, limiting the definition of a climate relict to only the Pleistocene seemingly runs contrary to Jeannel (1943) and is not supported in light of several recent genetic analyses of subterranean-adapted taxa. In some cases, these workers advanced CRH with limited knowledge concerning the potential time scales within which climate relicts could evolve. Several species, which meet the criteria of CRH, have been identified as diverging during the Miocene, Pliocene, and even Holocene (Contreras-Díaz et al. 2007; Faille et al. 2010; Osikowski et al. 2017; Latella, Chap. 4). These researchers considered several species to be "climate relicts," and they attributed CRH as the evolutionary paradigm.

Figure 2.3. (*opposite*) Seven evolutionary models identified in the literature to explain divergence and evolution of troglobiontic biodiversity. Symbols are (E) epigean/ancestral, (T) troglophilic, and multiple hypogean species (H1 through H3). An X over E1 indicates an extinction event, while a backslash indicates the epigean species may or may not go extinct. [A] Climatic relict hypothesis: ancestral species goes extinct, while the colonizing population ultimately evolves allopatrically into a subterranean-adapted species. [B] Adaptive shift hypothesis: an epigean species/population is extant with a subterranean-adapted sibling species/population (where parapatric divergence has occurred). [C] Divergence with gene flow occurs when a hypogean population is diverging from the adjacent epigean population despite ongoing gene flow. This may also be interpreted as an adaptive shift in progress. [D] Subterranean sympatric divergence where H1 colonizes and then diverges into two (H2 and H3) or more species. [E] Subterranean allopatric divergence: migration barriers arise (*denoted by an X*) and vicariance occurs. [F] Allopatric speciation with multiple subterranean colonization events resulting in multiple troglobiontic sibling species. This outcome may give rise to two or more distinct species or clades. [G] Subterranean dispersal with sympatric speciation.

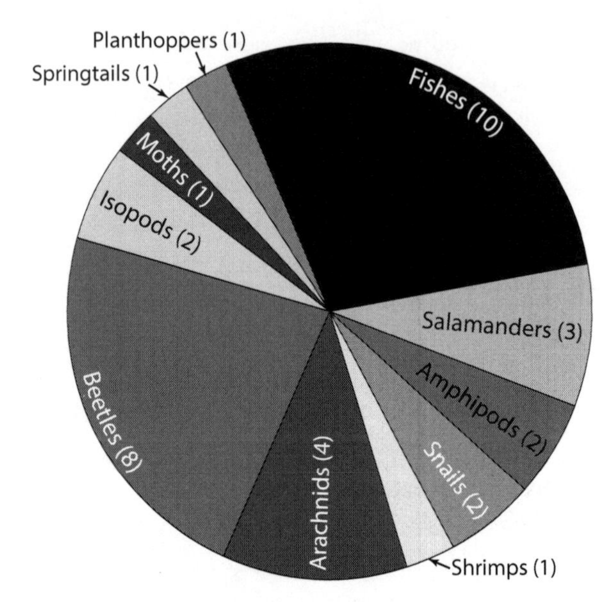

Figure 2.4. Thirty-five papers examined by study taxon. For beetles, these include studies on six terrestrial and two aquatic lineages/complexes.

Because ancestral species are hypothesized to go extinct due to climatic shifts, and ancestral species can only be presumed extinct (as additional sampling may result in detecting ancestral sibling species), CRH can only be reasonably deduced using the best available evidence. For example, Leys et al. (2003) examined 60 species of subterranean diving beetles of the subfamily Hydroporinae from the calcrete aquifers of inland arid Australia. They reported that species occurring within the same calcrete drainage systems did not group together phylogenetically, which is consistent with several coexisting, widespread ancestral species (all of which are extinct) that simultaneously colonized the subterranean realm between the late Miocene and early Pliocene. They evoked CRH to explain this phenomenon. Interestingly, Leray et al. (2019), in their study of *Ptomaphagus* cave beetles (Coleoptera: Leiodidae) of the southeastern United States, discounted CRH, as the colonization events commenced ~6 million years ago (mya), predating the Pleistocene. The authors applied a more rigid interpretation of CRH in which divergence is limited to the Pleistocene. However, these data support that climatic shifts before (and perhaps after) the Pleistocene also contributed to diversification of species. The findings of Leray et al.

Table 2.1. Summary of the 35 studies reviewed. Evolutionary models identified were [A] climatic relict hypothesis (with allopatric speciation), [B] adaptive shift hypothesis (with parapatric speciation), [C] divergence with gene flow, [D] subterranean sympatric speciation, [E] subterranean allopatric speciation, [F] allopatric speciation via multiple subterranean colonization events, [G] subterranean dispersal with sympatric speciation, and [H] a combination of two or more processes.

Study	Taxa	Habitat	A	B	C	D	E	F	G	H
Arnedo et al. (2007)	Spiders	Terrestrial				X	X			X
Bendik et al. (2013)	Salamanders	Terrestrial		X				X		X
Bradic et al. (2012)	Fishes	Aquatic			X					
Coghill et al. (2014)	Fishes	Aquatic			X			X		X
Contreras-Díaz et al. (2007)	Beetles	Terrestrial	X	X						X
Derkarabetian et al. (2010)	Harvestman	Terrestrial						X		
Faille et al. (2010)	Beetles	Terrestrial	X							
Faille et al. (2014)	Beetles	Terrestrial						X		
Faille, Tänzler, and Toussaint (2015)	Beetles	Terrestrial					X			
Faille, Bourdeau, et al. (2015)	Beetles	Terrestrial					X			
Gorički and Trontelj (2006)	Salamanders	Aquatic						X		
Hart et al. (2020)	Fishes	Aquatic						X		
Hedin and Thomas (2010)	Harvestman	Terrestrial						X		
Herman et al. (2018)	Fishes	Aquatic			X					
Katz et al. (2018)	Springtails	Terrestrial					X			
Konec et al. (2015)	Isopods	Aquatic		X						
Kruckenhauser et al. (2011)	Fishes	Aquatic					X			
Langille et al. (2020)	Beetles	Aquatic			X	X				X
Leray et al. (2019)	Beetles	Terrestrial	X				X			X
Leys et al. (2003)	Beetles	Aquatic	X							
Medeiros et al. (2009)	Moths	Terrestrial			X					
Mendes et al. (2019)	Fishes	Aquatic			X					
Niemiller et al. (2008)	Salamanders	Aquatic			X					
Niemiller, McCandless, et al. (2013)	Fishes	Aquatic					X			
Osikowski et al. (2017)	Gastropods	Aquatic	X					X		X
Plath et al. (2010)	Fishes	Aquatic			X					
Rivera et al. (2002)	Isopods	Terrestrial		X						
Santos (2006)	Shrimps	Aquatic					X			
*Schilthuizen et al. (2005, 2012)	Snails	Aquatic			X					
Segherloo et al. (2018)	Fishes	Aquatic				X				
Strecker et al. (2012)	Fishes	Aquatic				X				
Villacorta et al. (2008)	Amphipods	Terrestrial	X	X						X
Wessel et al. (2013)	Planthoppers	Terrestrial						X		
Zakšek et al. (2019)	Amphipods	Aquatic				X	X			X
Zhang and Li (2013)	Spiders	Terrestrial	X				X			X

* These two studies were combined as the latter study was a continuation of the first.

(2019) were similar to Zhang and Li (2013), who identified CRH as the driver for subterranean divergence in *Nesticella* spiders where climatic shifts occurred throughout the period from the late Miocene through the Pleistocene. When considering models of diversification in subterranean fauna, we recommend against a narrow interpretation (i.e., a Pleistocene-centric view) of CRH.

The Messinian salinity crisis was posited as being responsible for at least two CRH-driven events that transpired during the late Miocene (from 5.96 to 5.33 mya). Regionally characterized as a period of higher temperatures and increased aridity, this "crisis" resulted in a period of desiccation of the Mediterranean Sea (Krijgsman et al. 1996). Based on standard mitochondrial rate estimates of a clade consisting of three genera of trechine beetles (family Carabidae) from the Pyrenees Mountains, Spain, Faille et al. (2010) suggested the origin of this clade evolved during the mid-late Miocene (~10 mya), when the climate was more temperate than today. They inferred this may be an example of CRH. Additionally, in a study of troglobiontic aquatic gastropod lineages (family Hydrobiidae) from Bulgaria, Osikowski et al. (2017) forwarded a conservative case for CRH. Divergence time of these lineages ranged from 7 ± 1.0 to 6.75 ± 0.8 mya based upon Bayesian and penalized-likelihood analysis of sequence data, respectively. These dates coincide with the Messinian salinity crisis.

Regarding a more traditional interpretation of CRH, Zhang and Li (2013) analyzed DNA sequence data of *Nesticella* spiders (from the Yunnan-Guizhou Plateau, southern China) using Bayesian inference and maximum-likelihood techniques. They identified a kaleidoscopic diversification event occurring throughout the Pleistocene, and they further indicated this radiation event was most likely due to climatic shifts characteristic of that epoch.

As troglobiontic species were increasingly discovered and described from the tropics (e.g., Howarth 1972, 1973; Peck 1974), and the effects of the glacial-interglacial period were not as pronounced in equatorial regions, CRH was an implausible explanation for troglomorphies of these subterranean-dwelling taxa. Therefore, the adaptive shift hypothesis (ASH; refer to Howarth 1973, 2019) was posited. Adaptive shifts occur when an epigean species colonizes the subterranean environment, establishes a viable population, and ultimately evolves into a troglobiontic form. Under this paradigm, the hypogean population initially adapted to the cave environment with ongoing gene flow with the epigean population (Pinho and Hey 2010). Ultimately, the hypogean species becomes parapatric with the extant sibling species.

For the Hawaiian genus *Littorophiloscia* (family Halophilosciidae), Rivera et al. (2002) affirmed an adaptive shift occurred from marine supralittoral to subterranean terrestrial habitats. This was based upon the analysis of COI sequence data using a maximum-likelihood approach, whereby a reproductive relationship was identified between the extant epigean and hypogean species. Another example from the same study is the monophyletic terrestrial isopod genus *Hawaiioscia*, which contained four troglobiontic species where each species occurred on a different Hawaiian island. Rivera et al. (2002) concluded that these species diverged from a widespread ancestral epigean species (or perhaps a group of closely related species) via multiple and independent adaptive shifts.

Moreover, the evolutionary history of six *Eurycea* species (a complex of spring- and cave-dwelling salamanders) across seven counties on the southeastern Edwards Plateau, Texas, suggested shallow genetic divergences among several species, where gene flow and nested relationships across morphologically disparate cave and spring forms were all consistent with ASH (Bendik et al. 2013); furthermore, cave invasions were identified as recent, with many troglobiontic phenotypes arising independently. Similarly, the degree of phenotypic change was examined between two independent cave colonizations of the stygobiontic isopod *Asellus aquaticus* (Linnaeus, 1758) from a Romanian sulfidic aquifer and a sinking stream in Slovenia (these populations are ~1,200 km apart); both ancestral epigean populations intergrade with adjacent hypogean populations (Konec et al. 2015). Phylogenetic analysis provided support that subterranean colonization events were independent, while microsatellite analysis showed no evidence of gene flow between epigean and subterranean populations—providing support for two separate cases of ASH.

On the Canary Islands, two possible cases of both CRH and ASH were reported. The first was related to island age. Contreras-Díaz et al. (2007) posited the low genetic divergence between epigean and hypogean sibling species across 42 *Trechus* species (Trechine: Carabidae) was suggestive of multiple adaptive shifts on the younger islands. *Trechus detersus* Wollaston, 1864 also occurs on Lanzarote—one of the oldest, most eroded, and easternmost islands of the Canarian archipelago. Analysis of DNA sequence data showed no clear close relationship between the extant epigean relative, while a close sibling relationship was identified within a lineage distributed across the western-most island cluster of Tenerife, La Gomera, La Palma, and El Hierro. Given that Lanzarote is now much drier and that no sibling epigean species are known to occur there, this pattern is suggestive of a climatic relict scenario (Contreras-Díaz et al. 2007). In another example, in a study of epigean and troglobiontic *Palmorchestia* sp. (terrestrial Amphipoda) from La Palma, Villacorta et al. (2008) espoused support for both hypotheses based upon mtDNA phylogenies and results from molecular clock analyses. However, the authors later stated that a shift to drier conditions during the Pleistocene prompted multiple colonizations by an epigean species, and the subsequent extinction of this population was more congruent with CRH than ASH. Finally, Villacorta et al. (2008) emphasized lacunae in their sampling and the need to revise the taxonomy of this species, which may at least partially explain why clear patterns were indiscernible regarding the evolution of *Palmorchestia* on the Canarian island of La Palma.

Divergence with Gene Flow

Divergence with gene flow may be considered an intermediate model of connectivity and gene flow along a continuum between epigean and hypogean populations. This paradigm ranges from pure allopatric speciation with no geographic overlap to pure sympatric speciation with complete geographic overlap (e.g., syntopy) and potentially high levels of gene flow. Niemiller et al. (2008) identified divergence with gene exchange between epigean and hypogean populations, while Juan et al. (2010) considered this process as parapatric speciation with an ecological shift prompted by resource segregation. Four species within the Tennessee cave salamander *Gyrinophilus* complex have genetic signatures indicative of continuous or recurrent gene flow during divergence of the hypogean population from the epigean population (Niemiller et al. 2008). Schilthuizen et al. (2005, 2012) provided another example through examining genetic relationships between the hypogean and epigean snail complex (Gastropoda: Hydrocenidae: *Georissa*) from Sabah, Malaysian Borneo; they found the troglobiontic species (*Georissa filiasaulae* Haase & Schilthuizen, 2007) diverged from its epigean sibling species (*Georissa saulae* (van Benthem-Jutting, 1966)) without isolation and with ongoing unidirectional gene flow from surface to cave.

A study of eight populations of a cave-dwelling fish, the Atlantic molly (*Poecilia mexicana* Steindachner, 1863), pre- and post-catastrophic flood rendered similar results. These different populations selected for non-sulfidic surface, sulfidic surface, and sulfidic cave habitats. When pre- and post-flooding genetic diversity of these eight populations were compared, genotypes where not physically displaced in the cave populations post-flooding. This result provided support that cave-dwelling populations appeared largely unaffected by flood-induced epigean colonists. Thus, habitat rather than geographic distance was considered the primary driver for genetic divergence across the four habitats for both epigean and subterranean-dwelling populations (Plath et al. 2010). Additionally, differences across 11 hypogean and 10 epigean populations of the Mexican blind cavefish complex *Astyanax mexicanus* revealed ongoing gene flow from epigean populations, despite the fact that hypogean populations had diverged and were stygobiontic (Bradic et al. 2012). In a similar study, when analyzing 47 sequenced complete genomes from three cave populations and two epigean populations of *A. mexicanus*, Herman et al. (2018) reported evidence of gene exchange between epigean and stygobiontic phenotypes. Their study further demonstrated that even weak selection pressures were sufficient to retain stygobiontic traits despite admixture with epigean populations and a small population size (Herman et al. 2018). Similarly, Mendes et al. (2019), using a molecular genetic phylogeny, found high similarity and metapopulation connectivity between hypogean and epigean populations of the catfish species *Ancistrus cryptophthalmus* Reis, 1987 (a depigmented cavefish with reduced eyes) and *Ancistrus* sp. from central Brazil. Their results evinced these populations should be considered a single species.

Somewhat conversely, a moth from Hawai'i Island, *Schrankia howarthi* Davis & Medeiros, 2009, has established epigean, troglophilic, and dark zone populations. The dark zone population is "generally smaller, indistinctly patterned, and predominantly grey," behaviorally blind, and weakly volant with reduced eyes and wings (Medeiros et al. 2009). The authors posited that hybridization between the twilight and dark zone populations has impeded the deep cave variant from evolving into a separate troglobiontic species (Medeiros et al. 2009).

It remains unclear how often hypogean morphotypes evolve, via divergence with gene flow, into a distinct troglobiontic species, or to what extent gene flow impedes adaptation to the cave environment. In all but one case reported here, epigean and hypogean forms were observed between lineages and clades rather than between distinct sibling species. As Niemiller et al. (2008) implied, these examples likely represent divergence in progress. Moreover, the studies on *Schrankia* and *Astyanax* provide evidence that mating behavior may play an important role in species divergence.

Subterranean Sympatric Divergence

The evolutionary process of subterranean sympatric divergence occurs when an epigean species colonizes a subterranean system, becomes established, and becomes troglobiontic. The epigean population may or may not go extinct. The new troglobiontic phenotype evolves into multiple populations to fill different available niches. Arnedo et al. (2007) provided taxonomic evidence of sympatric speciation driven by niche partitioning of dysderid cave spiders—sibling species were both different sizes and displayed distinctive cheliceral morphologies.

Another example was revealed through molecular analysis and involved the most extensively studied cavefish group, *Astyanax* of northern Mexico. Strecker et al. (2012) examined two distinct morphotypes within a large subterranean stream pool (60 m in length), which is partially exposed to the surface via a skylight. The two morphotypes, an eyed population and a population with highly reduced eyes, select the illuminated and dark sections of the pool, respectively. Distinct microsatellite genotypes and the lack of admixture with other nuclear genotypic clusters suggested this is in-progress sympatric speciation (Strecker et al. 2012).

This model of speciation can also occur in a troglobiontic population. However, few studies to date have offered strong support for this mode of divergence. Segherloo et al. (2018) examined speciation of two blind cave barbs—*Garra typhlops* (Bruun & E. W. Kaiser, 1944) and *Garra lorestanensis* (Mousavi-Sabet & Eagderi, 2016), which occur syntopically in a single cave in the Zagros Mountains, Iran. They intimated that sympatric divergence occurred following a single colonization event of an ancestral source population. Mitochondrial and nuclear genomic phylogenies show that the two cave barbs are monophyletic, suggesting reproductive isolation; however, reproductive isolation appears to be incomplete because at least one hybridized individual was observed. Although historical demographic analysis supported a sympatric model over allopatric divergence followed by secondary contact (Segherloo et al. 2018), these results cannot completely rule out other allopatric or parapatric hypotheses.

In another recent study, Zakšek et al. (2019) examined sympatry among a clade of eight closely related species of *Niphargus* amphipods in the Dinaric Karst. Two species, *Niphargus croaticus* (Jurinac) and *Niphargus subtypicus* Sket, 1960, are sibling taxa with an extensive zone of sympatry and co-occur at several sites. Both species belong to different ecomorphs, which likely reflects differences in locomotion, habitat, prey, and possibly trophic position (Trontelj et al. 2012; Delić et al. 2016; Zakšek et al. 2019), and also likely reduces competition. However, phylogeographic analyses suggest that *N. croaticus* and *N. subtypicus* evolved in allopatry, with present-day sympatry reflecting post-speciation dispersal and secondary contact.

Langille et al. (2020) reported that a sibling species triplet of diving beetles (genus *Paroster*: Dytiscidae) from Western Australia underwent divergent or disruptive selection, as evidenced by their differences in body sizes, which simultaneously led to reproductive incompatibilities and assortative mating. Importantly, they share a variety of additional deleterious mutations in the long wavelength opsin 1 (*lwop*) gene, which indicated this radiation event occurred from a stygobiontic ancestor species. However, the authors averred the drivers of this phenomenon are presently unknown and that a population-level comparative genomic study should be conducted (Langille et al. 2020).

Subterranean Allopatric Divergence

In subterranean allopatric divergence, a troglobiontic species disperses and a barrier arises separating two populations (Phase 2 speciation, *sensu* Holsinger 2000). Hypogean populations can become isolated via hydrologic and geologic mechanisms, such as stratigraphic discontinuities, hydrologic flow changes due to stream capture, and/or loss of dispersal conduits due to erosion, siltation, and collapse. This diversification pattern has been observed for both stygobiontic and troglobiontic taxa. For aquatic species, subterranean allopatric speciation has been identified in some cavefish species where gene flow occurs among cave populations. For *Garra barreimiae* Fowler & Steinitz, 1956, a ray-finned fish species endemic in the southeastern Arabian Peninsula, Kruckenhauser et al. (2011) suggested that three clades underwent long-term isolation as their habitats became separated by a watershed—one clade was an epigean species, while the other two were stygobiontic. For two anchialine shrimps *Halocaridina rubra* Holthuis, 1963 lineages on Hawai'i Island, genetic analysis suggests isolation occurred between 2.13 and 1.17 mya (Santos 2006). Similarly, Niemiller, McCandless, et al. (2013), using sequence data, identified a divergence event where two populations of the northern cavefish *Amblyopsis spelaea* were isolated by the incision of the Ohio River; the population on the southern side of the river was later described as a separate species from the northern population (Chakrabarty et al. 2014).

For terrestrial taxa, subterranean allopatric divergence has been documented for several groups, including beetles, springtails, and spiders. Faille, Tänzler, and Toussaint (2015) and Faille, Bourdeau, et al. (2015) averred that two troglobiontic trechine lineages (*Aphaenops* and *Geotrechus*) from the Pyrenees diverged allopatrically due to rivers incising the karst massif and forming valleys. Leray et al. (2019) used mtDNA sequence data to examine diversification patterns of subterranean-adapted fungus beetles (genus *Ptomaphagus*) in the southeastern United States. They revealed that the "Southern Cumberlands" lineage began diversifying at the end of the Miocene (~6 mya) extending through the Pliocene and Pleistocene—this was due to downcutting via erosion and stream incision, which extensively fragmented the karst substrate. This fragmentation was postulated to have isolated cavernicolous populations, giving rise to 12 distinct species (Leray et al. 2019). Using mitochondrial and nuclear markers, Katz et al. (2018) identified a vicariance-driven divergence event of a troglobiontic *Pygmarrhopalites* sp. (Collembola: Arrhopalitidae), which coincided with the incision of the Mississippi and Kaskaskia Rivers. These subsequent valley incisions resulted in the formation of two distinct phylogenetic lineages on either side of each river.

Zhang and Li (2013) provided two examples where geologic barriers were implicated in vicariance events of *Nesticella* spiders of the Yunnan-Guizhou Plateau, southern China. The geologic strike-slip fault along the Red River shear zone (~22 mya) caused an offshore rift resulting in an approximately 1,000 km discontinuity between Indochina and South China (Searle 2006). This faulting event resulted in the divergence of one *Nesticella* group from the other two regional clades (Zhang and Li 2013).

Langille et al. (2020) examined a sibling species pair of diving beetles (genus *Paroster*: Dytiscidae) that occur in adjacent aquifers in Western Australia. These aquifers were once a contiguous unit about 3 mya (Leijs et al.

2012). A shared deleterious mutation of the *lwop* phototransduction gene was identified, indicating their ancestral species was stygobiontic—thus providing direct evidence of allopatric subterranean speciation.

Allopatric Speciation with Multiple Subterranean Colonizations

With allopatric speciation with multiple subterranean colonizations, a presumably wide-ranging ancestral species colonizes the subterranean realm multiple times across the landscape. Because subterranean populations are isolated from one another, allopatric speciation results. This process has been documented in numerous taxonomic groups, including arachnids, beetles, terrestrial amphipods, salamanders, and cavefishes.

Oromí et al. (1991) and Arnedo et al. (2007) observed varying degrees of troglomorphy via comparative systematics of *Dysdera* spiders of the Canary Islands. Using phylogenetic analyses, Arnedo et al. (2007) further suggested these variations were likely attributed to multiple independent subterranean colonizations of the epigean sibling species over time. Employing a combined phylogenetic-morphometric approach, Faille et al. (2014) indicated the *Trechus fulvus* group (Jeannel, 1927) from the Betic-Rifean region, between southeast Iberia and north Morocco, represented multiple subterranean colonization events by a single ancestral surface species. Additionally, Villacorta et al. (2008) affirmed the present distribution of *Palmorchestia hypogaea* Stock & Martin, 1988 (a troglobiontic amphipod species from La Palma, Canary Islands) was driven by multiple colonizations of an epigean species. Additionally, subterranean adaptation appears to have evolved independently at least three times in phalangodid harvestmen in the eastern United States (Hedin and Thomas 2010) and at least three times in travunioid harvestmen in the western United States (Derkarabetian et al. 2010).

For the *Eurycea* species complex of the Edwards Plateau, Texas, phylogenetic analyses showed that cave invasions were recent and many troglobiontic morphologies arose independently (Bendik et al. 2013). Using an mtDNA sequencing approach, Goricki and Trontelj (2006) identified no common haplotypes across several geographic groups of European cave salamander *Proteus anguinus* Laurenti, 1768. As such, their results were concordant with the independent hydrographic units occurring across several impermeable geologic substrates, supporting the necessity of multiple independent colonization events of the ancestral epigean salamander species. In northeastern Mexico, Strecker et al. (2003, 2012) identified three clades representing at least two (potentially three) separate colonizations of epigean populations of *Astyanax mexicanus* into the cave environment. Coghill et al. (2014) reported at least four independent colonization events of *A. mexicanus* based on analyses of single nucleotide polymorphisms and ancestral character state reconstructions; they also inferred their dataset would be useful in examining gene flow between epigean and hypogean populations.

At least three cave lineages of amblyopsid cavefishes from the southeastern United States independently accumulated unique loss-of-function (LOF) mutations over the last 10.3 million years (Niemiller 2011; Niemiller, Fitzpatrick, et al. 2013). Cave lineages do not share LOF mutations in the eye photoreceptor gene *rhodopsin*. At least some LOF mutations would be expected to be shared if cave lineages were derived from a single cave colonizing ancestor. Eye histological data (Eigenmann 1909; Niemiller and Poulson 2010) supported independent colonization events, as the number and degree of degenerate loss of eye structures differed among cave lineages. More recent phylogenomic work in amblyopsid cavefishes was also consistent with multiple, subterranean colonization events (Hart et al. 2020).

Subterranean Dispersal with Sympatric Speciation

Subterranean dispersal with sympatric speciation occurs when a subterranean-adapted species disperses underground to colonize new habitats. As the new populations become established sequentially further from the ancestral population, the new populations diverge. We know of only example of this phenomenon—the Hawaiian cave planthopper clade (*Oliarus polyphemus* Fennah, 1973) from lava tube caves on the volcanoes of Mauna Kea, Hualālai, Mauna Loa, and Kīlauea on Hawai'i Island. This animal was initially identified as a single species, but Wessel et al. (2013) found this clade contains at least seven distinct species. An mtDNA-based molecular phylogeny supported the likelihood of monophyly with a single founder event into caves (Wessel et al. 2013). The oldest surviving troglobiontic species is on Hualālai volcano, with different derived and younger species occurring both on Mauna Loa and Kīlauea. Since the maximum age of the surviving ancestral population (on Hualālai) is ~25,000 years, the six derived species represent the fastest speciation rate among all animals—all while dispersing and speciating underground (Wessel et al. 2013).

Future Research

The evolution and adaptation of subterranean organisms has long been of interest to students of biology (Darwin 1859; Poulson and White 1969; Mammola et al. 2019, 2020). While practitioners have collectively made considerable progress in advancing our understanding of the evolution of troglomorphy and speciation of subterranean fauna, many questions remain. In fact, in a recent exercise aimed toward identifying the 50 highest priority research questions in subterranean biology, Mammola et al. (2020) reported that 21 of these questions were related to the adaptation, origin, and evolution of subterranean fauna. Our aim is not to rehash this, nor do we desire to reexamine the content of other recent reviews and discussions (e.g., Barr and Holsinger 1985; Holsinger 2000; Juan et al. 2010; Culver and Pipan 2015; Mammola et al. 2020). Instead, we wish to expound upon a few topics deemed particularly important to the adaptation, evolution, and speciation of subterranean organisms, outline the current limitations in addressing these questions, and offer some recommendations for future research.

Numerous previous researchers who examined the evolution of subterranean fauna designed their studies as a competing models framework between CRH and ASH (e.g., Rivera et al. 2002; Arnedo et al. 2007; Villacorta et al. 2008; Medeiros et al. 2009; Bribiesca-Contreras et al. 2019; Mendes et al. 2019). While this approach certainly represents a useful starting point, the models driving adaptation and diversification are often quite complex. Firstly, at least seven models driving troglomorphy have been identified (this summary; Juan et al. 2010). Secondly, divergence with gene flow (*sensu* Niemiller et al. 2008) may be interpreted as an intermediate adaptive shift (i.e., speciation in progress). Furthermore, nearly one-third of the studies (10 of 35 papers) examined here involved multiple evolutionary models—elucidating the intricacies of subterranean adaptation. By applying a broader view concerning the drivers of troglomorphy and diversification, future studies may provide additional support for the currently recognized models, reveal variations of existing models, and perhaps lead to identification of novel paradigms.

There is a growing corpus of evidence to support the adaptive shift model. Forty percent of the studies examined here provided evidence for adaptive shifts or divergence with gene flow. As this is not a complete review on evolutionary models driving troglomorphy, we recognize there are likely studies on adaptive shifts not discussed herein. Moreover, all the models discussed here (involving the transition from epigean to hypogean environments) exhibited some degree of an adaptive shift (including CRH). Subsequently, we suspect this process is probably a rather common phenomenon in subterranean systems. We look forward to more evidence testing these evolutionary models as studies on the evolution of subterranean fauna continue.

Our understanding of subterranean-adapted organisms and diversification paradigms has advanced principally from studies of a few well-known species and groups, such as *Astyanax mexicanus*, *Asellus aquaticus*, amblyopsid cavefishes, and dytiscid beetles. However, identification of more universal evolutionary processes governing troglomorphy will require expanding upon the phylogenetic breadth of species studied. In particular, studies on these popular taxonomic groups are expected to continue, while other groups, including flatworms, spiders, pseudoscorpions, mites, millipedes, and springtails, will likely linger behind. We recommend a slight shift in focus so that molecular phylogenies may be generated for these underrepresented groups. Comparative analysis of a robust dataset, representing a broad spectrum of taxonomic groups, may facilitate further refinements in common patterns of subterranean diversity and diversification. Through such an effort, we further anticipate an emergence of a deeper understanding concerning the patterns and models driving species diversity, the prevalence of cryptic diversity, and diversification processes.

Moreover, the range and parameters of subterranean habitats inhabited by each study species should be incisively characterized (e.g., Culver et al. 2010; Hutchins et al. 2014; Pérez-Moreno et al. 2017; Howarth and Wynne, Chap. 1). Relating evolutionary processes to habitat will strengthen our ability to derive more meaningful inference concerning how the environment can influence and shape the evolutionary dynamics of subterranean animals.

Although past discussions of subterranean adaptation and evolution focused primarily on caves (e.g., most of the papers discussed in this chapter), a plethora of terrestrial and aquatic subterranean environments occur across all caverniferous substrates—including the MSS, epikarst, calcrete aquifers, and lava tubes (refer to Howarth and Wynne, Chap. 1, for an overview of subterranean habitats). Additionally, within these environments, the distribution and occurrence of microhabitats are patchily distributed (Chapman 1983; Weinstein and Slaney 1995; Zagmajster et al. 2008; Pellegrini and Ferreira 2013; Wynne et al. 2019). Microhabitats are influenced by proximity and connectivity to the surface, energy and nutrient input, pore size and connectivity, and temperature and moisture, and vary considerably across the subterranean biome. Consequently, the relative importance of various microevolutionary processes, modes of speciation, and evolution of troglobiontic organisms may also differ among these various subterranean environments and habitats. Examining how biotic and abiotic factors influence evolutionary processes in these other subterranean substrates could both bolster and refine our understanding of how fauna becomes adapted to the subterranean environment.

As Juan et al. (2010) and others have stated, our understanding of the drivers of adaptation and diversification will continue to advance through identifying and studying model systems in which surface and cave populations/species are of recent origin. This will enable us to gain insights into early speciation processes operating during colonization and transition from epigean to hypogean habitats, including the roles of selection, drift, mutation, and gene flow. Additionally, this approach will enable us to further codify many of the evolutionary models discussed herein. Studying model organisms that are evolving underground offers important insights into both the later stages of cladogenesis and the evolution of troglomorphy. Importantly, obtaining a better understanding concerning the interplay of evolutionary processes could provide us with insights concerning how anthropogenic climate change and other detrimental human activities influence adaptation to hypogean habitats, as well as the fate of subterranean-restricted/troglobiontic populations and species.

ACKNOWLEDGMENTS

The authors are grateful to Carlos Juan Clar, Arnaud Faille, Francis G. Howarth, and Aron D. Katz for providing insightful comments leading to the improvement of this chapter.

REFERENCES

Ahearn, Gregory A., and Francis G. Howarth. 1982. "Physiology of Cave Arthropods in Hawaii." *Journal of Experimental Zoology* 222:227–238.

Arnedo, Miquel A., Pedro Oromí, Cesc Múrria, Nuria Macías-Hernández, and Carles Ribera. 2007. "The Dark Side of an Island Radiation: Systematics and Evolution of Troglobitic Spiders of the Genus *Dysdera* Latreille (Araneae: Dysderidae) in the Canary Islands." *Invertebrate Systematics* 21:623–660.

Ashmole, N. Philip. 1993. "Colonization of the Underground Environment in Volcanic Islands." *Mémoires de Biospéologie* 20:1–11.

Assmann, Thorsten, Achille Casale, Claudia Drees, Jan C. Habel, Andrea Matern, and Andreas Schuldt. 2010. "Review: The Dark Side of Relict Species Biology: Cave Animals as Ancient Lineages." In *Relict Species*, edited by Jan Christian Habel and Thorsten Assmann, 91–103. Berlin, Heidelberg: Springer.

Barr, Thomas C., Jr. 1967. "Observations on the Ecology of Caves." *American Naturalist* 101:475–491.

Barr, Thomas C., Jr. 1968. "Cave Ecology and the Evolution of Troglobites." In *Evolutionary Biology*, edited by Theodosius Dobzhansky, Max K. Hecht, and William C. Steere, 35–102. Boston: Springer.

Barr, Thomas C., Jr., and John R. Holsinger. 1985. "Speciation in Cave Faunas." *Annual Review of Ecology and Systematics* 16:313–337.

Barr, Thomas C., Jr., and Robert A. Kuehne. 1971. "Ecological Studies in the Mammoth Cave System of Kentucky: II. The Ecosystem." *Annales de Spéléologie* 26:47–96.

Bartlow, Andrew W., and Salvatore J. Agosta. 2020. "Phoresy in Animals: Review and Synthesis of a Common but Understudied Mode of Dispersal." *Biological Reviews* https://doi.org/10.1111/brv.12654.

Beier, M. 1948. "Phoresie und Phagophilie bei Psudoscorpionen." *Ost.* Zoo. *Zeit.* 1:441–497.

Bendik, Nathan F., Jesse M. Meik, Andrew G. Gluesenkamp, Corey E. Roelke, and Paul T. Chippindale. 2013. "Biogeography, Phylogeny, and Morphological Evolution of Central Texas Cave and Spring Salamanders." *BMC Evolutionary Biology* 13:201.

Benoit, Joshua B., Jay A. Yoder, Lawrence W. Zettler, and Horton H. Hobbs. 2004. "Mycoflora of a Trogloxenic Cave Cricket, *Hadenoecus cumberlandicus* (Orthoptera: Rhaphidophoridae), from Two Small Caves in Northeastern Kentucky." *Annals of the Entomological Society of America* 97:989–993.

Bilandžija, Helena, Helena Ćetković, and William R. Jeffery. 2012. "Evolution of Albinism in Cave Planthoppers by a Convergent Defect in the First Step of Melanin Biosynthesis." *Evolution & Development* 14: 196–203.

Binns, E. S. 1982. "Phoresy as Migration—Some Functional Aspects of Phoresy in Mites." *Biological Reviews* 57:571–620.

Bradic, Martina, Peter Beerli, Francisco J. García-de León, Sarai Esquivel-Bobadilla, and Richard L. Borowsky. 2012. "Gene Flow and Population Structure in the Mexican Blind Cavefish Complex (*Astyanax mexicanus*)." *BMC Evolutionary Biology* 12:9.

Bribiesca-Contreras, Guadalupe, Tania Pineda-Enríquez, Francisco Márquez-Borrás, Francisco A. Solís-Marín, Heroen Verbruggen, Andrew F. Hugall, et al. 2019. "Dark Offshoot: Phylogenomic Data Sheds Light on the Evolutionary History of a New Species of Cave Brittle Star." *Molecular Phylogenetics and Evolution* 136:151–163.

Chakrabarty, Prosanta, Jacques A. Prejean, and Matthew L. Niemiller. 2014. "The Hoosier Cavefish, a New and Endangered Species (Amblyopsidae, *Amblyopsis*) from the Caves of Southern Indiana." *ZooKeys* 412:41–57.

Chapman, Phillip R. J. 1983. "Species Diversity in a Tropical Cave Ecosystem." *Proceedings of the University of Bristol Speleological Society* 16:201–213.

Christiansen, Kenneth. 2012. "Morphological Adaptations." In *Encyclopedia of Caves*, 2nd ed., edited by David C. Culver and William B. White, 517–528. Amsterdam: Elsevier, Academic Press.

Coghill, Lyndon M., C. Darrin Hulsey, Johel Chaves-Campos, Francisco J. García de Leon, and Steven G. Johnson. 2014. "Next Generation Phylogeography of Cave and Surface *Astyanax mexicanus*." *Molecular Phylogenetics and Evolution* 79:368–374.

Contreras-Díaz, Hermans G., Oscar Moya, Pedro Oromí, and Carlos Juan. 2007. "Evolution and Diversification of the Forest and Hypogean Ground-Beetle Genus *Trechus* in the Canary Islands." *Molecular Phylogenetics and Evolution* 42:687–699.

Culver, David C. 1982. *Cave Life: Evolution and Ecology*. Cambridge, MA: Harvard University Press.

Culver, David C., and Tanja Pipan. 2014. *Shallow Subterranean Habitats: Ecology, Evolution, and Conservation*. Oxford: Oxford University Press.

Culver, David C., and Tanja Pipan. 2015. "Shifting Paradigms of the Evolution of Cave Life." *Acta Carsologica* 44:415–425.

Culver, David C., John R. Holsinger, Mary C. Christman, and Tanja Pipan. 2010. "Morphological Differences among Eyeless Amphipods in the Genus *Stygobromus* Dwelling in Different Subterranean Habitats." *Journal of Crustacean Biology* 30:68–74.

Ćurčić, Božidar P. M., Walter Sudhaus, Rajko N. Dimitrijević, Slobodan E. Makarov, and Vladimir T. Tomić. 2008. "*Rhabditophanes schneideri* (Rhabditida) Phoretic on a Cave Pseudoscorpion." *Journal of Invertebrate Pathology* 99:254–256.

Ćurčić, B. P. M., B. S. Ilić, T. Rađa, S. Makarov, V. T. Tomić, and R. N. Dimitrijević. 2011. "On Two New Cave-Dwelling and Relict Pseudoscorpions of the Genus *Chthonius* C. L. Koch (Chthonidae, Pseudoscorpiones) from Bosnia." *Archives of Biological Sciences* 63:847–854.

Darwin, Charles. 1859. *On the Origin of Species by Means of Natural Selection, or the Preservation of Favoured Races in the Struggle of Life*. London: John Murray.

Dattagupta, Sharmishtha, Irene Schaperdoth, Alessandro Montanari, Sandro Mariani, Noriko Kita, John W. Valley, and Jennifer L. Macalady. 2009. "A Novel Symbiosis between Chemoautotrophic Bacteria and a Freshwater Cave Amphipod." *The ISME Journal* 3:935–943.

Deharveng, Louis. 2004. "Diversity Patterns in the Tropics." In *Encyclopedia of Caves*, edited by David C. Culver and William B. White, 166–170. Amsterdam: Elsevier, Academic Press.

Deharveng, Louis, Cyrille A. D'Haese, and Anne Bedos. 2008. "Global Diversity of Springtails (Collembola; Hexapoda) in Freshwater." *Hydrobiologia* 595:329–338.

Del-Claro, Kleber, and Everton Tizo-Pedroso. 2009. "Ecological and Evolutionary Pathways of Social Behavior in Pseudoscorpions (Arachnida: Pseudoscorpiones)." *Acta Ethologica* 12: 13–22.

Delić, Teo, Peter Trontelj, Valerija Zakšek, and Cene Fišer. 2016. "Biotic and Abiotic Determinants of Appendage Length Evolution in a Cave Amphipod." *Journal of Zoology* 299:42–50.

Derkarabetian, Shahan, David B. Steinmann, and Marshal Hedin. 2010. "Repeated and Time-Correlated Morphological Convergence in Cave-Dwelling Harvestmen (Opiliones, Laniatores) from Montane Western North America." *PLOS ONE* 5:e10388.

Dole-Olivier, M. J., F. Castellarini, N. Coineau, D. M. P. Galassi, P. Martin, N. Mori, A. Valdecasas, and J. Gibert. 2009. "Towards an Optimal Sampling Strategy to Assess Groundwater Biodiversity: Comparison across Six European Regions." *Freshwater Biology* 54:777–796.

Ducarme, Xavier, and Philippe Lebrun. 2004. "Spatial Microdistribution of Mites and Organic Matter in Soils and Caves." *Biology and Fertility of Soils* 39: 457–466.

Ducarme, Xavier, Henri M. André, Georges Wauthy, and Philippe Lebrun. 2004. "Comparison of Endogenic and Cave Communities: Microarthropod Density and Mite Species Richness." *European Journal of Soil Biology* 40:129–138.

Dumnicka, Elzbieta, Tanja Pipan, and David C. Culver. 2020. "Habitats and Diversity of Subterranean Macroscopic Freshwater Invertebrates: Main Gaps and Future Trends." *Water* 12:2170.

Ebermann, Ernst, and Jose G. Palacios-Vargas. 1989. "*Imparipes (Imparipes) tocatlphilus* n. sp. (Acari, Tarsonemina, Scutacaridae) from Mexico and Brazil—First Record of Ricinuleids as Phoresy Hosts for Scutacarid Mites." *Acarologia* 29:347–354.

Eigenmann, Carl H. 1909. *Cave Vertebrates of America: A Study in Degenerative Evolution*. Washington, DC: Carnegie Institution of Washington.

Faille, Arnaud, Carmelo Andújar, Floren Fadrique, and Ignacio Ribera. 2014. "Late Miocene Origin of an Ibero-Maghrebian Clade of Ground Beetles with Multiple Colonizations of the Subterranean Environment." *Journal of Biogeography* 41:1979–1990.

Faille, Arnaud, Charles Bourdeau, Xavier Belles, and Javier Fresneda. 2015. "Allopatric Speciation Illustrated: The Hypogean Genus *Geotrechus* Jeannel, 1919 (Coleoptera: Carabidae: Trechini), with Description of Four New Species from the Eastern Pyrenees (Spain)." *Arthropod Systematics & Phylogeny* 73:439–455.

Faille, Arnaud, Ignacio Ribera, Louis Deharveng, Charles Bourdeau, Lionel Garnery, Eric Quéinnec, et al. 2010. "A Molecular Phylogeny Shows the Single Origin of the Pyrenean Subterranean Trechini Ground Beetles (Coleoptera: Carabidae)." *Molecular Phylogenetics and Evolution* 54:97–106.

Faille, Arnaud, Rene Tänzler, and Emmanuel F. A. Toussaint. 2015. "On the Way to Speciation: Shedding Light on the Karstic Phylogeography of the Microendemic Cave Beetle *Aphaenops cerberus* in the Pyrenees." *Journal of Heredity* 106:692–699.

Fenolio, Danté B., G. O. Graening, Bret A. Collier, and Jim F. Stout. 2006. "Coprophagy in a Cave-Adapted Salamander: The Importance of Bat Guano Examined through Nutritional and Stable Isotope Analysis." *Proceedings of the Royal Society B: Biological Sciences* 273:439–443.

Ferreira, Rodrigo Lopes, and Rogério Parentoni Martins. 1999. "Trophic Structure and Natural History of Bat Guano Invertebrate Communities, with Special Reference to Brazilian Caves." *Tropical Zoology* 12:231–252.

Ferreira, Rodrigo Lopes, Xavier Prous, and Rogério Parentoni Martins. 2007. "Structure of Bat Guano Communities in a Dry Brazilian Cave." *Tropical Zoology* 20:55–74.

Fitzpatrick, Benjamin M., James A. Fordyce, and S. Gavrilets. 2008. "What, If Anything, Is Sympatric Speciation?" *Journal of Evolutionary Biology* 21:1452–1459.

Francke, Oscar F., and Gabriel A. Villegas-Guzmán. 2006. "Symbiotic Relationships between Pseudoscorpions (Arachnida) and Packrats (Rodentia)." *The Journal of Arachnology* 34: 289–298.

Gers, Charles. 1998. "Diversity of Energy Fluxes and Interactions between Arthropod Communities: From Soil to Cave." *Acta Oecologica* 19:205–213.

Gnaspini, Pedro, and Eleonora Trajano. 2000. "Guano Communities in Tropical Caves." In *Ecosystems of the World: Subterranean Ecosystems*, edited by Horst Wilkens, David C. Culver, and William F. Humphreys, 251–268. Amsterdam: Elsevier.

Goricki, Špela, and Peter Trontelj. 2006. "Structure and Evolution of the Mitochondrial Control Region and Flanking Sequences in the European Cave Salamander *Proteus anguinus*." *Gene* 378:31–41.

Gould, Stephen Jay, and Elisabeth S. Vrba. 1982. "Exaptation—a Missing Term in the Science of Form." *Paleobiology* 8:4–15.

Grandcolas, Philippe, Romain Nattier, and Steve Trewick. 2014. "Relict Species: A Relict Concept?" *Trends in Ecology and Evolution* 29:655–663.

Hahn, Hans J., 2002. "Methods and Difficulties of Sampling Stygofauna—an Overview." In *Field Screening Europe 2001: Proceedings of the Second International Conference on Strategies and Techniques for the Investigation and Monitoring of Contaminated Sites*, edited by Wolfgang Breh, Johannes Gottlieb, Heinz Hötzl, Frieder Kern, Tanja Liesch, and Reinhard Niessne, 201–205. Dordrecht: Springer.

Hart, Pamela B., Matthew L. Niemiller, Edward D. Burress, Jonathan W. Armbruster, William B. Ludt, and Prosanta Chakrabarty. 2020. "Cave-Adapted Evolution in the North American Amblyopsid Fishes Inferred Using Phylogenomics and Geometric Morphometrics." *Evolution* 74:936–949.

Hedin, Marshal. 1997. "Molecular Phylogenetics at the Population/Species Interface in Cave Spiders of the Southern Appalachians (Araneae: Nesticidae: *Nesticus*)." *Molecular Biology and Evolution* 14:309–324.

Hedin, Marshal, and Steven M. Thomas. 2010. "Molecular Systematics of Eastern North American Phalangodidae (Arachnida: Opiliones: Laniatores), Demonstrating Convergent Morphological Evolution in Caves." *Molecular Phylogenetics and Evolution* 54:107–121.

Herman, Adam, Yaniv Brandvain, James Weagley, William R. Jeffery, Alex C. Keene, Thomas J. Y. Kono, et al. 2018. "The Role of Gene Flow in Rapid and Repeated Evolution of Cave-Related Traits in Mexican Tetra, *Astyanax mexicanus*." *Molecular Ecology* 27:4397–4416.

Hichem, Khammar, Hadjab, Ramzi, and Merzoug Djemoi. 2019. "Biodiversity and Distribution of Groundwater Fauna in the Oum-El-Bouaghi Region (Northeast of Algeria)." *Biodiversitas Journal of Biological Diversity* 20:3553–3558.

Hobbs, Horton H., III. 1992. "Caves and Springs." In *Biodiversity of the Southeastern United States Aquatic Communities*, edited by Courtney T. Hackney, S. Marshall Adams, and William H. Martin, 59–131. London: Wiley.

Hoch, Hannelore, and Francis G. Howarth. 1999. "Multiple Cave Invasions by Species of the Planthopper Genus *Oliarus* in Hawaii (Homoptera: Fulgoroidea: Cixiidae)." *Zoological Journal of the Linnean Society* 127:453–475.

Holdhaus, K. 1933. "Die europäische Höhlenfauna in ihren Beziehungen zur Eiszeit." *Zoogeographica* 1:1–53.

Holsinger, John R. 2000. "Ecological Derivation, Colonization, and Speciation." In *Ecosystems of the World: Subterranean Ecosystems*, edited by Horst Wilkens, David C. Culver, and William F. Humphreys, 399–416. Amsterdam: Elsevier.

Houck, M. A., and Barry M. O'Connor. 1991. "Ecological and Evolutionary Significance of Phoresy in the Astigmata." *Annual Review of Entomology* 36:611–636.

Howarth, Francis G. 1972. "Cavernicoles in Lava Tubes on the Island of Hawaii." *Science* 175:325–326.

Howarth, Francis G. 1973. "The Cavernicolous Fauna of Hawaiian Lava Tubes, 1. Introduction." *Pacific Insects* 15:139–151.

Howarth, Francis G. 1980. "The Zoogeography of Specialized Cave Animals: A Bioclimatic Model." *Evolution* 34:394–406.

Howarth, Francis G. 1982. "Bioclimatic and Geologic Factors Governing the Evolution and Distribution of Hawaiian Cave Insects." *Entomologia Generalis* 8:17–26.

Howarth, Francis G. 1983. "Ecology of Cave Arthropods." *Annual Review of Entomology* 28:365–389.

Howarth, Francis G. 1993. "High-Stress Subterranean Habitats and Evolutionary Change in Cave-Inhabiting Arthropods." *The American Naturalist* 142:S65–S77.

Howarth, Francis G. 2019. "Adaptive Shifts." In *Encyclopedia of Caves*, 3rd ed., edited by William B. White, David C. Culver, and Tanja Pipan, 47–55. Amsterdam: Elsevier, Academic Press.

Howarth, Francis G., and Oana T. Moldovan. 2018. "The Ecological Classification of Cave Animals and Their Adaptations." In *Cave Ecology*, edited by Oana T. Moldovan, Ľubomír Kováč, and Stuart Halse, 41–67. New York: Springer Publishing Company.

Howarth, Francis G., and Fred D. Stone. 1990. "Elevated Carbon Dioxide Levels in Bayliss Cave, Australia: Implications for the Evolution of Obligate Cave Species." *Pacific Science* 44:207–218.

Howarth, Francis G., and Fred D. Stone. 2020. "Impacts of Invasive Rats on Hawaiian Cave Resources." *International Journal of Speleology* 49:35–42.

Howarth, Francis G., Matthew J. Medieros, and Fred D. Stone. 2020. "Hawaiian Lava Tube Cave Associated Lepidoptera from the Collections of Francis G. Howarth and Fred D. Stone." *Bishop Museum Occasional Papers* 129:37–54.

Hunter, Rebecca L., Michael Scott Webb, Thomas M. Iliffe, and Jaime R. Alvarado Bremer. 2008. "Phylogeny and Historical Biogeography of the Cave-Adapted Shrimp Genus *Typhlatya* (Atyidae) in the Caribbean Sea and Western Atlantic." *Journal of Biogeography* 35:65–75.

Hutchins, Benjamin T., Benjamin F. Schwartz, and Weston H. Nowlin. 2014. "Morphological and Trophic Specialization in a Subterranean Amphipod Assemblage." *Freshwater Biology* 59:2447–2461.

Jasinska, Edyta J., Brenton Knott, and Arthur J. McComb. 1996. "Root Mats in Ground Water: A Fauna-Rich Cave Habitat." *Journal of the North American Benthological Society* 15:508–519.

Jeannel, R. 1943. *Les Fossiles Vivants des Cavernes*. Gallimard.

Juan, Carlos, Michelle T. Guzik, Damià Jaume, and Steven J. B. Cooper. 2010. "Evolution in Caves: Darwin's 'Wrecks of Ancient Life' in the Molecular Era." *Molecular Ecology* 19:3865–3880.

Juberthie, Christian, Bernard Delay, and Michael Bouillon. 1980. Extension du Milieu Souterrain Superficiel en Zone Non-Calcaire: Description d'un Nouveau Milieu et de son Peuplement par les Coleopteres Troglobies. *Mémoires de Biospéologie* 7:19–52.

Kano, Yasunori, and Tomoki Kase. 2004. "Genetic Exchange between Anchialine Cave Populations by Means of Larval Dispersal: The Case of a New Gastropod Species *Neritilia cavernicola*." *Zoologica Scripta* 33:423–437.

Katz, Aron D., Steven J. Taylor, and Mark A. Davis. 2018. "At the Confluence of Vicariance and Dispersal: Phylogeography of Cavernicolous Springtails (Collembola: Arrhopalitidae, Tomoceridae) Codistributed across a Geologically Complex Karst Landscape in Illinois and Missouri." *Ecology and Evolution* 8:10306–10325.

Konec, M., S. Prevorčnik, S. M. Sarbu, R. Verovnik, and P. Trontelj. 2015. "Parallels between Two Geographically and Ecologically Disparate Cave Invasions by the Same Species, *Asellus aquaticus* (Isopoda, Crustacea)." *Journal of Evolutionary Biology* 28:864–875.

Kozel, Peter, Tanja Pipan, Stefano Mammola, David C. Culver, and Tone Novak. 2019. "Distributional Dynamics of a Specialized Subterranean Community Oppose the Classical Understanding of the Preferred Subterranean Habitats." *Invertebrate Biology* 138:e12254.

Krijgsman, W., M. Garcés, C. G. Langereis, R. Daams, J. Van Dam, A. J. Van der Meulen, et al. 1996. "A New Chronology for the Middle to Late Miocene Continental Record in Spain." *Earth and Planetary Science Letters* 142:367–380.

Kruckenhauser, Luise, Elisabeth Haring, Robert Seemann, and Helmut Sattmann. 2011. "Genetic Differentiation between Cave and Surface-Dwelling Populations of *Garra barreimiae* (Cyprinidae) in Oman." *BMC Evolutionary Biology* 11:1–15.

Ladle, Richard J., João V. L. Firmino, Ana C. M. Malhado, and Armando Rodríguez-Durán. 2012. "Unexplored Diversity and Conservation Potential of Neotropical Hot Caves." *Conservation Biology* 26:978–982.

Langille, Barbara L., Josephine Hyde, Kathleen M. Saint, Tessa M. Bradford, Danielle N. Stringer, Simon M. Tierney, et al. 2020. "Evidence for Speciation Underground in Diving Beetles (Dytiscidae) from a Subterranean Archipelago." *Evolution* https://doi.org/10.1111/evo.14135.

Leijs, Remko, Egbert H. Van Nes, Chris H. Watts, Steven J. B. Cooper, William F. Humphreys, and Katja Hogendoorn. 2012. "Evolution of Blind Beetles in Isolated Aquifers: A Test of Alternative Modes of Speciation." *PLOS ONE* 7:e34260.

Leray, Vincent L., Jason Caravas, Markus Friedrich, and Kirk S. Zigler. 2019. "Mitochondrial Sequence Data Indicate 'Vicariance by Erosion' as a Mechanism of Species Diversification in North American *Ptomaphagus* (Coleoptera, Leiodidae, Cholevinae) Cave Beetles." *Subterranean Biology* 29:35–57.

Leys, Remko, Chris H. S. Watts, Steve J. B. Cooper, and William F. Humphreys. 2003. "Evolution of Subterranean Diving Beetles (Coleoptera: Dytiscidae Hydroporini, Bidessini) in the Arid Zone of Australia." *Evolution* 57:2819–2834.

Liu, Sai, Jianling Li, Kun Guo, Haili Qiao, Rong Xu, Jianmin Chen, et al. 2016. "Seasonal Phoresy as an Overwintering Strategy of a Phytophagous Mite." *Scientific Reports* 6:25483.

Liu, Weixin, Sergei Golovatch, Thomas Wesener, and Mingyi Tian. 2017. "Convergent Evolution of Unique Morphological Adaptations to a Subterranean Environment in Cave Millipedes (Diplopoda)." *PLOS ONE* 12:e0170717.

Mammola, Stefano, Isabel R. Amorim, Maria E. Bichuette, Paulo A. V. Borges, Naowarat Cheeptham, Steven J. B. Cooper, et al. 2020. "Fundamental Research Questions in Subterranean Biology." *Biological Reviews* 95:1855–1872.

Mammola, Stefano, Elena Piano, Pedro Cardoso, Philippe Vernon, David Domínguez-Villar, David C. Culver, et al. 2019. "Climate Change Going Deep: The Effects of Global Climatic Alterations on Cave Ecosystems." *The Anthropocene Review* 6:98–116.

Medeiros, Matthew J., Don Davis, Francis G. Howarth, and Rosemary Gillespie. 2009. "Evolution of Cave Living in Hawaiian *Schrankia* (Lepidoptera: Noctuidae) with Description of a Remarkable New Cave Species." *Zoological Journal of the Linnean Society* 156:114–139.

Mendes, Izabela Santos, Francisco Prosdocimi, Alex Schomaker-Bastos, Carolina Furtado, Rodrigo Lopes Ferreira, Paulo Santos Pompeu, et al. 2019. "On the Evolutionary Origin of Neotropical Cavefish *Ancistrus cryptophthalmus* (Siluriformes, Loricariidae) Based on the Mitogenome and Genetic Structure of Cave and Surface Populations." *Hydrobiologia* 842:157–171.

Mitchell, Rodger. 1970. "An Analysis of Dispersal in Mites." *The American Naturalist* 104:425–431.

Niemiller, Matthew L. 2011. "Evolution, Speciation, and Conservation of Amblyopsid Cavefishes." PhD diss., University of Tennessee, Knoxville.

Niemiller, Matthew L., and Thomas L. Poulson. 2010. "Subterranean Fishes of North America: Amblyopsidae." *Biology of Subterranean Fishes*, edited by Eleonora Trajano, Maria Elina Bichuette, and B. G. Kapoor, 169–280. Enfield, NH: Science Publishers.

Niemiller, Matthew L., Benjamin M. Fitzpatrick, and Brian T. Miller. 2008. "Recent Divergence with Gene Flow in Tennessee Cave Salamanders (Plethodontidae: Gyrinophilus) Inferred from Gene Genealogies." *Molecular Ecology* 17:2258–2275.

Niemiller, Matthew L., Benjamin M. Fitzpatrick, Premal Shah, Lars Schmitz, and Thomas J. Near. 2013. "Evidence for Repeated Loss of Selective Constraint in *Rhodopsin* of Amblyopsid Cavefishes (Teleostei: Amblyopsidae)." *International Journal of Organic Evolution* 67:732–748.

Niemiller, Matthew L., James R. McCandless, R. Graham Reynolds, James Caddle, Thomas J. Near, Christopher R. Tillquist, et al. 2013. "Effects of Climatic and Geological Processes during the Pleistocene on the Evolutionary History of the Northern Cavefish, *Amblyopsis spelaea* (Teleostei: Amblyopsidae)." *Evolution* 67:1011–1025.

Niemiller, Matthew L., Thomas J. Near, and Benjamin M. Fitzpatrick. 2012. "Delimiting Species Using Multilocus Data: Diagnosing Cryptic Diversity in the Southern Cavefish, *Typhlichthys subterraneus* (Teleostei: Amblyopsidae)." *International Journal of Organic Evolution* 66:846–866.

Nitzu, E., A. Nae, R. Băncilă, I. Popa, A. Giurginca, and R. Plăiaşu. 2014. "Scree Habitats: Ecological Function, Species Conservation and Spatial-Temporal Variation in the Arthropod Community." *Systematics and Biodiversity* 12:65–75.

Oromí, P., J. L. Martín, A. L. Medina, and I. Izquierdo. 1991. "The Evolution of the Hypogean Fauna in the Canary Islands." In *The Unity of Evolutionary Biology*, edited by E. C. Dudley, 380–395. Portland, OR: Dioscorides Press.

Osikowski, Artur, Sebastian Hofman, Dilian Georgiev, Aleksandra Rysiewska, and Andrzej Falniowski. 2017. "Unique, Ancient Stygobiont Clade of Hydrobiidae (Truncatelloidea) in Bulgaria: The Origin of Cave Fauna." *Folia Biologica (Kraków)* 65:79–93.

Pacioglu, Octavian, Nicoleta Ianovici, Mărioara N. Filimon, Adrian Sinitean, Gabriel Iacob, Henrietta Barabas, et al. 2019. "The Multifaceted Effects Induced by Floods on the Macroinvertebrate Communities Inhabiting a Sinking Cave Stream." *International Journal of Speleology* 48:167–177.

Palacios-Vargas, José G., Daniela Cortés-Guzmán, and Javier Alcocer. 2018. "Springtails (Collembola, Hexapoda) from Montebello Lakes, Chiapas, Mexico." *Inland Waters* 8:264–272.

Pallarés, Susana, Raquel Colado, Toni Pérez-Fernández, Thomas Wesener, Ignacio Ribera, and David Sánchez-Fernández. 2019. "Heat Tolerance and Acclimation Capacity in Subterranean Arthropods Living under Common and Stable Thermal Conditions." *Ecology and Evolution* 9: 13731–13739.

Peck, Stewart B. 1974. "The Invertebrate Fauna of Tropical American Caves, Part II: Puerto Rico, an Ecological and Zoogeographic Analysis." *Biotropica* 6:14–31.

Peck, Stewart B. 1980. "Climatic Change and the Evolution of Cave Invertebrates in the Grand Canyon, Arizona." *NSS Bulletin* 42:53–60.

Peck, Stewart B., and J. Judson Wynne. 2013. "*Ptomaphagus parashant* Peck and Wynne, New Species (Coleoptera: Leiodidae: Cholevinae: Ptomaphagini): The Most Troglomorphic Cholevine Beetle Known from Western North America." *The Coleopterists Bulletin* 67:309–317.

Pellegrini, Thais Giovannini, and Rodrigo Lopes Ferreira. 2013. "Structure and Interactions in a Cave Guano-Soil Continuum Community." *European Journal of Soil Biology* 57:19–26.

Pérez-Moreno, Jorge L., Gergely Balázs, Blake Wilkins, Gábor Herczeg, and Heather D. Bracken-Grissom. 2017. "The Role of Isolation on Contrasting Phylogeographic Patterns in Two Cave Crustaceans." *BMC Evolutionary Biology* 17:247.

Pinho, Catarina, and Jody Hey. 2010. "Divergence with Gene Flow: Models and Data." *Annual Review of Ecology and Evolutionary Systematics* 51:215–230.

Pipan, Tanja, and Anton Brancelj. 2001. "Ratio of Copepods (Crustacea: Copepoda) in Fauna of Percolation Water in Six Karst Caves in Slovenia." *Acta Carsologica* 9:257–265.

Pipan, Tanja, and David C. Culver. 2012. "Convergence and Divergence in the Subterranean Realm: A Reassessment." *Biological Journal of the Linnean Society* 107:1–14.

Pipan, Tanja, Mary C. Christman, and David C. Culver. 2020. "Abiotic Community Constraints in Extreme Environments: Epikarst Copepods as a Model System." *Diversity* 12:269.

Plath, Martin, Bernd Hermann, Christiane Schröder, Rüdiger Riesch, Michael Tobler, Francisco J. García de León, et al. 2010. "Locally Adapted Fish Populations Maintain

Small-Scale Genetic Differentiation Despite Perturbation by a Catastrophic Flood Event." *BMC Evolutionary Biology* 10:256.

Poinar, George O., Jr., Bozidar P. M. Curcic, and James C. Cokendolpher. 1998. "Arthropod Phoresy Involving Pseudoscorpions in the Past and Present." *Acta Arachnologica* 47:79–96.

Poulson, Thomas L. 1963. "Cave Adaptation in Amblyopsid Fishes." *American Midland Naturalist* 70:257–290.

Poulson, Thomas L., and William B. White. 1969. "The Cave Environment." *Science* 165:971–981.

Proudlove, Graham S. 2006. *Subterranean Fishes of the World*. Moulis, France: International Society for Subterranean Biology.

Racovitza, Emil G. 1907. Éssai sur les Problèmes Biospéleologiques. *Archives du Zoologie Experimentale et Generale* 6:371–488.

Rendoš, Michal, Dana Miklisová, Ľubomír Kováč, and Andrej Mock. 2020. "Dynamics of Collembola (Hexapoda) in a Forested Limestone Scree Slope, Western Carpathians, Slovakia." *Journal of Cave and Karst Studies* 82:18–29.

Rivera, Malia Ana J., Francis G. Howarth, Stefano Taiti, and George K. Roderick. 2002. "Evolution in Hawaiian Cave-Adapted Isopods (Oniscidea: Philosciidae): Vicariant Speciation or Adaptive Shifts?" *Molecular Phylogenetics and Evolution* 25:1–9.

Rizzo, Valeria, David Sánchez-Fernández, Javier Fresneda, Alexandra Cieslak, and Ignacio Ribera. 2015. "Lack of Evolutionary Adjustment to Ambient Temperature in Highly Specialized Cave Beetles." *BMC Evolutionary Biology* 15:10.

Santos, Scott R. 2006. "Patterns of Genetic Connectivity among Anchialine Habitats: A Case Study of the Endemic Hawaiian Shrimp *Halocaridina rubra* on the Island of Hawaii." *Molecular Ecology* 15:2699–2718.

Schäffer, Sylvia, and Stephan Koblmüller. 2020. "Unexpected Diversity in the Host-Generalist Oribatid Mite *Paraleius leontonychus* (Oribatida, Scheloribatidae) Phoretic on Palearctic Bark Beetles." *PeerJ* 8:e9710.

Schilthuizen, Menno, A. S. Cabanban, and Martin Haase. 2005. "Possible Speciation with Gene Flow in Tropical Cave Snails." *Journal of Zoological Systematics and Evolutionary Research* 43:133–138.

Schilthuizen, Menno, Elise M. J. Rutten, and Martin Haase. 2012. "Small-Scale Genetic Structuring in a Tropical Cave Snail and Admixture with its Above-Ground Sister Species." *Biological Journal of the Linnean Society* 105:727–740.

Searle, Michael P. 2006. "Role of the Red River Shear Zone, Yunnan and Vietnam, in the Continental Extrusion of SE Asia." *Journal of the Geological Society* 163:1025–1036.

Segherloo, Iraj Hashemzadeh, Eric Normandeau, Laura Benestan, Clément Rougeux, Guillaume Coté, Jean-Sébastien Moore, et al. 2018. "Genetic and Morphological Support for Possible Sympatric Origin of Fish from Subterranean Habitats." *Scientific Reports* 8:1–13.

Sendra, Alberto, and Ana Sofia P.S. Reboleira. 2012. "The World's Deepest Subterranean Community—Krubera-Voronja Cave (Western Caucasus)." *International Journal of Speleology* 41:221–230.

Sarbu, Serban M., Thomas C. Kane, and Brian K. Kinkle. 1996. "A Chemoautotrophically Based Cave Ecosystem." *Science* 272:1953–1955.

Shaw, Peter, Mark Dunscombe, and Anne Robertson. 2011. "Collembola in the Hyporheos of a Karstic River: An Overlooked Habitat for Collembola Containing a New Genus for the UK." *Soil Organisms* 83:507–514.

Silva, Marconi, Rogério Parentoni Martins, and Rodrigo Lopes Ferreira. 2011. "Trophic Dynamics in a Neotropical Limestone Cave." *Subterranean Biology* 9:127–138.

Simon, Kevin S., Tanja Pipan, and David C. Culver. 2007. "A Conceptual Model of the Flow and Distribution of Organic Carbon in Caves." *Journal of Cave and Karst Studies* 69:279–284.

Sket, Boris. 2008. "Can We Agree on an Ecological Classification of Subterranean Animals?" *Journal of Natural History* 42:1549–1563.

Souza, Rodrigo Antônio Castro, Leopoldo Ferreira de Oliveira Bernardi, and Rodrigo Lopes Ferreira. 2018. "First Record of a Phoretic Mite (Histiostomatidae) on a Cave Dwelling Cricket (Phalangopsidae) from Brazil." *Neotropical Biology and Conservation* 13:171–176.

Stone, Fred D., Francis G. Howarth, Hannelore Hoch, and Manfred Asche. 2012. "Root Communities in Lava Tubes." In *Encyclopedia of Caves*, 2nd ed., edited by William B. White and David C. Culver, 658–664. Amsterdam: Elsevier, Academic Press.

Strecker, Ulrike, Bernhard Hausdorf, and Horst Wilkens. 2003. "Genetic Divergence between Cave and Surface Populations of *Astyanax* in Mexico (Characidae, Teleostei)." *Molecular Ecology* 12:699–710.

Strecker, Ulrike, Bernhard Hausdorf, and Horst Wilkens. 2012. "Parallel Speciation in *Astyanax* Cave Fish (Teleostei) in Northern Mexico." *Molecular Phylogenetics and Evolution* 62:62–70.

Titus, Timothy N., J. Judson Wynne, Murzy D. Jhabvala, Glen E. Cushing, Peter Shu, and Nathalie A. Cabrol. 2011. "Cave Detection Using Oblique Thermal Imaging, Abstract #8024." *First International Planetary Caves Workshop, Carlsbad, New Mexico.*

Tobin, Benjamin W., Benjamin T. Hutchins, and Benjamin F. Schwartz. 2013. "Spatial and Temporal Changes in Invertebrate Assemblage Structure from the Entrance to Deep-Cave Zone of a Temperate Marble Cave." *International Journal of Speleology* 42:203–214.

Trajano, Eleonora, and Marcelo R. de Carvalho. 2017. "Towards a Biologically Meaningful Classification of Subterranean Organisms: A Critical Analysis of the Schiner-Racovitza System from a Historical Perspective, Difficulties of its Application and Implications for Conservation." *Subterranean Biology* 22:1–26.

Trontelj, Peter, Andrej Blejec, and Cene Fišer. 2012. "Ecomorphological Convergence of Cave Communities." *International Journal of Organic Evolution* 66:3852–3865.

Urry, Lisa A., Michael L. Cain, Steven A. Wasserman, Peter V. Minorsky, and Jane B. Reece. 2016. *Campbell Biology, 11th edition, Study Guide*. New York: Pearson.

Villacorta, Carlos, Damià Jaume, Pedro Oromí, and Carlos Juan. 2008. "Under the Volcano: Phylogeography and Evolution of the Cave-Dwelling *Palmorchestia hypogaea* (Amphipoda, Crustacea) at La Palma (Canary Islands)." *BMC Biology* 6:1–14.

Weinstein, Phillip, and David Slaney. 1995. "Invertebrate Faunal Survey of Rope Ladder Cave, Northern Queensland: A Comparative Study of Sampling Methods." *Journal of the Australian Entomological Society* 34:233–236.

Welbourn, W. Calvin. 1976. "Preliminary Report on the Fauna of the Earth Cracks." In *Wupatki National Monument Earth Cracks*, 32–41. Yellow Springs, Ohio: Cave Research Foundation.

Wessel, Andreas, Hannelore Hoch, Manfred Asche, Thomas von Rintelen, Björn Stelbrink, Volker Heck, et al. 2013. "Founder Effects Initiated Rapid Species Radiation in Hawaiian Cave Planthoppers." *Proceedings of the National Academy of Sciences* 110:9391–9396.

Wilkens, Horst. 2001. "Convergent Adaptations to Cave Life in the *Rhamdia laticauda* Catfish Group." *Environmental Biology of Fishes* 62:251–261.

Wilkens, Horst, and Ulrike Strecker. 2003. "Convergent Evolution of the Cavefish *Astyanax* (Characidae, Teleostei): Genetics Evidence from Reduced Eye-Size and Pigmentation." *Biological Journal of the Linnean Society* 80:545–554.

Witt, Jonathan D. S., Doug L. Threloff, and Paul D. N. Hebert. 2006. "DNA Barcoding Reveals Extraordinary Cryptic Diversity in an Amphipod Genus: Implications for Desert Spring Conservation." *Molecular Ecology* 15:3073–3082.

Wynne, J. Judson, Francis G. Howarth, Stefan Sommer, and Brett G. Dickson. 2019. "Fifty Years of Cave Arthropod Sampling: Techniques and Best Practices." *International Journal of Speleology* 48:33–48.

Zagmajster, Maja, David C. Culver, and Boris Sket. 2008. "Species Richness Patterns of Obligate Subterranean Beetles (Insecta: Coleoptera) in a Global Biodiversity Hotspot—Effect of Scale and Sampling Intensity." *Diversity and Distributions* 14:95–105.

Zakšek, Valerija, Teo Delić, Cene Fišer, Branko Jalžić, and Peter Trontelj. 2019. "Emergence of Sympatry in a Radiation of Subterranean Amphipods." *Journal of Biogeography* 46:657–669.

Zakšek, Valerija, Boris Sket, Sanja Gottstein, Damjan Franjević, and Peter Trontelj. 2009. "The Limits of Cryptic Diversity in Groundwater: Phylogeographic of Cave Shrimp *Troglocaris anophthalmus* (Crustacea: Decapoda: Atyidae)." *Molecular Ecology* 18:931–946.

Zhang, Yuanyuan, and Shuqiang Li. 2013. "Ancient Lineage, Young Troglobites: Recent Colonization of Caves by *Nesticella* Spiders." *BMC Evolutionary Biology* 13:1–10.

Zhang, Yuanyuan, and Shuqiang Li. 2014. "A Spider Species Complex Revealed High Cryptic Diversity in South China Caves." *Molecular Phylogenetics and Evolution* 79:353–358.

3

Biology and Ecology of Subterranean Mollusca

Jozef Grego

Introduction

The subterranean radiation of Mollusca has taken place in both main subterranean habitats: freshwater saturated underground crevices and caves, which are inhabited by species called stygobionts, and terrestrial subterranean habitats with species called troglobionts (Ruffo 1957). Subterranean species of Mollusca appear to be absent from the areas covered by Pleistocene glaciation. Presently, there is no species list of hypogean Mollusca available, and the estimates of global numbers vary considerably. Culver (2012) estimated 350 worldwide stygobionts, including about 100 troglobionts; Bole and Velkovrh (1986) listed 364 stygobiontic Mollusca species, while Prié (2019) provided the most recent estimate of over 450 stygobiontic taxa. Based upon my calculations, there are at least 1,069 taxa (including species and subspecies) of hypogean mollusks, including 747 subterranean freshwater and 326 subterranean terrestrial species. Freshwater species are more diverse at genus level with 171 genera compared to 78 terrestrial genera. Conversely, hypogean terrestrial fauna is more diverse at the family level with 32 subterranean families versus 18 freshwater families.

Most hypogean freshwater Mollusca of Europe (78%) are known from the Dinaric Karst region (approximately 60,000 km^2), which may be considered a global hotspot. This region supports 30% of freshwater Mollusca global diversity. The second-most diverse region is the Ponto-Caspian region (i.e., the Caucasus Mountains spanning between the Black Sea and the Caspian Sea), which supports 8% of global subterranean freshwater species. The remaining regions are eastern Asia, including Japan with 4% of hypogean freshwater mollusk species; North America at 5%; Australia, New

Zealand, and Papua New Guinea at 3%; and the least studied regions, North Africa and South America, presently representing 1% each of known global subterranean Mollusca (Fig. 3.1).

Global terrestrial hypogean species differs somewhat from freshwater species. While all of Europe boasts 58% of species diversity, the Dinaric Karst alone comprises 25% of global diversity. Additionally, South and East Asia (i.e., Laos, Myanmar, and Japan) represent 33% of the known terrestrial species (Fig. 3.2). It is important to emphasize these numbers represent our current level of knowledge—as data from most parts of the world are either lacking or incomplete.

To better organize and characterize subterranean mollusks, I have provided some basic concepts on habitats and evolutionary categories. Many surface (or *epigean*) Mollusca are occasionally found in subterranean

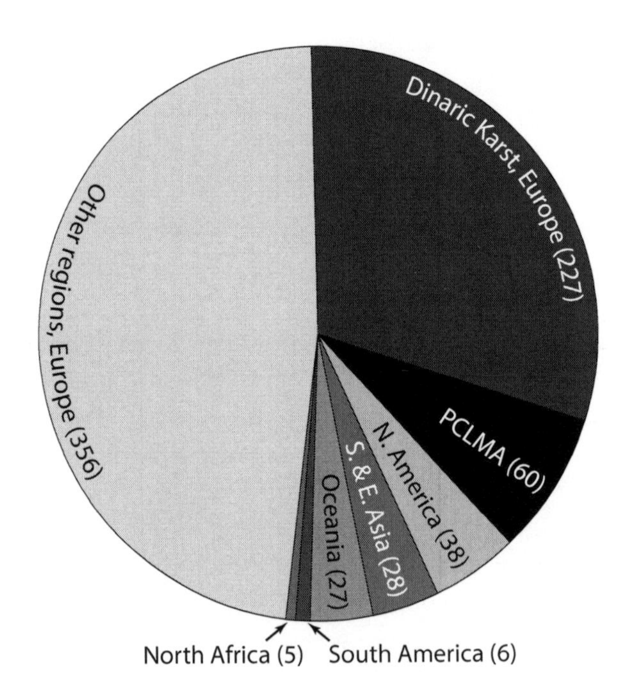

Figure 3.1. Global diversity of hypogean freshwater Mollusca representing the 747 known taxa. PCLMA encompasses the Ponto-Caspian, Levant, and middle Asian regions and includes Cyprus, Turkey, Israel, southern Russia, Georgia, Iran, Turkmenistan, and Uzbekistan. Oceania consists of Australia, New Zealand, and Papua New Guinea, while South and East Asia represent Laos, Myanmar, and Japan.

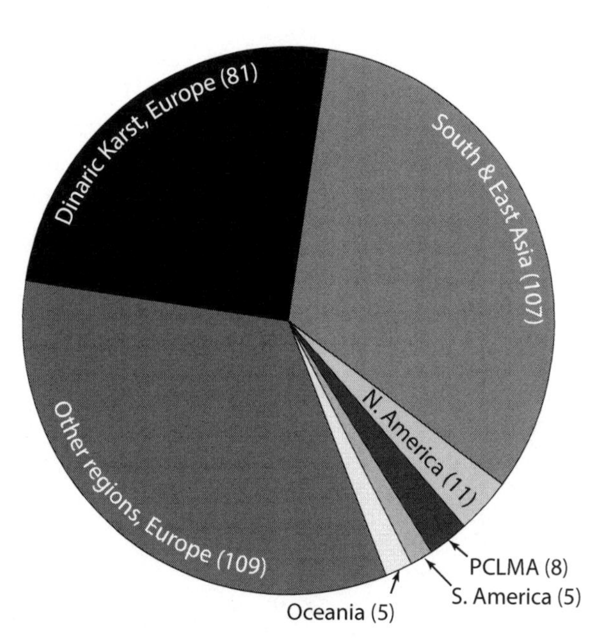

Figure 3.2. Global diversity of hypogean terrestrial Mollusca depicting the 326 known taxa. PCLMA and Oceania consist of the same countries/regions denoted in the caption for Fig. 3.1.

habitats. These species occasionally fall, or are washed, into these subterranean habitats (i.e., they become *accidental* inhabitants). Other surface species (*troglophiles* for terrestrial and *stygophiles* for freshwater habitats) can complete their life cycle underground but are not adapted to the subterranean environment. Subterranean-adapted species inhabiting the deep cave zone (see Howarth 1980) include *troglobionts* (terrestrial subterranean-adapted species) and *stygobionts* (aquatic subterranean-adapted species) (refer to Howarth and Wynne, Chap. 1, for additional information). Additionally, *hypogean freshwater* and *hypogean terrestrial* are used as ecological groups to further define the aforementioned evolutionary categories.

From the perspective of mollusks, hypogean freshwater habitats include caves and phreatic channels, as well as a wide range of small water-saturated rock crevices, debris interstices, and alluvial gravel interstices (i.e., areas larger than 1 mm in size). Alluvial gravel interstices can occur along cave streams and are considered one of the most important dispersal pathways for specialized hypogean freshwater gastropod species (Richling et al. 2016).

Hypogean freshwater snails can also be found at spring heads, which also support crenobiont (spring-dwelling) Mollusca species specialized to the spring zone (these species occur within flowing or semistagnant spring water). These spring habitats represent the transition between the epigean and hypogean freshwater environments connecting groundwater habitats to the surface. Many crenobiont and freshwater benthic (bottom-dwelling) species have life cycles and estivation periods similar to species inhabiting hypogean habitats (Grego, unpublished data). While some authors have referred to these animals as stygophiles (Delicado 2018), these intermediate gastropod species may have reduced eyes and pigmentation and may occur in spring habitats. Examples of this intermediate group include *Sadleriana cavernosa* Radoman, 1978 and the genera *Belgrandiella*, *Agrafia*, *Tschernomorica*, and *Daphniola*.

Hypogean terrestrial gastropods occur in habitats ranging from the *milieu souterrain superficiel* (MSS; Juberthie et al. 1980; Racovitza 1983) to the deep recesses of caves. Mammola et al. (2016) divided the MSS into four types: (1) *colluvial MSS* (as defined by Juberthie et al. 1980), (2) *bedrock MSS* (Racovitza 1983), (3) *alluvial MSS* (Ortuño et al. 2013), and (4) *volcanic MSS* (Oromí et al. 1986). Most hypogean gastropods are found in colluvial and bedrock MSS habitats and are largely absent from alluvial and volcanic MSS—especially when conditions are acidic. I introduce a fifth MSS type, the *macroporous rock MSS*. This MSS type represents spaces inside porous/cavernous secondary carbonate bedrock and is characterized by a network of small pore spaces and fissures penetrated by tree roots. The hypogean gastropod genera selecting this MSS type are primarily *Agardhiella* and *Balcanodiscus* (Grego, unpublished data); both genera are known to the south Dinaric Alps.

My overall goals of this chapter are to synthesize the information on global subterranean Mollusca, highlight their vulnerability and how they are threatened by anthropogenic activities, and identify future research needs.

Drivers of Subterranean Mollusca Diversity

In general, the factors driving colonization and adaptation of hypogean species is the highly stable deep cave zone, which is characterized by constant temperature, low to no airflow, and a near water-saturated atmosphere (Howarth 1983; Howarth and Wynne, Chap. 1). Grego et al. (2017)

suggests there are comparatively fewer molluscan competitors, predators, and parasites in this zone. Because deep cave zones often have lower nutrient availability when compared to the surface, most subterranean air-breathing gastropod species occur in habitats containing plant roots. For example, plant roots, vegetation detritus, and microorganisms (including fungi) provide the nutrient resources in the MSS (Rendoš et al. 2012), which are critical to supporting gastropod populations. Plant roots penetrate the cave environment via cracks and fissures in rock, while other organic material is transported underground through the horizontal and vertical movement of water. The least understood nutrient source for gastropods is autochthonous sources (i.e., nutrients generated from soil auxotrophic bacteria, auxotrophic fungi, chemolithotrophic bacteria, and in some cases, chemoautotrophic bacteria; Dudich 1930, 1932); however, autochthonous nutrients are important as these chemoautotrophic food webs include mollusk species (Sarbu et al. 1996; Falniowski et al. 2008; Fatemi et al. 2019).

Due to nutrient limitations and adaptation to a nutrient limited environment, subterranean faunas may have longer lifespans compared to their epigean relatives (e.g., Voituron et al. 2011; Puljas et al. 2014). Longer lifespan, coupled with longer maturation period and longer reproduction cycle, and comparatively small populations with limited distributional ranges (Eme et al. 2018), can make hypogean fauna highly vulnerable to human disturbance (Fišer et al. 2013; Raschmanová et al. 2018; Mammola et al. 2019). Based upon my research, this applies to most hypogean Gastropoda as well.

Mollusca inhabiting both freshwater and terrestrial subterranean habitats have significantly different taxonomy, biology, ecology, habitat, and natural history. Therefore, I am treating my discussion on these Mollusca groups and their habitats separately.

Hypogean Freshwater Mollusca

Taxonomy

At least 747 hypogean freshwater molluscan species are known. Historically, 98% of the global hypogean freshwater fauna was treated as one truncatelloid megafamily, Hydrobiidae, due to similar shell morphologies and anatomy. Recently, Hydrobiidae was split into eight different families according to both molecular analysis and zoogeography (Wilke et al. 2009). Similar revisions have been applied to its hypogean freshwater relatives, including Amnicolidae, Cochliopidae, Hydrobiidae, Lithoglyphidae, Moitessieriidae, Pomatiopsidae, and Tateidae. Higher taxonomy of truncatelloid gastropod families is under continuous reevaluation and revision, with genera reassigned according to results of molecular analysis (Hofman et al. 2018). Incidentally, representatives of nontruncatelloid gastropod genera are less common in subterranean habitats (2% of global fauna) and include genera such as *Theodoxus*, *Acroloxus*, *Emmericia*, and *Valvata* in the north Dinaric region and *Physella* in North America. The list of worldwide hypogean freshwater families is presented in Table 3.2 at the end of the chapter.

Hypogean freshwater Bivalvia are represented by three species of Dreissenidae in the Dinaric Alps in southeastern Europe (Bilandžija et. al. 2013) and six species of Sphaeriidae from southwestern Caucasus (Starobogatov 1962), Crimea, Japan, Laos, and Mexico (Yucatán) (Grego, unpublished data). When adequate nutrients are available, epigean truncatelloids, including Physidae and Sphaeriidae (occurring as troglophiles), may establish permanent populations in cave streams. Examples of troglophilic Mollusca include (1) the widespread *Physella acuta* (Draparnaud, 1805) from Stemler Cave, Illinois (Weck and Taylor 2016); (2) the alien species *Potamopyrgus antipodarum* (J. E. Gray, 1843) from New Zealand, which recently colonized Frasassi Cave in Italy (Bodon et al. 2009); (3) *Pisidium casertanum* (Poli, 1791) in Silická Ľadnica Ice Cave, Slovakia; (4) *Bythinella austriaca* (Frauenfeld, 1857), which also occurs in Silická Ľadnica Ice Cave, Slovakia; and (5) *Bythinella pannonica* (Frauenfeld, 1865) from Drienovecká Cave, Slovakia (Kováč et al. 2014).

Many taxonomically and geographically distant hypogean freshwater gastropod species inhabiting similar habitats evolved to have similar shell shapes (Grego 2018). Conversely, many different morphotypes previously considered as different species (occurring within a 500 km radius) represent one species, which is likely due to recent postglacial subterranean colonization (Richling et al. 2016). Thus, shell morphology differences are driven more by the environmental factors (discussed in the habitat section) than by the taxonomic position. Accordingly, I recognize nine shell-shape morphological groups, which are discussed at the genus level: (1) flat discoid shells with angled or sinuated aperture, occasionally carinated (e.g., *Hadziella*, *Phreatodrobia*, *Dalmatella*, and *Daudebardiella*); (2) flat discoid to flat pyramidal (valvatiform) shells (e.g., *Hauffenia*, *Kerkia*, *Bracenica*, *Gocea*, *Hausdorfenia*, and *Coahuilix*); (3) elongate to elongate-conical shells (e.g.,

Iglica, *Paladilhiopsis*, *Bythiospeum*, *Pseudoiglica*, *Thamkhondonia*, *Imeretiopsis*, and *Pterides*); (4) elongate-conical thin shells with expanded aperture margins (e.g., *Plagigeyeria*, *Lanzaia*, *Caucasogeyeria*, and *Pseudotricula*); (5) globose shells with expanded aperture (e.g., *Dabriana*, *Tricula*, and *Antroselates*); (6) oval to oval-conical thick shells with expanded aperture and ribbed surface (e.g., *Travunijana* and *Emmericiella*); (7) limpet-form gastropod shells (e.g., *Acroloxus*, *Ancylus*, and *Ferissia*); (8) pyramidal shells (e.g., *Antrobia*, *Pontohoratia*, *Motsametia*, *Horatia*, *Phreatomascogos*, *Islamia*, and *Daphniola*); and (9) uncoiled species, with open scalaroid coiling or with detached last whorl (e.g., *Phreatoceras*).

Despite increased sampling efforts and improved sampling methods over the past decades, live-collected hypogean freshwater specimens are still rarely used in taxonomic analysis; subsequently, our understanding of their taxonomy and phylogeny based on molecular data remains elusive (Szarowska 2006; Beran et al. 2014, 2016; Richling et al. 2016; Rysiewska et al. 2017; Angyal et al. 2018; Grego et al. 2018; Hofman et al. 2018; Osikowski 2018; Grego, Glöer, et al. 2019).

Due to often inaccessible hypogean habitats (Culver and Pipan 2014), typically only a few empty shells are collected from different localities (e.g., caves, wells, springs, or river deposits). In many cases, taxonomy of many species was based only on the empty shell morphology from the type locality (Culver 2012). Taxonomically, the most important truncatelloid shell morphology features were first proposed by Davis et al. (1992) and later revised by Hershler and Ponder (1998). While this approach is useful to hypogean relatives, shell morphology alone does not reflect all the characters important for describing new species and assigning them to their correct taxonomic position. Incidentally, many new genera had been established solely on the shell characters. The highly variable intraspecific shell morphology can be misleading when identifying specimens to the species level and many synonyms for the same species may have been incorrectly created (Richling et al. 2016).

Zoogeography

Our understanding of the zoogeography of hypogean freshwater Mollusca species is restricted to the hydrologically and speleologically most studied regions including the Balkans (Grego et al. 2017; Osikowski et al. 2017), Alps, southern and western Germany (Richling et al. 2016), Pyrenees and Iberian Peninsula (Arconada and Ramos 2011), Apennines, Italy (Bodon and Cianfanelli 2012), Carpathians (Ložek and Brtek 1964), Carpathian Basin (Angyal et al. 2018) and the Caucasus (Vinarski et al. 2014; Chertoprud et al. 2020; Grego et al. 2020), Asia Minor (Odabashi and Arlarslan 2015; Yildirim et al. 2017), Central Asia (Starobogatov 1962), Japan (Kuroda and Habe 1957; Kuroda 1963; Habe 1965), the United States (Culver 2012; Gladstone 2018; Gladstone, Perez, et al. 2019), parts of Mexico (Hershler 1985; Czaja et al. 2017, 2019; Grego, Angyal, and Beltrán 2019), Australia (Clark et al. 2003), Tasmania (Ponder 1992; Ponder et al. 2005), and New Zealand (Climo 1974). Europe supports most (78%) of the known hypogean freshwater fauna. The main hotspots of stygobiontic gastropod biodiversity are the Pyrenees Mountains (Darwall et al. 2014) and Dinaric Mountains (Zagmajster et al. 2010; Sket 2012), including the Skadar Lake basin (Pešić and Glöer 2013). The Dinaric region is the most diverse, representing 30% of the worldwide fauna (Osikowski et al. 2017). The Ponto-Caspian region (northern Turkey, the Caucasus, Iran, and Central Asia) has high potential; although largely understudied, it supports 8% of global hypogean freshwater mollusk diversity. The United States and Mexico are well studied locally, representing 5% of global diversity (Hershler 1985; Culver 2012; Gladstone et al. 2018; Czaja et al. 2019; Grego, Angyal, and Beltrán 2019), while East Asia (Grego 2018) and Japan (Kuroda and Habe 1957; Habe 1965) host 4%. Australia (Ponder 1992, Clark et al. 2003, Ponder et al. 2005), New Zealand (Climo 1974), and Papua New Guinea (Bernasconi 1995) contribute 3% to the global diversity of hypogean freshwater species. Africa (Morocco, Algeria, and Egypt; Bole and Velkovrh 1986; Ghamizi et al. 1998; Ghamizi and Boulal 2017) and South America (Brazil, Ecuador, and Colombia; Hershler and Velkovrh 1993; Simone 2012) represent the least studied regions yet retain a high potential of new hypogean species. Currently, both regions contain ~1% each of known global diversity.

Distributional patterns of stygobiontic mollusks are provided in Figure 3.3. Most recorded gastropod species are based on empty shells collected from caves, at surface springs near underground habitats, or from river deposits. In most cases, species are known from only the type locality. Because of this, most species are tentatively considered short-range endemics.

While hypogean freshwater Gastropoda appear more common globally, stygobiontic Bivalvia are known only from the Dinaric Mountains (genus *Congeria*; Bilandžija et al. 2013), the southwestern Caucasus region

Figure 3.3. Global distribution of subterranean-dwelling Gastropoda (*white circles*) and Bivalvia (*black triangles*).

(genus *Pisidium* [*Euglesa*]; Starobogatov 1962), and Japan (genus *Neopisidium*; Odhner 1921). However, given the lack of zoogeographic data from the surface waters of the Western Caucasus, it remains unclear whether the two subterranean-dwelling *Pisidium* (*Euglesa*) species are stygobionts or stygophiles—as is more clearly the case with the stygophilic *Pisidium* sp. from Silická Ľadnica Ice Cave, Slovakia. Epigean *Pisidium* species could inhabit various habitats, including moist soil or crevices; thus, their presence in a cave environment is not surprising.

Seven families comprise the majority of known hypogean Gastropoda species. The truncatelloid family Hydrobiidae dominates freshwater subterranean habitats of European, North African, Ponto-Caspian, and Central Asian regions. Moitessieriidae likely range from northern Africa through southern Europe from the Iberian Peninsula to the Balkans (including Greece) and eastward to southern Iran, whereas Cochliopidae occurs in subterranean habitats of North and South America, the Sultanate of Oman, and Romania. Amnicolidae and Lithoglyphidae occur in phreatic waters of North America. Pomatiopsidae dominates hypogean waters of Southeast Asia, Japan, and most likely China, while a single species is known from Brazil. Members of the family Tateidae are known from Australia and New Zealand and may also occur in Papua New Guinea.

Subterranean Gastropoda adapted to sulfidic environments are known from only five cave localities (Fatemi et al. 2019). Only four anchialine or brackish gastropod species have been recorded from Yucatán, Mexico (Rubio et al. 2016), Cebu Island, Philippines (Kano and Kase 2004), Turkmenistan (Vinarski and Kantor 2016), and the Sultanate of Oman (J. Šteffek, pers. comm., March 25, 2009).

Habitats and Ecology

The occurrence of specialized hypogean freshwater Mollusca (both Gastropoda and Bivalvia) in various subterranean freshwater habitats is generally well recognized, but much remains to be learned concerning environmental factors that influence their biology, evolutionary biology, habitat selection, food preferences, reproduction cycles, and parasitology (Fig. 3.4; Grego et al. 2017). Among the best-studied groups are subterranean truncatelloids, which are found in a wide range of subterranean habitats. Although most of the subterranean truncatelloids are referred to as cave dwellers, they are not exclusively restricted to cave habitats. Based upon two

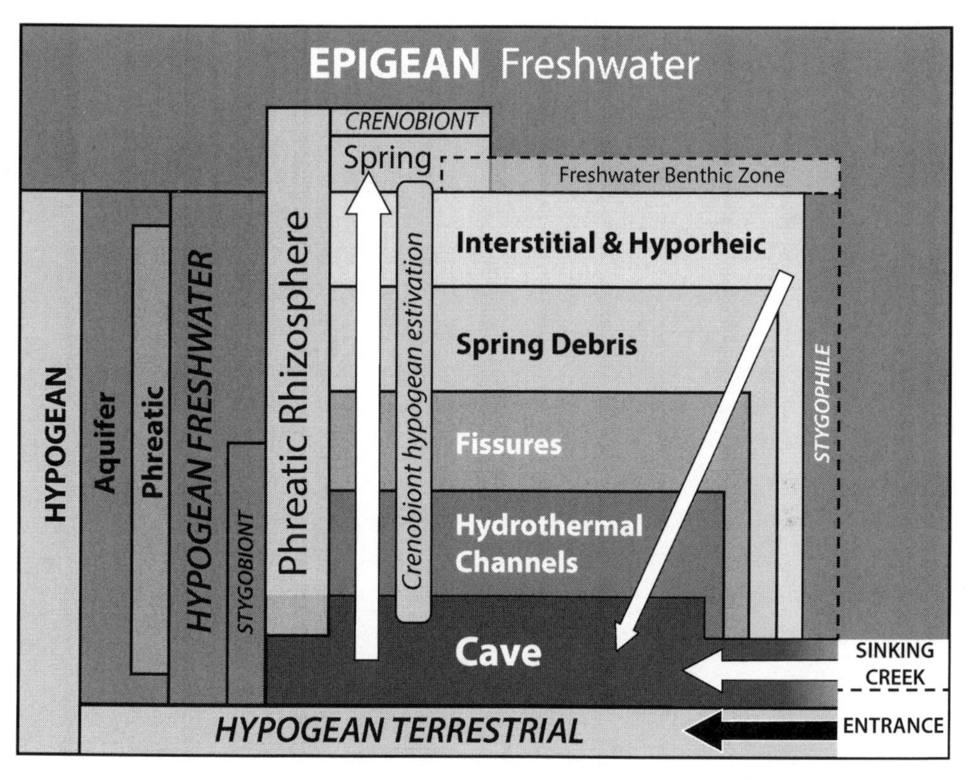

Figure 3.4. Subterranean freshwater habitats and related evolutionary categories (*in italics*). Nutrient pathways are illustrated with arrows. Surface water enters the subterranean habitat through sinking creeks (a flowing body of water that "disappears" into a cave entrance). Freshwater rhizosphere can supply nutrients via plant roots and decaying organic material. Freshwater benthic zone represents the lowest level of a body of water such as lake, spring, or stream. An intermediate zone can occur between the epigean (a surface body of freshwater) and hypogean habitats (e.g., interstitial and hyporheic zone). Estivation of crenobionts (spring-dwelling species) occurring in the hypogean region is typical for many spring-dwelling Mollusca during drought periods.

decades of research from the Carpathian Mountains, the Carpathian Basin, Dinaric Alps, Greece, the Caucasus, and southeastern Asia, the primary habitat of most hypogean truncatelloid species appears to be the groundwater-saturated phreatic and hyporheic zones; these zones consist of interstitial spaces among coarse sand and gravel deposits within the alluvium (Orghidan 1955; Grego, Glöer, et al. 2019). While some phreatic truncatelloid gastropods (e.g., *Paladilhiopsis* and *Bythiospeum*) occur in larger cavities and caves, they are most commonly encountered in the alluvial gravel deposits accessible only via springs and human-made wells—however, they are typically detected as empty shells. Few true stygobiontic genera (e.g., *Plagigeyeria*, *Pseudotricula*, and *Dabriana*) select for open spaces within caves versus the interstitial habitats—these genera have been found within subterranean streambeds and in open cavities. Species within these genera have developed thicker, mostly ribbed shells with expanded apertures enabling them to exploit habitats with rapidly flowing water (Grego et al. 2017).

Subterranean freshwater Mollusca species inhabit a range of hypogean freshwater habitats. Individuals are regularly washed out from the subsurface from subterranean streams into epigean environments. As they are typically unable to survive on the surface, this is evidenced by an accumulation of empty shells occurring at the spring bottom. However, more recent thanatocoenoses frequently occurred along cave stream bottoms, where few to no live individuals are encountered. This may indicate the range extent of stygobiontic populations in near-surface upstream microhabitats of the cave system. As with the distributions of most subterranean-specialized species, many hypogean freshwater Mollusca species do not inhabit a given cave system evenly, rather they are found only where suitable habitat occurs. These microhabitats are typically associated with food availability, such as pockets of accumulated flood detritus, plant roots, or chemolithotrophic bacteria biofilms (Dudich 1930, 1932; Grego et al. 2017).

Recent research (e.g., Šteffek and Grego 2005; Grego and Šteffek 2010; Grego et al. 2020) has led us to recognize a specific hypogean freshwater habitat, which I call the "phreatic rhizosphere." This habitat is characterized by phreatic spring debris (Fig. 3.5; Jasinska et al. 1996) or phreatic alluvial gravel (Fig. 3.6) interspersed with fibrous tree roots and exposed to hypergenic water flow. This habitat often hosts numerous populations of valvatiform-shelled gastropods of various genera and families. The valvatiform discoid, or flat pyramidal shelled, species are most typical in this habitat. European genera occurring within this habitat include *Hauffenia*, *Kerkia*, *Bracenica*, and *Daphniola*, as well as *Pontohoratia* and *Imeretiopsis* in the Caucasus (the region between the Black and Caspian Seas). These genera were found alive in the root masses within the spring zone. However, in adjacent caves, we found no live individuals beyond the spring zone with tree roots, and empty shells

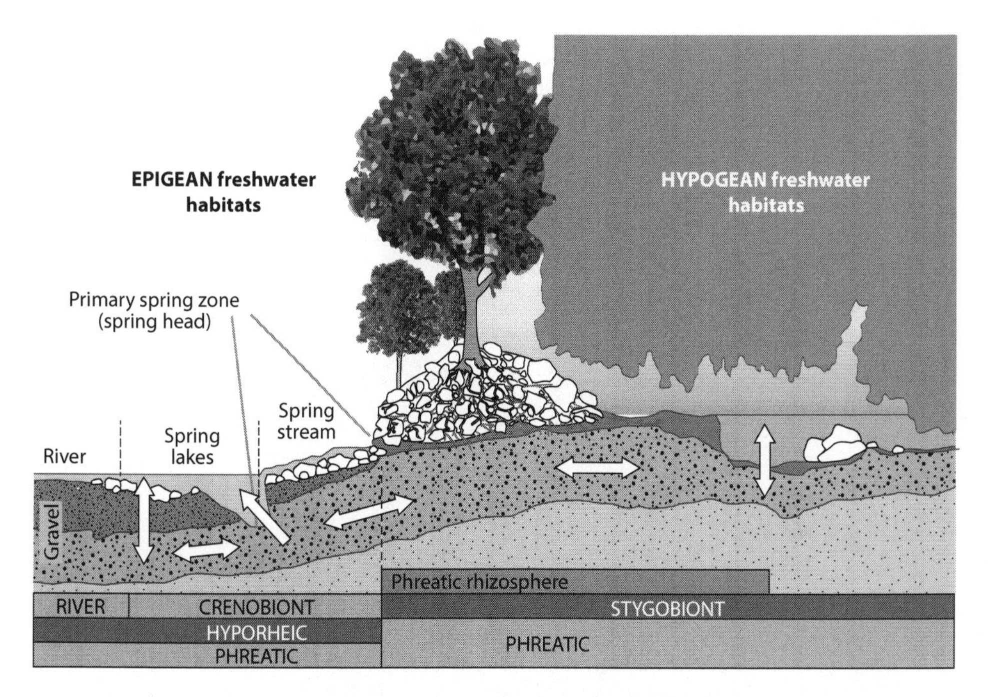

Figure 3.5. Phreatic rhizosphere habitat with hypogean migration routes of subterranean freshwater gastropods (*arrows*). Tree roots penetrate the spring zone, cave streams, and pools to provide both live and decaying organic material (directly or indirectly as fungi and bacteria mats) for hypogean freshwater Mollusca. Gastropod species occurring within this habitat can migrate into the cave system, as well into the interstices within the alluvial gravel and further down into hyporheic zone.

were scarce. Furthermore, the density of empty shells decreased with the distance from the spring head (Grego, unpublished data).

Plant roots support the growth of symbiotic bacteria and fungi, which can represent 20% to 40% of the carbon fixed by photosynthesis (Badri and Vivanco 2009). Some hypogean gastropods found in the phreatic rhizosphere may feed on either roots or microbial mats. Future molecular investigations of live gastropod intestinal content could provide evidence to support rhizosphere food specialization and the importance of this nutrient source in hypogean habitats.

Several hypogean freshwater truncatelloids may be found in the phreatic rhizosphere. These include *Hauffenia kissdalmae* Erőss & Petró, 2008 in

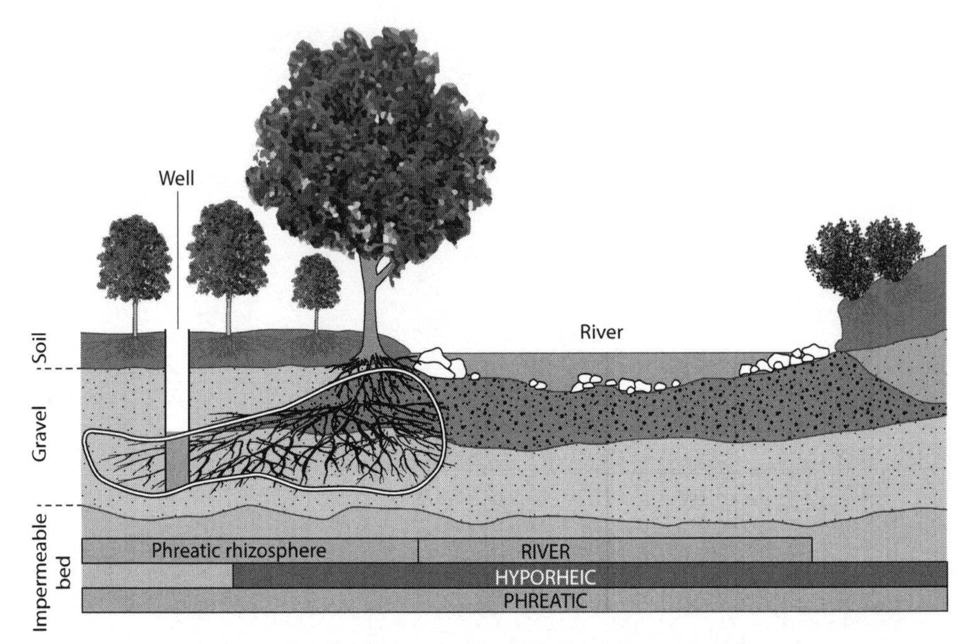

Figure 3.6. Phreatic rhizosphere habitat in alluvial gravel. Tree roots occur within the water-saturated gravel interstices. The roots represent a major allochthonous nutrient source for subterranean freshwater Mollusca.

the volcanic (andesite) Börzsöny Mountains, Hungary; *Paladilhiopsis oshanovae* (Pintér, 1968) in the alluvial gravel of the Danube River, Hungary; *Hauffenia* sp. in the late-Miocene Bretka formation sandstone in the Rimavská Basin, Slovakia; *Alzoniella slovenica* (Ložek & Brtek, 1964) in the Outer Carpathian Flysch, Slovakia and Czech Republic; and *Paladilhiopsis carpathicum* (Soós, 1940) on the same flysch formation of Mount Hoverla in Ukraine (Glöer et al. 2015).

Most surface-dwelling crenobionts require hyporheic and hypogean habitats as refugia during seasonal or drought periods (Schmid-Araya 2000). I observed the live crenobiont, *Bythinella austriaca*, in Hučiaca Cave several hundred meters from the spring entrance and detected the same species inside several closed well reservoirs within the alluvium of Slovenský Kras Mountains, Slovakia (Grego, unpublished data). Additionally, the population of *B. austriaca* in the Silická Ľadnica Ice Cave, Slovakia, may represent a permanent subterranean population persisting 100 m beneath the surface (Grego, unpublished data). I suggest crenobiont gastropods use the hyporheic and shallow hypogean habitats to lay their eggs during reproduction, as this habitat has more stable water conditions and may offer protection from epigean predators.

Our knowledge concerning the habitat requirements of stygobiontic Mollusca is largely based on limited field observations and indirectly based on characterizing the groundwater habitats where these animals are likely to occur. Combinations of these variables contribute to hypogean freshwater Mollusca ecology and habitat selection, and likely influence community composition and structure (Grego et al. 2017). Here, I provide a summary of what I consider the five primary habitat requirements for stygobiontic mollusks.

ABSENCE OF SUNLIGHT. Primary producers in spring ecosystems (Cyanobacteria, Algae, and auxotrophic Labyrinthulomycetes) typically occur from the entrance to twilight zones. Their presence gives rise to a community of primary consumers (e.g., crenobiont gastropod genera, including *Bythinella*, *Belgrandiella*, and *Agrafia*) and secondary consumers (e.g., Odonata nymphs and Dytiscidae beetle larvae). Subsequently, this facilitates the evolution of a wide range of diurnal food competitors, predators, and parasites. Conversely, the deep cave zone is characterized by a lower number of predators, parasites, and food competitors, which may have driven freshwater gastropods species to inhabit and ultimately adapt to hypogean habitats. The only potential predators of hypogean Mollusca species are cave salamanders and cavefishes.

FOOD AVAILABILITY. As caves are often considered allochthonous habitats, nutrients are transported via various mechanisms from the surface into the subterranean environment. This represents the primary food resource for most cave-dwelling gastropod species. For example, there are indications that some hypogean valvatiform-shelled hydrobioid species (Carpathian populations of *Hauffenia*, western Balkan *Islamia*, Greek *Daphniola*, and Caucasian *Pontohoratia*) feed on the dense, soft, tiny plant roots penetrating the interstitial water among debris within the hyporheic spring zone (Grego et al. 2017) and likely within the phreatic rhizosphere as well.

Additionally, some hypogean freshwater gastropods feed on organic material produced by autochthonous chemolithotrophic archaebacteria (Dudich 1930, 1932; Engel 2012) and fungi (de la Torre and Gomez-Alarcon 1994). The latter represents a resource that requires further examination in relation to its importance to specialized stygobiontic Mollusca (which I herein provide). While hydrogen sulfide is toxic for most organisms (Kelley et al. 2016), chemoautotrophic sulfide-based cave ecosystems can host relatively rich and diversified communities, including crustaceans (Peterson et al. 2013; Por 2014), Hemiptera (Tobler et al. 2013), and less commonly cavefishes (Riesch et al. 2010; Roach et al. 2011; Mousavi-Sabet et al. 2016) and gastropods (Grossu and Negrea 1989; Bodon et al. 2009). While sulfide-based ecosystems are typically associated with deep-sea hydrothermal vents (Deming and Baross 1993), subterranean sulfide ecosystems remain poorly understood (Engel 2007). This ecosystem was first described by Sarbu et al. (1996) at Movile Cave, Romania, and represents the most intensively studied sulfide cave ecosystem—which incidentally supports a subterranean-adapted gastropod (Sarbu 2000). Today, nearly 20 sulfide cave ecosystems have been cataloged (Por 1963, 2007, 2011; Arroyo et al. 1997; Latella et al. 1999; Engel et al. 2001; Holmes et al. 2001; Jaume et al. 2001; Porter et al. 2002; Schabereiter-Gurtner 2002; Engel 2007; Macalady et al. 2007; Falniowski and Sarbu 2015; Mousavi-Sabet et al. 2016). Of these, only six support stygobiontic gastropod species. These include *Heleobia dobrogica* (Grossu & Negrea 1989) from Movile Cave (Falniowski et al. 2008), *Islamia sulfurea* Bodon & Cianfanelli, 2012 from Frasassi Cave (Bodon et al. 2009), *Physella spelunca* Turner & Clench, 1974 from Lower Kane Cave in Wyoming (Porter et al. 2002; Wethington and Guralnick 2004), *Iglica hellenica* Falniowski & Sarbu, 2015 and *Daphniola magdalenae* Falniowski & Sarbu, 2015 from Melissotrypa Cave in Greece, and *Trogloiranica tashanica* (Fatemi et al. 2019) from south Zagros Mountains, Iran. Given the specialized and rich invertebrate biodiversity of the sulfidic caves and the estimation of 10% to 20% of worldwide caves with sulfidic speleogenesis (Palmer 2007; Vaxevanopoulos 2009), high subterranean diversity may have persisted in most of those former-sulfidic caves after the hydrogen sulfide ducts became inactive. Moreover, the former-sulfidic cave Tham Nam Dôn in Khammouane, Laos, supported significantly higher stygobiontic gastropod diversity (at least 12 species) than its neighboring Pha Soung Cave, which lacked any traces of the sulfide genesis (four species; Grego 2018). It is also likely the sulfide environment induced the evolution of stygobiontic *Congeria* species (Harzhauser and Mandic 2004). If confirmed, the importance of these sulfidic ecosystems and the bacterial biofilm food source in the generation of the recent stygobiontic diversity would be significantly increased. Subsequently, the high subterranean diversity of these unique subterranean habitats could be better understood.

GAS SATURATION IN WATER. Dissolved oxygen is essential to gastropods for respiration and oxidative phosphorylation reactions in mitochondria. Oxygen is also important for auxotrophic species, and chemolithotrophic and chemoautotrophic processes depend on oxidative reactions related to autochthonous food generation. Dissolved CO_2 is the main carbon source for autotrophic food bacteria. Both CO_2 and O_2 saturation in water are strongly dependent on temperature (i.e., saturation in water increasing by lower temperature). Thus, subterranean temperature has a direct effect on optimal water saturation range for aquatic gastropods, as well as for their bacterial food resource. Furthermore, CO_2 saturation, coupled with dissolved hydrocarbonate equilibrium, determine the corrosive (dissolving) versus accumulative (precipitating) character of the karstic waters (Zupan-Hajna 2015).

Six subterranean gastropod species are adapted to waters with high carbon dioxide and hydrogen sulfide content with low oxygen saturation (Falniowski et al. 2008; Fatemi et al. 2019). Hydrogen sulfide is toxic to oxidative phosphorylation in mitochondria (Kelley et al. 2016). Therefore, atmospheric oxygen demand driven mainly by phosphorylation is much less essential for animals capable of living in low oxygen environments (Sarbu et al. 1996). Although poorly understood, these sulfide extremophiles have found other pathways to supply their oxygen demand. They may be using similar mechanisms as anaerobic sulfate-reducing archaebacteria (SRA; Dillon et al. 2007) that use sulfate as the terminal electron acceptor and release hydrogen sulfide and oxygen (Muyzer and Stams 2008). This mechanism is likely provided by such microorganisms in the sulfide environment or in the digestive system of the extremophile organisms—which aids in the access of limited molecular oxygen in an hypoxygenic environment. Additionally, chemoautotrophic SRA feeding on slowly soluble gypsum may represent the primary source of hydrogen sulfide for chemoautotrophic thiobacteria in this environment. This underscores the complicated nature of these food webs, and how they may provide food and oxygen when oxygen saturation levels are drastically below the surface levels.

Methane also occurs in caves (e.g., Movile Cave; Sarbu 2000) and could serve as a carbon source for microorganisms (genus *Methanomonas*) synthesizing protein-rich biofilms (Monteiro et al. 1982). In the case of *Methanomonas* sp. and thiobacteria ecosystems, the dissolved gas concentrations may directly influence the chemoautotrophic food web.

WATER CHEMISTRY. Dissolved ions (Fe^{2+} and Mn^{2+}) are important for reactions driving the growth of autochthonous food biomass (Dudich 1930, 1932; de la Torre and Gomez-Alarcon 1994; Engel 2012), as well as key metabolic microelements. In addition to metabolic function, calcium content drives the development and maintenance of the gastropod shell. A broad range of other water-soluble ions are essential as microelements for both molluscan growth and food generation. Water chemistry (i.e., dissolved ions) can stabilize pH and hydrocarbonate equilibrium in freshwater environments (Zupan-Hajna 2015). As most subterranean Mollusca species are narrowly specialized to a stable chemical habitat, a pH level above 7.9 can cause strong carbonate precipitation, which can cover the entire shell and operculum (i.e., the door). Hypogean freshwater species can adapt to such extremes, where precipitates of inorganic compounds (usually calcium, iron, and manganese) thickly cover their shells. This can make their shell shape indistinguishable from grains of sand—save for their aperture and the precipitate-covered operculum (Grego, unpublished data). In some sulfate caves, when pH levels drop to between 2 and 3, these acidic conditions may dissolve the shell's inorganic material. Evolutionarily, the shells of some gastropod species have evolved to have a thick periostracum (a corneous protein layer), which protects the shell against corrosion (Grego, unpublished data). Subsequently, even low-level anthropogenic pollution or other environment changes may significantly alter water chemistry, which may cause irreversible damage both to the ecosystem (Mammola et al. 2019) and mollusks.

WATER FLOW VELOCITY AND HYDRODYNAMICS. Water flow velocity and turbulence influences oxygen and CO_2 saturation and distribution. Water flow rate is also important for species motility and can drive the evolution of gastropod shell shape. Stagnant water has lower oxygen saturation and could have higher CO_2 content originating from decaying organic material. Low flow rates allow better active animal motility both for locating food and reproduction, while high flow rates can restrict or stop motility. High flow also presents challenges for the adhesive strength of gastropod musculature and slime, which are important for their attachment to the substrate (i.e., suction effect). High water flow, which is typically seasonal, can result in removing them from their suitable habitat. When this occurs, it can result in mass mortality events. This may represent the main reason for massive and recent thanatocoenoses accumulations observed in many caves and especially at spring zones (Grego, unpublished data).

As stygobiontic species have adapted to flowing water in cave streams, it is likely flow velocity has influenced shell-shape evolution and may have influenced several adaptive strategies reflected in the shell morphology (Fig. 3.7). From general hydrodynamic principles, I suggest the main evolutionary adaptive strategies to exist in high water stream velocity include (1) maximal horizontal swinging liberty angle (i.e., the angle of flexible shell joint swinging around the animal attached to the substrate), (2) maximal span of adhesion surface where the foot surface is supported by expanded aperture, (3) minimal lifting force or the flat shape reducing hydrodynamic lift of the "wing" profile, and (4) minimal frontal hydrodynamic resistance area (i.e., shell cross section is perpendicular to water flow direction).

In general, more slender, elongated, or flat shell shapes (which convey lower frontal hydrodynamic resistance) may have evolved in species occurring in waters with permanently higher velocity, whereas more robust, globose shell shapes may have adapted to habitats with still or slow-flowing waters, where adhesion surface and strength of mucus with suction effect are sufficient to keep the animal attached to the substrate (Grego et al. 2017). Evolutionary convergence of shell shape across several unrelated global stygobiontic families inhabiting similar habitats (with similar flow velocities) indirectly supports this (Grego 2018). Starting with more insulated habitats, smaller species living in interstitial areas (within gravel or sand) have elongated but proportionally shorter shells (e.g., genera such as *Iglica*, *Moitessiera*, *Stygobium*, and *Tricula*); this adaptation permits the animals to move and disperse within the smaller interstitial spaces. Elongated shell shape (e.g., genera including *Iglica* from the Balkans and Greece and *Pseudoiglica* and *Thamkhondonia* from Laos) has lower frontal hydrodynamic resistance by angling the pointed tip of the shell (swinging) in the flow direction, thus avoiding the shear stress of water turbulence and reducing the likelihood of becoming dislodged from the substrate (Grego 2018). A similar strategy, but with even

1
2
3
4
5
6
7
8
1 mm
9
10
11
12
13
14
15
16
17
18
19
20
21
22
23
24
25
26
27
28
29
30
31
32
33
34
35

Figure 3.7. (*opposite*) Examples of hypogean freshwater gastropod species from Mexico, Morocco, and Georgia (Caucasus). [1] *Phreatomascogos gregoi* Czaja & Estrada-Rodríguez, 2019, Sabinas River, Coahuila, Mexico; [2] *Balconorbis sabinasense* Czaja, Cardoza-Martínez & Estrada-Rodriguez, 2019, Rio Sabinas, Coahuila, Mexico; [3] *Coahuilix landyei* Hershler, 1985, Cuatro Ciénegas, Coahuila, Mexico; [4] *Coahuilix parrasense* Czaja, Estrada-Rodrigue, Romero-Mendez, Avila-Rodriguez, Meza-Sanchez & Covich, 2017, Parras de la Fuente, Coahuila, Mexico; [5] *Paludiscala thompsoni* Czaja, Estrada-Rodríguez, Romero-Méndez, Ávila-Rodríguez, Meza-Sanchez & Covich, 2017, Viesca, Coahuila, Mexico; [6] *Juturnia* sp., Viesca, Coahuila, Mexico (courtesy of A. Czaja); [7] *Paludiscala* sp., Viesca, Coahuila, Mexico; [8] *Phreatoceras taylori* Hershler & Longley, 1986, Parras de la Fuente, Coahuila, Mexico; [9] *Pyrgophorus cenoticus* Grego, Angyal & Bertrán, 2019, Cenote Xoch, Yucatán, Mexico; [10] *Pyrgophorus coronatus* (L. Pfeiffer, 1840), Cenote Xoch, Yucatán, Mexico; [11] *Mexicenotica xochi* Grego, Angyal & Bertrán, 2019, Cenote Xoch, Yucatán, Mexico; [12] *Hausdorfenia shareula* Grego & Mumladze, 2020, Tsivtskala 2 Cave Spring, Georgia; [13] *Paludiscala caramba* Taylor, 1966, Cuatro Ciénegas, Coahuila, Mexico; [14] *Milesiana* sp., Bin El Ouidane, Morocco; [15] *Pontohoratia mapeli* Grego & Mumladze, 2020, Kanti, Mukhuri, Georgia; [16] *Pontohoratia vinarskii* Grego & Mumladze, 2020, Letsurtume Cave, Georgia; [17] *Hausdorfenia pseudohauffenia* Grego & Mumladze, 2020, Krikhula Spring, Georgia (aberrant form); [18] *Caucasopsis letsurtsume* Grego & Mumladze, 2020, Letsurtsume Cave, Georgia; [19] undescribed genus and species, Maghara Cave, Georgia; [20] *Imeretiopsis nakeralaensis* Grego & Mumladze, 2020, Nakerala Spring, Georgia; [21] *Imeretiopsis cameroni* Grego & Mumladze, 2020, Iazoni Cave, Georgia; [22] cf. *Iglica soussenisis* Ghamizi & Boulal, 2017, Souss plains, Agadir, Morocco (courtesy of V. Héros and M. Caballer [MNHN, project E-RECOLNAT: ANR-11-INBS-0004]); [23] *Pontohoratia pichkhaiai* Grego & Mumladze, 2020, Shisha Spring, Georgia; [24] *Caucasogeyeria* cf. *gloeri*, Shurubumu Spring, Georgia; [25] *Caucasogeyeria gloeri* Grego & Mumladze, 2020, Turchusmtha Spring, Georgia; [26] *Hausdorfenia pseudohauffenia* Grego & Mumladze, 2020, Krikhula Spring, Georgia; [27] undescribed genus and species, Kinchkhaperdi, Georgia; [28] *Motsametia borutzkii* (Shadin, 1933), Iazoni Cave, Georgia; [29] *Caucasogeyeria gloeri* Grego & Mumladze, 2020, Turchu Gamosadivari Cave, Georgia; [30] *Sitnikovia* sp., Shareula Cave, Georgia; [31] *Sitnikovia ratschuli* Chertoprud, Palatov & Vinarski 2020, Kidobana Cave, Georgia; [32] *Kartvelobia sinuata* Grego & Mumladze, 2020, Turchusmtha Spring, Georgia; [33] *Imeretiopsis prometheus* Grego & Palatov 2020, Prometheus Cave, Georgia; [34] *Caucasopsis letsurtsume* Grego & Mumladze, 2020, Letsurtsume Cave, Georgia (robust form); [35] *Caucasogeyeria ignidona* Grego & Palatov, 2020, Prometheus Cave, Georgia. Images courtesy of Jozef Grego unless otherwise noted.

more reduced frontal resistance area, has evolved in species of the genera (*Phreatoceras*), which has either open shell coiling (Fig. 3.7, #6) or ephemeral coiling (Fig. 3.7, #8). Gastropods with globose shells (e.g., *Dabriana* sp.) would have difficulties remaining attached at high water velocity. The high frontal resistance surface is compensated by a larger adhesion area (i.e., expanded aperture with larger foot). Species with globose shells (e.g., *Dabriana, Horatia, Pontohoratia, Motsametia*, and *Tricula* spp.) are adapted to calmer waters, such as semistagnant cave pools.

Conversely, species living in habitats with alternating water velocities (calm waters with frequent or occasional floods) have more robust and solid shell shapes with thicker shell walls that are supported by broad, thick apertures (sometimes characterized as a limpet-like shape). The expanded aperture frequently has folds and sinuations along the margin, and the animal has an enlarged foot to increase adhesion and suction surface—enabling it to remain attached to the substrate during periods of high water velocity. Examples include the following genera and species: *Plagigeyeria* sp., *Tricula davisi* Grego, 2018, *Tricula lenahani* Grego, 2018, and *Pseudotricula eberhardi* Ponder, 1992.

The best shell morphology adaptation for high water velocity occurs in limpet-shaped gastropod genera, such as *Acroloxus* sp. (Plate 3.1, #3). Species within this genus can firmly attach themselves to the surface with a large adhesion and suction surface; concomitantly, this adaptation also substantially reduces motility. This is perhaps the main reason these forms are not very successful in the cave environment—as no more than five hypogean species are known globally. "Swinging angle liberty" is a similar strategy where the shell is pivoted in the flow direction. This strategy has been observed in species with elongated shells (e.g., *Paladilhiopsis* spp.), as well as the flat discoid (valvatiform) shelled species (e.g., the genera *Hauffenia, Kerkia*, and *Phreatodrobia*). The foot is attached and the shell swings in stream by exposing the lowest frontal hydrodynamic resistance area of the flat shell and reducing the reflux turbulence behind the shell.

Additionally, discoid or valvatiform-shaped species mostly inhabit the phreatic rhizosphere. Their flat shape may enable them to successfully navigate plant roots, while the widely open umbilicus minimizes the possibility of trapping sand grains in the umbilicus and thus limiting mobility. Other valvatiform-shelled gastropod species with angled or sinuated aperture applied the strategy of being flatly attached to the substrate with minimal swinging (e.g., *Hadziella* and *Coahuilix*).

The majority of subterranean freshwater gastropod species have small shell dimensions ranging from 0.8 to 3.0 mm. Smaller shell size better protects the animal against damage in harsh, turbulent waters because the laminar film covering the shell surface better protects proportionally smaller objects. Conversely, when immobile shells are attached to a hard substrate, hypogean freshwater bivalve species of the genus *Congeria* have evolved larger shell sizes up to 15 mm. The high shell morphological diversity of global epigean freshwater Gastropoda is illustrated in Figures 3.7 through 3.9 and Plates 3.1 through 3.3, with stygobiontic limpets (Gastropoda) and Bivalvia examples displayed on Plate 3.1.

Dispersal and Reproductive Biology

Alluvial gravel interstices and hyporheic habitats should be considered as the main path for the dispersal and migration of phreatic species. Richling et al. (2016) revealed a close and young molecular relationship among morphologically different *Bythiospeum* populations (previously recognized as 23 independent species) in areas north of the Alps in Germany. Genetic relatedness of remote *Bythiospeum* populations ranging from 400 to 500 km can be explained by the relatively recent postglacial colonization and ultimate dispersal of only one confirmed refugia in the foothills of the Northern Alps (Richling et al. 2016). The colonization pathway was connected via phreatic habitats within the hyporheic alluvial gravel sediments of glacial river valleys and basins. Presently, there are few records of hypogean freshwater gastropod species from nonglaciated areas with absent alluvial migration pathways (which are usually represented by south to north oriented river deposits).

Compared to northern Europe, in hypogean habitats south of the main Pleistocene glaciation impact, Mollusca had a longer and different evolutionary history combined with their limited dispersal capabilities through poorly developed alluvial beds. Subsequently, this resulted in greater diversity throughout the Balkan and Iberian Peninsulas. The diverse hypogean truncatelloid fauna in the Dinaric Alps and southern Balkan regions is likely represented by many Tertiary relict species. Epigean ancestors colonized the hypogean environment prior the late Miocene Messinian salinity crisis

1
2
3
4
5
6
7
8
9
10
11
12
13
14
15
16
17
18
19
20
21
22
23
24
25
26
27
28
1 mm

(MSC), which resulted in the nearly complete desiccation of the Mediterranean Sea Basin between 5.96 and 5.33 million years ago (mya; Gargani and Rigollet 2007; Murphy et al. 2009). Surface population extinctions during MSC were followed by additional Pleistocene extinctions. Subterranean-dwelling populations of Mollusca that survived the abovementioned extinctions likely evolved via allopatry as climatic relicts (refer to Wynne

Figure 3.8. (*opposite*) Examples of European hypogean freshwater gastropod species of the genera *Plagigeyria*, *Travunijana*, *Zeteana*, *Lanzaia*, *Orientalina*, *Devetakia*, *Balcanica*, *Palaeospeum*, *Narentiana* (*Zavalia*), *Sardopaladilhia*, and *Moitessiera*. [1] *Plagigeyeria zetaprotogona* Schütt, 1960, Viška Vrela, Montenegro; [2] *Plagigeyeria zetaprotogona* Schütt, 1960, Glava Zete Spring, Montenegro; [3] *Plagigeyeria pageti* Schütt, 1961, Cave Sopot, Risan, Montenegro; [4] *Plagigeyeria montenegrina* Bole, 1961, Vrijesko Vrelo, Bandiči, Montenegro; [5] *Zeteana ljiljanae* Glöer & Pešić, 2014, Glava Zete, Montenegro; [6] *Plagigeyeria montenegrina* Bole, 1961, Obodska Pećina Cave, Montenegro; [7] *Islamia azarum* Boeters & Rolán, 1988, Manantial la Fontana, Borondes, Spain; [8] *Orientalina* cf. *troglobia* (Bole, 1961), Ombla Spring, Komolac, Croatia; [9] *Palaospeum bessoni rebenacquensis* Boeters & Bertrand, 2001, Spring of Nez, Rébénacq, France; [10] *Narentiana* (*Zavalia*) cf. *vjetrenicae* Radoman, 1973, Vruljak 1 Cave, Gorica, Hercegovina; [11] *Sardopaladilhia mariannae* Rolán & Ortí, 2003, Sueras, Castellón, Spain; [12] *Moitessieria foui* Boeters, 2003, Molino de Espasa, Tarragona, Spain; [13] *Plagigeyeria zetatridyma* Schütt, 1960, Glava Zete, Montenegro; [14] *Balcanica yankovi* Georgiev, 2011, Izvora Cave, Sutari, Bulgaria; [15] *Plagigeyeria zetatridyma* Schütt, 1960, Viška Vrela, Montenegro; [16] *Plagigeyeria angyaldorkae* Grego, 2020, Jakšenica Spring, Lukende, Hercegovina; [17] *Devetakia pandurskii* Georgiev & Glöer, 2011, Devetàshka Cave, Bulgaria; [18] *Lanzaia brandti* (Schütt, 1968), Ombla Spring, Komolac, Croatia; [19] *Plagigeyeria gladilini* Kuščer, 1936, Drini i Bardhë Spring, Radac, Kosovo; [20] *Plagigeyeria steffeki* Grego, Glöer, Fehér & Erőss, 2017, Krumë Spring, Albania; [21] *Plagigeyeria mostarensis* Kuščer, 1933, Izvor Buna Spring, Blagaj, Hercegovina; [22] *Plagigeyeria pseudocostellina* Grego, 2020 (Kuščer, 1933), Izvor Buna Spring, Blagaj, Hercegovina; [23] *Plagigeyeria* sp., Nikšićko Polje, Montenegro; [24] *Plagigeyeria zetaprotogona* Schütt, 1960, Vitoja Spring, Drume, Montenegro; [25] *Belgrandia* cf. *torifera* Schütt, 1961, Peć Mlini Cave, Tihajlina, Hercegovina; [26] *Travunijana robusta* (Schütt, 1961), Vruljak 1 Cave, Gorica, Hercegovina; [27] *Travunijana robusta asculpta* (Schütt, 1972), Ombla Spring, Komolac, Croatia; [28] *Travunijana ovalis* (Kuščer, 1933), Izvor Bunica Spring, Hodbina, Hercegovina. Images courtesy of Jozef Grego.

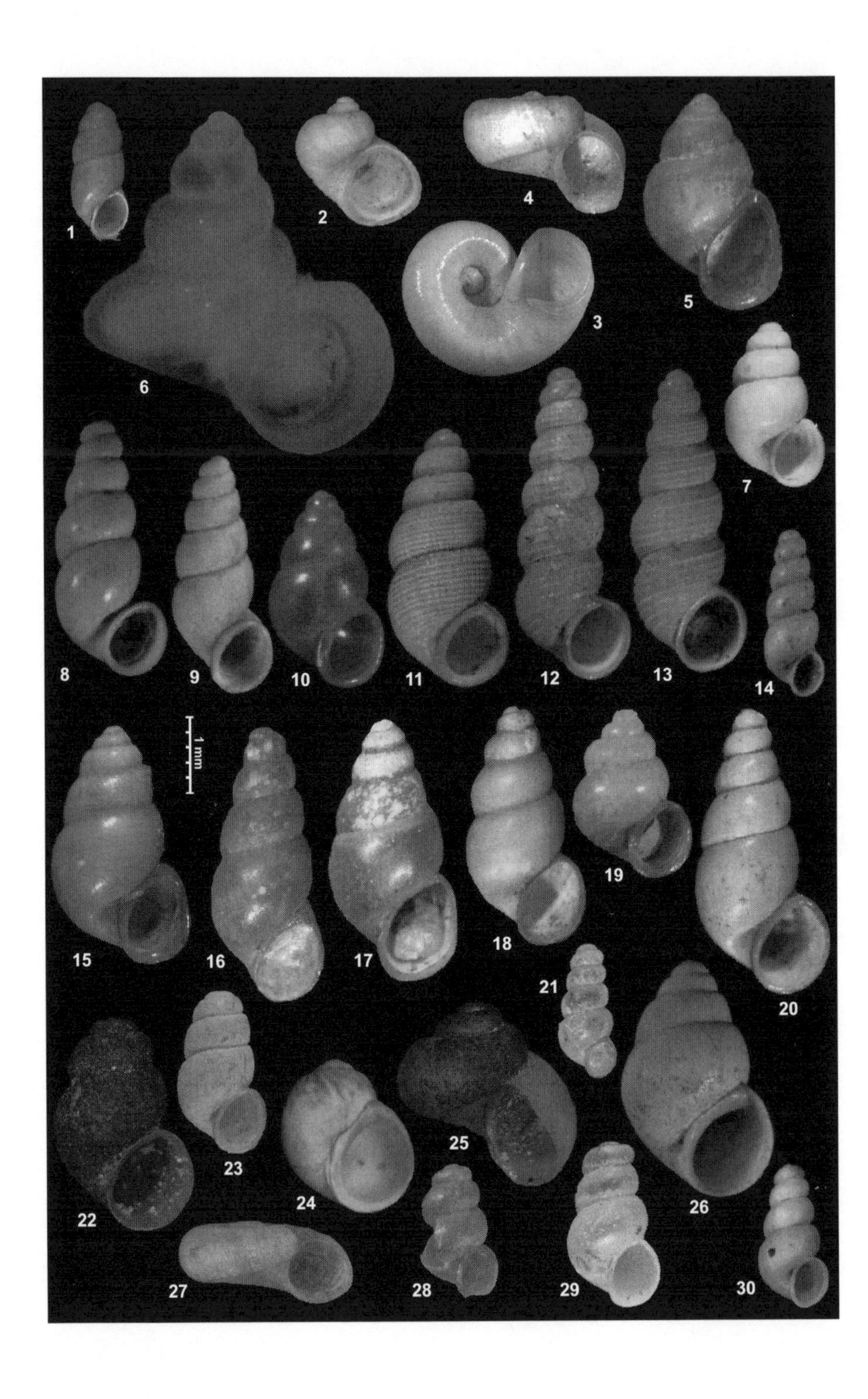

1 mm

Figure 3.9. (*opposite*) Examples of hypogean freshwater gastropod species from New Zealand, Croatia, Iran, Brazil, Laos, and the United States. [1] *Leptopyrgus manneringi* (Climo, 1974), Waitomo, North Island, New Zealand; [2] *Horatia* cf. *knorri* Schütt, 1961, Konavlonska Ljuta, Croatia; [3] *Horatia* sp., Badžula Spring, Croatia; [4] *Horatia sp.*, Žrnovnica Spring, Croatia; [5] *Trogloiranica tashanica* Fatemi, Malek-Hosseini, Falniowski, Hofman, Kuntner & Grego, 2019, Tashan Cave, Zagros Mountains, Iran; [6] *Spiripockia punctata* Simone, 2012, Lapa dos Peixes Cave, Serra Ramalho, Bahia, Brazil (courtesy of R. L. Simone); [7] *Tricula phasoungensis* Grego, 2017, Pha Song Cave, Khammouane, Laos; [8] *Pseudoiglica pseudoiglica* Grego, 2017, Tham Nam Don Cave, Khammouane, Laos; [9] *Pseudoiglica kameniari* Grego, 2017, Tham Nam Don Cave, Khammouane, Laos; [10] *Tricula spelaea* Grego, 2017, Tham Nam Don Cave, Khammouane, Laos; [11] *Thamkhondonia vacquiei* Grego, 2017, Tham Nam Don Cave, Khammouane, Laos; [12] *Thamkhondonia moureti* Grego, 2017, Tham Nam Don Cave, Khammouane, Laos; [13] *Thamkhondonia smidai* Grego, 2017, Tham Nam Don Cave, Khammouane, Laos; [14] *Pseudoiglica olsavskyi* Grego, 2017, Tham Nam Don Cave, Khammouane, Laos; [15] *Tricula lenahani* Grego, 2017, Tham Nam Don Cave, Khammouane, Laos; [16] *Pseudoiglica phonsavanica* Grego, 2017, Ban Nadom, Xiangkhouang, Laos; [17] *Tricula viengthongensis* Grego, 2017, Spring cave at Vien Thong, Bolikhamsai, Laos; [18] *Tricula reischuetzorum* Grego, 2017, Na Li Cave, Khammouane, Laos; [19] *Tricula bannaensis* Grego, 2017, Tham Nam Don Cave, Khammouane, Laos; [20] *Tricula valenasi* Grego, 2017, Tham Nam Don Cave, Khammouane, Laos; [21] *Fontigens turritella* Hubricht, 1976, McClung Cave, West Virginia (courtesy of J. Gerber); [22] *Fontigens antroecetes* (Hubricht, 1940), Stemler Cave, Illinois; [23] *Fontigens tartarea* Hubricht, 1963, Organ Cave, West Virginia (courtesy of J. Gerber); [24] *Potamotithus troglobius* Simone & Moracchioli, 1994, Iporanga, Areias Cave, São Paulo, Brazil (courtesy of R. L. Simone); [25] *Amnicola stygius* Hubricht, 1971, Soehl Cave, Missouri; [26] *Tricula davisi* Grego, 2017, Tham Nam Don Cave, Khammouane, Laos; [27] *Phreatodrobia micra* (Pilsbry & Ferriss, 1906), Manitou Cave, Alabama (courtesy of J. Gerber); [28] *Fontigens holsingeri* Hubricht, 1976, Harman Cave, West Virginia (courtesy of J. Gerber); [29] *Fontigens cryptica* Hubricht, 1963, Bethlehem, Indiana (courtesy of J. Gerber); [30] *Fontigens proserpina* (Hubricht, 1940), Rockwoods Reservation, Missouri. Images courtesy of Jozef Grego unless otherwise noted.

et al., Chap. 2, for a complete discussion on this topic). Genomic studies of hypogean Mollusca date their separation from epigean species between 5.96 and 5.33 mya and provide support for MSC-induced hypogean fauna evolution in the Mediterranean region (Osikowski et al. 2017; Osikowski 2018; Prevorčnik et al. 2019; Sands et al. 2019).

It is also possible some of their epigean ancestors originated from the marine or brackish environment that infiltrated submarine freshwater karst springs and gradually adapted to the freshwater environment (Osikowski 2018). For example, the superfamily Trucatelloidea is adapted to marine, brackish, and freshwater habitats. The large and morphologically diverse karstic areas of the Dinaric region had a high potential to host a diversity of subterranean refugial species—combined with a reduced extent of alluvial gravel conducive to stygobiontic gastropod dispersal. Moreover, alluvial gravel beds in the Dinaric Karst are highly fragmented, discontinuous, and unsuitable for longer distance dispersal of phreatic gastropods. In most cases, the Dinaric alluvium is characterized by narrow karst ridges created by gorges and rivers within small and isolated karstic basins. Larger accumulations of alluvial gravel suitable for eventual stygobiontic species dispersal occurs only at the periphery of the Dinaric Alps within the Sana, Bosna, Sava, Drin, and Neretva river valleys, the Lower Neretva Delta, and the Skadar Lake basin (Stevanović et al. 2014). Thus, these phreatic species had a much longer period of isolation with lower dispersal potential; as a result, this likely facilitated an impressive species radiation, compared to other regions in central and northern Europe (Grego, Glöer, et al. 2019).

Few hypogean freshwater gastropod genera are subterranean specialized to the degree to be considered stygobionts. The genus *Plagigeyeria*, which is characterized by broadened apertures and thicker shells (e.g., see Fig. 3.10, rows 4 and 6) is one example (Schütt 1972; Grego 2020). Shell shape aids the animal in resisting high water flow during the spring season. However, its flaring aperture is not conducive for inhabiting gravel interstices, which also results in low vagility and dispersal. Subsequently, species with this shell morphology likely evolved as endemic to a single cave system or aquifer (Schütt 1972; Grego 2020).

For bivalves, little is known concerning their dispersal. Motility of adult Bivalvia is dramatically reduced (e.g., *Pisidium*) to absent (e.g., *Congeria*). Most epigean Bivalvia (e.g., Unionidae and Margaritiferidae) use a glochidial parasite larval stage, whereby they disperse by phoretically attaching them-

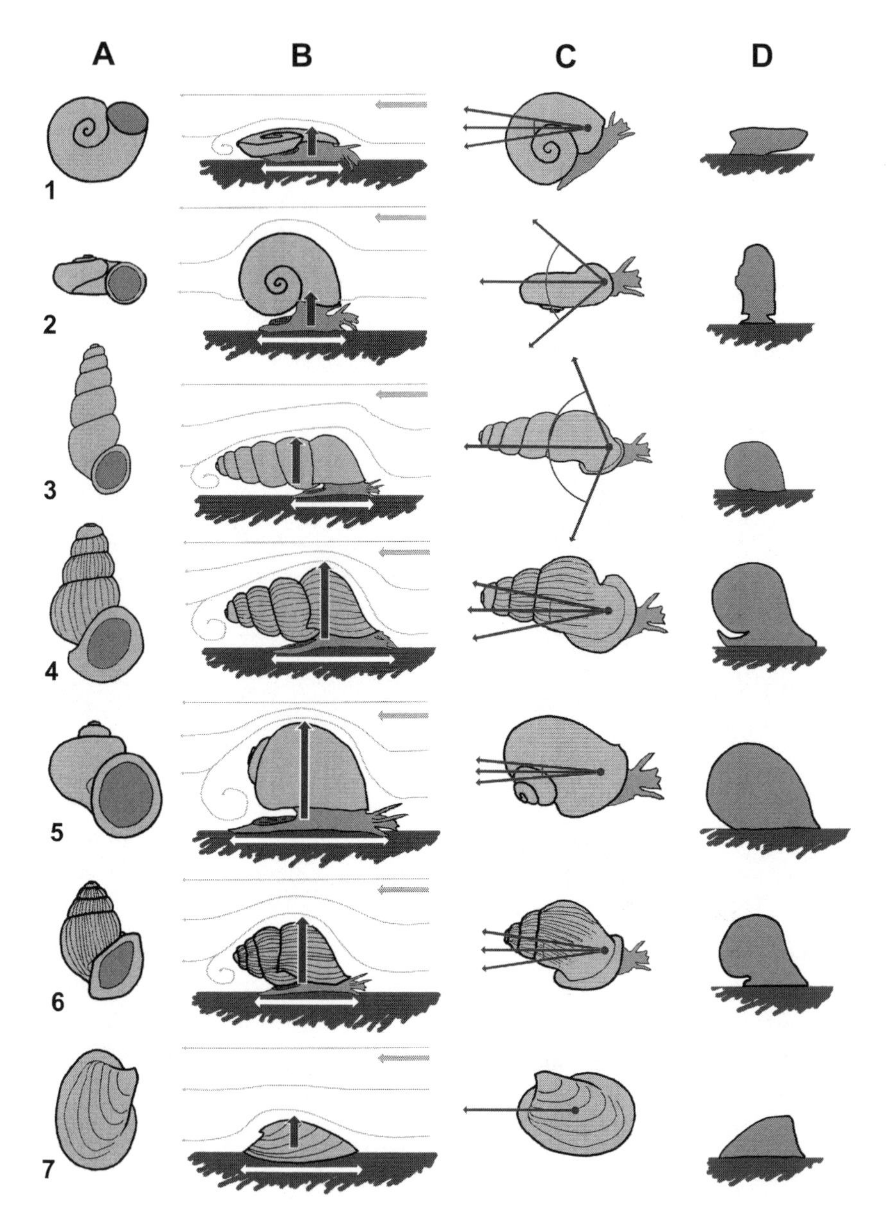

Figure 3.10. Stygobiontic gastropod shell morphology based upon adaptations to subterranean hydrodynamic conditions. [A] General shell morphology with examples of different gastropod genera representing different adaptive strategies of (1) *Hadziella* sp., (2) *Hauffenia* sp., (3) *Paladilhiopsis* sp., (4) *Plagigeyeria steffeki*, (5) *Dabriana bosniaca*, (6) *Plagigeyeria zetaprotogona*, and (7) *Acroloxus tetensi*. [B] Hydrodynamic profile with hydrodynamic lifting forces (*black arrows*). The span of gastropod foot adhesion surface is represented by the *white double-arrows*. [C] Horizontal swinging liberty (*angle between the arrows*) to compensate for changes in flow direction. [D] Frontal hydrodynamic resistance area perpendicular to water flow direction.

selves to the fins and gills of fish (Schwartz and Dimock 2001). This evolutionary step has enabled these groups to colonize more distant and remote upstream areas. However, this glochidial stage has not been confirmed in the stygobiontic Bivalvia genera *Congeria* (Rada and Rada 2012; Morton and Puljas 2013; Glavaš et al. 2017) and *Pisidium* (Mouthon and Daufresne 2008; Pettinelli and Bicchierai 2009), which had to develop another, thus far unconfirmed, dispersal mechanism. There is one example where the epigean bivalve *Dreissena polymorpha* (Pallas, 1771) disperses via phoresy on epigean fishes, crayfish, and Odonata nymphs (Coughlan et al. 2017). Dispersal mechanisms of *D. polymorpha* may be useful for inferring possible dispersal mechanisms of other stygobiontic bivalves. The distributional range of the only Dinaric stygobiontic bivalve *Congeria* (Bilandžija 2013) overlaps with the distribution of a cave-obligate salamander *Proteus anguinus* Laurenti, 1768 (Balázs and Lewarne 2017). This salamander is the only suitable-sized potential host in *Congeria* habitat. It is likely that *Proteus* occasionally feeds on the *Congeria* and may serve to disperse *Congeria* larvae to upstream portions of the cave either by phoresy or parasitic pseudoglochidia.

In surface environments, juveniles of epigean bivalve species may disperse via phoresy on shorebirds or other terrestrial animals (Kappes and Haase 2012). However, upstream dispersal in cave systems and other hypogean environments is unknown. It is unclear whether phoresy is the dispersal method for juvenile *Pisidium* or whether a stygobiontic host is involved. The most likely upstream transporters of the *Pisidium* juveniles in springs could be small crustaceans (e.g., *Gammarus* spp.), which are abundant. Juveniles could attach to their legs for dispersal. In the cave populations of *Pisidium* (known from Slovakia, Crimea, and southwestern Caucasus), dispersal likely occurs via the stygobiontic crustacean species of the genera *Niphargus* and/or *Xiphocaridinella*.

The reproductive biology of subterranean truncatelloid gastropods and the possible sexual dimorphism of some species with high shell morphology variability is also virtually unknown. Phreatic animals produce a pronounced slime track during movement on sand within freshwater habitats (Grego, unpublished data), which adheres to sand grains along the movement path and may be an important mechanism for communicating with other individuals. In particular, the slime track may contain sex pheromones to aid in locating mates, and it may contain additional information, such as movement direction and slime age. Additionally, female truncatelloid gastropods lay small egg clusters on a solid surface in areas protected from water currents, such as behind rocks, between larger sand grains, or in other still-water areas within gravel (Grego, unpublished data). To my knowledge, no other information exists on the reproductive biology or life span of phreatic gastropods.

Hypogean Terrestrial Mollusca

Taxonomy

Subterranean terrestrial environments host a remarkable and largely undescribed molluscan assemblage. Such molluscan assemblages are typically closely related phylogenetically to the local (recent or fossil) epigean fauna. Thus, knowledge of the regional native epigean taxa is essential for understanding subterranean taxa. Importantly, many troglobiontic species represent ancient relict species and are taxonomically more closely related to fossil species that inhabited a given study region. In some cases, these fossil species date back to the Cenozoic Era. As a result, the study of local fossil fauna is also a relevant area of research for broadening our understanding of troglobiontic Mollusca.

To date, 322 terrestrial gastropod species occur in subterranean habitats worldwide. They belong to 74 genera and 32 families. The larger number of terrestrial families compared to their freshwater analogues suggests that in the more xeric parts of caves, air-breathing snails have not evolved effective dispersal capabilities. The most diverse subterranean family is Zonitidae, which, with its closely related families Strobilopsidae, Oxychilidae, and Pristilomatidae—in the aggregate—consists of at least 70 species. Also, the Tertiary period relict family Clausiliidae is composed of eight short-range endemic subterranean genera (including *Sciocochlea* and *Tsoukatosia*) and represents 29 different species. Impressively diverse, the genus *Zospeum* from Dinaric and Iberian caves and the family Agardhiellidae support 25 (Weigand 2013) and 24 species (Jochum et al. 2015), respectively. In Southeast Asia, the family of Hypselostomatidae represents at least 39 species, while Diplommatinidae has 35 species that dominate the well-developed karstic regions; both groups are likely associated with the karstic rhizosphere.

Worldwide subterranean habitats are likely inhabited by a large number of undescribed species with shell sizes ranging from 0.6 to 5 mm. Due to the scarcity of live individuals, it is difficult to assess their taxonomic position. Fortunately, empty shells of terrestrial species usually possess taxonomically important characters (such as denticles, folds, plicae, ribbing, and shell surface), which makes it possible to assign empty shells to higher taxonomic levels (e.g., Maassen 2008; Páll-Gergely 2014; Páll-Gergely et al. 2015, 2016, 2017, 2019; Inkhavilay et al. 2016). Interestingly, some hypogean terrestrial families (e.g., Clausiliidae, Agardhiellidae, and Diplommatinidae) retain typical family-specific shell morphologies and size long after subterranean isolation. For example, the hypogean Clausiliidae, an ancient relict lineage from the Tertiary period (Grego and Szekeres 2017) has identical shell morphologies to fossil relatives from the Miocene (Nordsieck 2007). Worldwide

Figure 3.11. (*opposite*) Examples of European hypogean terrestrial gastropod species from the families of *Argnidae*, *Strobilopsidae*, *Spelaeodiscidae*, and *Pristilomatidae*. [1] *Agardhiella crassilabris* (Grossu & Negrea, 1968), Tismana, Romania; [2] *Agardhiella armata* (Clessin, 1887), Băile Herculane, Romania; [3] *Agardhiella grossui* (Zilch, 1958), Zaton Cave, Romania; [4] *Agardhiella banatica* (Zilch, 1958), Cheile Jelărăului, above Baile Herculane, Romania; [5] *Agardhiella macrodonta* (Heese, 1916), Knjaževac, Serbia; [6] *Agardhiella caesa* (Westerlund, 1871), Cheile Sohodol, Romania; [7] *Agardhiella biarmata* (Boettger, 1880), Gorica, Trebinje, Hercegovina; [8] *Agardhiella truncatella* (L. Pfeiffer, 1846), Kärnten, Austria; [9] *Spelaeodiscus albanicus albanicus* (A. J. Wagner, 1914), Prekal, Albania; [10] *Spelaeodiscus albanicus edentatus* Pál-Gergely & P. L. Reischütz, 2018, Izvor Vitoja near Skadar Lake, Drume, Montenegro; [11] *Vitrea megistislavras* A. Reischütz & P. L. Reischütz, 2014, Megistis Lavras Monastery, Athos Peninsula, Macedonia, Greece; [12] *Virpazaria pageti* E. Gittenberger, 1969, Igalo, Montenegro; [13] *Virpazaria aspectulabeatidis* A. Reischütz, N. Reischütz & P. L. Reischütz, 2009, Besa at Lake Skadar, Montenegro; [14] *Virpazaria nicoleae* A. Reischütz & P. L. Reischütz, 2012, Izvor Vitoja near Skadar Lake, Drume, Montenegro; [15] *Klemmia sinistrorsa* E. Gittenberger, 1969, Vilina Pećina Cave, Montenegro; [16] *Klemmia magnicosta* E. Gittenberger, 1975, Jabukov Do Cave, Montenegro; [17] *Spelaeodiscus virpazarioides* Páll-Gergely & Fehér, 2018, near Rraps-Starjë road, Albania; [18] *Spelaeodiscus unidentatus acutus* Páll-Gergely & Fehér, 2018, Vau i Dejës, Albania; [19] *Virpazaria ripkeni pastorpueri* A. Reischütz, N. Reischütz & P. L. Reischütz, 2011, Drisht, Albania; [20] *Virpazaria dhorai* A. Reischütz, N. Reischütz & P. L. Reischütz, 2010, Makaj, Albania; [21] *Conulopolita retowskii* (Lindholm, 1914), Banoja near Sataplia Cave, Imereti, Georgia. Images courtesy of Jozef Grego.

1
2
3
4
5
6
7
8
9
10
11
12
13
14
15
16
17
18
19
20
21
5 mm

1a
1b
1c
1d
2
3a
3b
4
5a
5b
6
1 mm
500 µm
7
8
9
10a
10b
11
12
13
14
15
16
17
18
19

subterranean gastropod genera are listed in Table 3.1 with their shell morphology variability illustrated in Figs. 3.11 through 3.13 and Plate 3.4.

Zoogeography

Currently, most of our knowledge on subterranean terrestrial gastropods is from Europe and Southeast Asia. The highest diversity of hypogean species is known from Europe (with 59% of global fauna) and includes the Dinaric region (which representing 25% of worldwide taxa). This region supports the largest shelled representatives of true troglobiontic snails from the family Zonitidae, including *Paraegopis oberwimmeri* Klemm, 1965, *Aegopis spelaeus* A. J. Wagner, 1914, *Troglaegopis mosorensis* (Kuščer, 1933), and *Meledella werneri* Sturany, 1908. Additionally, a few Cenozoic Era relict species, including *Pholeoteras euthrix* Sturany, 1904 (Štamol et al. 1999) and *Spelaeoconcha paganetti* Sturany, 1901 (Maassen 1989), occur in the Dinaric Mountains. Within the Skadar Lake subregion of Montenegro and Albania, many short-range endemic species occur within the genera *Virpazaria*, *Klemmia*, *Spelaeodiscus*, *Agardhiella*, and *Platyla* (Reischütz and Reischütz

Figure 3.12. (*opposite*) Examples of global hypogean terrestrial gastropod species. [1, 2] *Opisthostoma mirabile* E. A. Smith, 1893, Gomantong Cave, Sabah, Malaysia; [3] *Opisthostoma lissopleuron* Vermeulen, 1994, Medai Cave, Sabah, Malaysia; [4] *Fadyenia* sp., Pomé Cave, Formón, Massif de la Hotte, Haiti; [5] *Plectostoma concinnum* (Fulton, 1901), Gomantong Cave, Sabah, Malaysia; [6] *Diplommatina* sp., Pitangueiras Cave, Bonito, Mato Grosso do Sul, Brazil (courtesy of R. L. Simone); [7] *Hypselostoma lacrima* Páll-Gergely & Hunyadi, 2015, Lenglei, Guangxi, China; [8] *Angustopila* sp., Ban Thathom, Xiangkhouang, Laos; [9] *Angustopila* sp., Lak Sao, Bolikhamsai, Laos; [10] *Opisthosoma brachyacrum lambi* Vermeulen, 1996, Laying Cave, Sabah, Malaysia; [11] *Angustopila* sp., Pu Phuong, Thanh Hoa, Vietnam; [12] *Angustopila* sp., Ha Long Bay, Quang Ning, Vietnam; [13] *Angustopila subelevata* Páll-Gergely & Hunyadi, 2015, Cong Troi Cave, Ha Giang, Vietnam; [14] *Tonkinospira* sp., Pu Phuong, Thanh Hoa, Vietnam; [15] *Tonkinospira* sp., Tham Xienliab Cave, Khammouane, Laos; [16] *Angustopila dominikae* Páll-Gergely & Hunyadi, 2015, Jialoe Cun, Guangxi, China; [17] *Krobylos laosensis* (Saurin, 1953), Pha Hia, Xiangkhouang, Laos; [18] *Pupisoma lignicola* (Stoliczka, 1871), Tenasserim, Moulmein, Malaysia; [19] *Dentisphaera maxema* Páll-Gergely & Jochum, 2017, Cong Troi Cave, Ha Giang, Vietnam. Images courtesy of Jozef Grego and Barna Páll-Gergely unless otherwise noted.

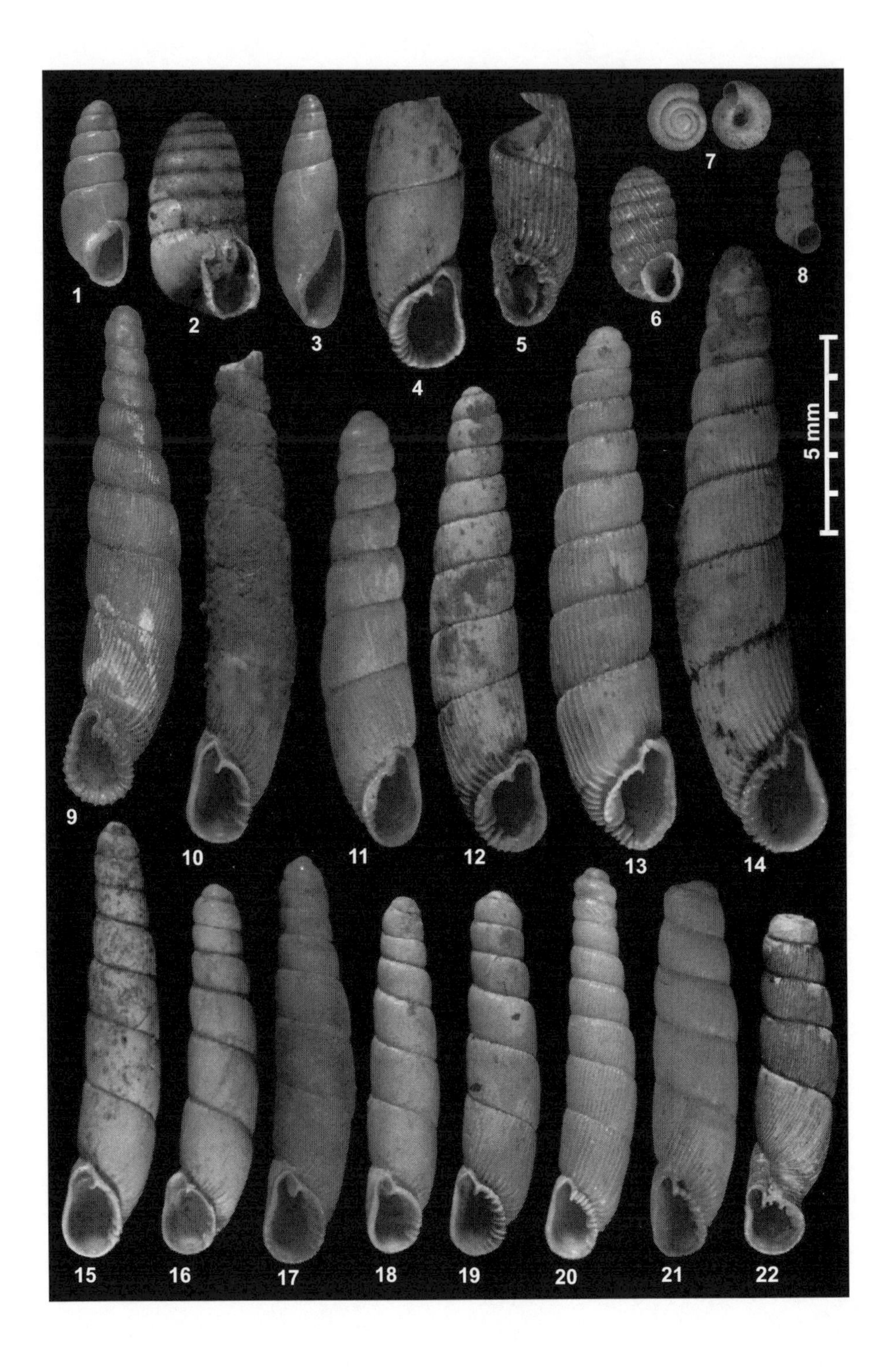
1
2
3
4
5
6
7
8
5 mm
9
10
11
12
13
14
15
16
17
18
19
20
21
22

2009; Páll-Gergely et al. 2018; Fehér et al. 2019). Southward to Albania and Greece, the relict clausiliid genera *Sciocochlea*, *Tsoukatosia*, and *Graecophaedusa*, with orculid *Speleodentorcula beroni* Gittenberger, 1985, have persisted since at least the Miocene (Nordsieck 2007). The eastern Balkans (i.e., Serbia, Macedonia, and Bulgaria) support the subterranean genera *Lindbergia* and *Gyralina*. The genera *Cryptazeca*, *Zospeum*, *Platyla*, *Acicula*, and *Bofilliella* are found in the Alps, Apennines, Pyrenees, and Iberian Peninsula. Along the southern coast of the Black Sea, the clausiliid genera *Cotyorica* and *Nothoserrulina* have been documented, while the eastern coast is typically characterized by the epigean genera *Truncatophaedusa* and *Troglolestes*.

Figure 3.13. (*opposite*) Examples of hypogean terrestrial gastropod species of families Clausiliidae, Ferussacidae, Orculidae, Vertiginidae, and Cyclophoridae. [1] *Spelaeoconcha paganettii* Sturany, 1901, Gorica, Trebinje, Hercegovina; [2] *Speleodentorcula beroni* E. Gittenberger, 1985, Skotini Cave near Tharounia, Evia, Greece; [3] *Cecilioides spelaea* A. J. Wagner, 1914, Gorica, Trebinje, Hercegovina; [4] *Tsoukatosia chistinae* A. Reischütz & P. L. Reischütz, 2003, Taigetos Mountains, Dirranio, Greece; [5] *Graecophaedusa sperrlei* Rähle, 1982, Pangaion Mountains, Macedonia, Greece; [6] *Pagodulina kaeufeli* Klemm, 1939, Vilina Pećina, Montenegro; [7] *Spelaeodiscus dejongi* E. Gittenberger, 1969, south of Virpazar, Montenegro; [8] *Pholeoteras euthrix* Sturany, 1904, cave above Uskoplje, Croatia; [9] *Cotyorica nemethi* Grego & Szekeres, 2017, between Perşembe and Mersin, Ordu, Turkey; [10] *Sciocochlea llogaraensis* A. Reischütz & P. L. Reischütz, 2009, Kudhës, Vlorë, Albania; [11] *Tsoukatosia arabatzis* A. Reischütz & P. L. Reischütz, 2014, Megistis Lavras Monastery, Athos Peninsula, Greece; [12] *Tsoukatosia nicoleae* A. Reischütz, & P. L. Reischütz, 2016, Karveli, Greece; [13] *Tsoukatosia subaii* Hunyadi & Szekeres, 2009, Fokida, Greece; [14] *Tsoukatosia liae excelsa* A. Reischütz, P. L. Reischütz & Szekeres, 2018, Agios Petros, Parnon Mountains, Peloponnese, Greece; [15] *Sciocochlea collasi* (Sturany, 1904), Barbati Cave, Corfu Island, Greece; [16] *Sciocochlea cryptica* Subai & Szekeres, 1999, Anthousa, Epirus, Greece; [17] *Sciocochlea cryptica acheron* A. Reischütz & P. L. Reischütz, 2004, Glyki, Epirus, Greece; [18] *Sciocochlea cryptica filiates* A. Reischütz & P. L. Reischütz, 2009, Filiates, Epirus, Greece; [19] *Sciocochlea nordsiecki* Subai, 1993, Dafni, Epirus, Greece; [20] *Serrulina senghanensis* Germain, 1933, Lunak, Gilan, Iran; [21] *Truncatophaedusa evae* Majoros, Nemeth, Szili-Kovacs, 1994, Sochi, Russia; [22] *Bofiliella subarcuata* (Bofill, 1897), Ermitons Cave, Olot, Girona, Spain. Images courtesy of Jozef Grego.

Table 3.1. Known genera of global hypogean terrestrial Gastropoda (with example species) and distribution by country or region, including Albania (AL); Australian regions: New South Wales (AU-NSW), Queensland (AU-QLD), Victoria (AU-VIC); Austria (AT); Belize (BZ); Bosnia and Hercegovina (BA); Brazil (BR); Bulgaria (BG); China (CN); Croatia (HR); European Union (EU); France (FR); Georgia (GE); Greece (GR); Haiti (HT); India (IN); Indonesia (ID); Iran (IR); Italy (IT); Laos (LA); Macedonia (MK); Malaysia (MY); Mexico (MX); Montenegro (ME); Myanmar (MM); New Zealand (NZ); Papua New Guinea (PG); Romania (RO); Russia (RU); Serbia (RS); Slovenia (SI); Spain (ES); the Sultanate of Oman (OM); Thailand (TH); Turkey (TR); United States of America (USA); Uzbekistan (UZ); and Vietnam (VN). Taxonomy is organized according to morphological homology. An asterisk (*) is used to denote troglophiles; all other species are troglobionts.

Taxonomy	Distribution	Example species
Gastropoda: Caenogastropoda		
Cyclophoridae		
Cochlostoma (Lovcenia)	AL, ME	*Cochlostoma (Lovcenia) erika* (A. J. Wagner, 1906)
Pholeoteras	BA, GR, HR	*Pholeoteras euthrix* Sturany, 1904
Alycaeidae		
Dicharax	ID, LA, TH	*Dicharax caudapiscis* Páll-Gergely & Hunyadi, 2017
Laotia	LA, VN	*Laotia pahiensis* Saurin, 1953
Diplommatinidae		
Diplommatina	LA, MM, MY, TH, VN	*Diplommatina boucheti* Maassen, 2007
Opisthostoma	MY	*Opisthostoma mirabile* E. A. Smith, 1893
Plectostoma	MY	*Plectostoma annandalei* (Sykes, 1903)
Habeas	BR	*Habeas data* Simone, 2013
Habeastrum	BR	*Habeastrum parafusum* Simone, 2018
Aciculidae		
Acicula	IT	*Acicula benoiti* (Bourguignat, 1864)
Platyla	AL, GR, HR, IT, ME, MK, RO, SI	*Platyla feheri* Subai, 2009
Renea	AL, ME	*Renea kobelti* (A. J. Wagner, 1910)
Gastropoda: Neritimorpha		
Helicinidae		
Fadyenia	HT	*Fadyenia* sp.
Hydrocenidae		
Georissa	ID, MY, PG	*Georissa pangianensis* Maassen, 2000

Table 3.1. (continued)

Taxonomy	Distribution	Example species
Gastropoda: Heterobranchia		
Ellobiidae		
Carychium	USA	*Carychium stygium* Call, 1897
Zospeum	AT, BA, ES, FR, HR, IT, ME, SI	*Zospeum spelaeum* (Rossmässler, 1839)
Gastropoda: Stylommatophora		
Strobilopsidae		
Klemmia	ME	*Klemmia sinistrorsa* Gittenberger, 1969
Spelaeodiscus	AL, ME, SI	*Spelaeodiscus albanicus* (A. J. Wagner, 1914)
Virpazaria	AL, HR, ME	*Virpazaria pageti* E. Gittenberger, 1969
Zonitidae		
Aegopis	BA, HR	*Aegopis spelaeus* A. J. Wagner, 1914
Balcanodiscus	BG, GR, RS	*Balcanodiscus frivaldskyanus* (Rossmässler, 1842)
Daudebardia	GE	*Daudebardia nivea* Schileyko, 1988
Meledella	HR	*Meledella werneri* Sturany, 1908
Paraegopis	ME	*Paraegopis oberwimmeri* Klemm, 1965
Thasiogenes	GR	*Thasiogenes difficilis* A. Riedel, 1988
Troglaegopis	HR	*Troglaegopis mosorensis* (Kuščer, 1933)
Oxychilidae		
Conulopolita	GE	*Conulopolita cavatica* (A. Riedel, 1966)
Oxychilus	Palearctic/ Pontic	**Oxychilus koutaisianus* (Mousson, 1863)
Vitrinoxychilus	GE	**Vitrinoxychilus subsuturalis* (O. Boettger, 1888)
Pristilomatidae		
Gyralina	AL, BA, GR, HR, ME, MK	*Gyralina velkovrhi* A. Riedel, 1985
Lindbergia	BG, GR, IT, TR	*Lindbergia spiliaenymphis* A. Riedel, 1959
Spelaeopatula	AL, HR, MK	*Spelaeopatula mljetica* (Pintér & A. Riedel, 1973)
Troglovitrea	RO	*Troglovitrea argintarui* Negrea & A. Riedel, 1968
Vitrea	BA, GR, MK	*Vitrea spelaea* (A. J. Wagner, 1914)
Argnidae		
Argna	IT	*Argna valsabina* (Spinelli, 1851)
Speleodentorcula	GR	*Speleodentorcula beroni* Gittenberger, 1985

(continued)

Table 3.1. (continued)

Taxonomy	Distribution	Example species
Agardhiellidae		
Agardhiella	AL, BA, BG, GR, HR, MK, RO, RS	*Agardhiella formosa* (Westerlund, 1887)
Vertiginidae		
Paraboysidia	LA, TH, VN	*Paraboysidia serpa* van Benthem-Jutting, 1950
Spelaeoconcha	BA, HR	*Spelaeoconcha paganettii* Sturany, 1901
Gastrocoptidae		
Gastrocopta	BR	*Gastrocopta sharae* Salvador, Cavalliari & Simone, 2017
Ferussaciidae		
Cecilioides	EU/ Ponto-Caspian	*Cecilioides spelaea* A. J. Wagner, 1914
Cryptazeca	ES	*Cryptazeca spelaea* Gómez, 1990
Clausiliidae		
Acrotoma	GE	**Acrotoma baryshnikovi* Likharev & Schileyko, 2007
Bofilliella	ES	*Bofilliella subarcuata* (Bofill, 1897)
Cotyorica	TR	*Cotyorica nemethi* Grego & Szekeres, 2017
Graecophaedusa	GR	*Graecophaedusa sperrlei* Rähle, 1982
Nothoserrulina	TR	*Nothoserrulina subterranea* Németh & Szekeres, 1995
Sciocochlea	AL, GR	*Sciocochlea collasi* C. Boettger, 1935
Serrulina	IR	*Serrulina senghanensis* Germain, 1933
Truncatophaedusa	RU	*Truncatophaedusa evae* Majoros, Nemeth & Szili-Kovács, 1994
Tsoukatosia	GR	*Tsoukatosia liae* Gittenberger, 2000
Helicodiscidae		
Lucilla	NZ	*Lucilla academia* (Climo, 1970)
Helicodiscus	USA	*Helicodiscus hadenoecus* Hubricht, 1962
Scolodontidae		
Prohappia	BR	*Prohappia besckei* (Dunker, 1847)
Hypselostomatidae		
Angustopila	LA, TH, VN	*Angustopila stochi* Páll-Gergely & Jochum, 2017
Dentisphaera	VN	*Dentisphaera maxema* Páll-Gergely & Jochum, 2017
Hypselostoma	CN, LA, TH, VN	*Hypselostoma edentata* (Panha & Burch, 1999)

Table 3.1. (continued)

Taxonomy	Distribution	Example species
Krobylos	CN LA, TH, VN	*Krobylos sinensis* Páll-Gergely & Hunyadi, 2015
Tonkinospira	LA, VN	*Tonkinospira tomasini* Páll-Gergely & Jochum, 2017
Valloniidae		
Pupisoma	BR, LA, TH, VN	*Pupisoma dioscoricola* (C. B. Adams, 1845)
Diapheridae		
Sinoennea	LA, MM, MY, TH	*Sinoennea lizae* Maassen, 2008
Subulinidae		
Lamellaxis	BZ	*Lamellaxis matola* Dourson & Caldwell, 2018
Lavajatus	BR	*Lavajatus moroi* Simone, 1918
Leptinaria	BZ	*Leptinaria doddi* Dourson & Caldwell, 2018
Leptopeas	BZ	*Leptopeas corwinii* Dourson & Caldwell, 2018
Opeas	IN, LA, MM, MY, TH	*Opeas cavernicola* Annetale & Chopra, 1924
Trigonochlamididae		
Lesticulus	GE	*Lesticulus nocturnus* Schileyko, 1988
Troglolestes	GE	*Troglolestes sokolovi* Liovushkin & Matiokin, 1965
Punctidae		
Pseudiotula	AU-QLD, AU-VIC	**Pseudiotula eurysiana* Stanisic, 2010
Charopidae		
Charopa	MY	*Charopa kanthanensis* Vermeulen & Marzuki, 2014
Elsothera	AU-NSW	*Elsothera sericatula* (L. Pfeiffer, 1849)

For Southeast Asia, at least 11 genera have been identified. This includes *Diplommatina*, *Plectostoma*, and *Opisthostoma* from Malaysia (Liew et al. 2014) and *Angustopila, Hypselostoma, Paraboysidia, Pupisoma, Tonkinospira, Dicharax*, *Sinoennea*, and *Laotia* reported from Myanmar, Thailand, Malaysia, Laos, Vietnam, and China (Yunnan and Guangxi; Páll-Gergely 2014; Páll-Gergely et al. 2019, 2020; Liew at et al. 2014).

Except for limited investigations in the United States (Culver 2012; Gladstone, Niemiller, et al. 2019) and Brazil (Simone 2018), the Americas are largely undocumented. In the Caribbean, a few undescribed species (e.g.,

Microsagda sp. and *Fadyenia* sp.) are known from caves of Macaya National Park, Haiti (Grego and Šteffek 2007), and four cave Subulinidae species have been documented in Belize (Dourson et al. 2018). Finally, in Australia, species from the genera *Calvigenia*, *Austrochloritis*, *Charopa*, *Laevidelos*, and *Gastrocopta* were identified as troglophiles (Slack-Smith 1993; Eberhard et al. 2014).

Habitat and Ecology

Researchers remain challenged to disentangle the distributions of terrestrial hypogean Mollusca species between soil/leaf-litter habitats and caves. I have provided a schematic that illustrates my interpretations concerning how this group uses the substratum (Fig. 3.14). As deposits of empty shells are often found inside caves, this group's biology and ecological function (i.e., epigean or hypogean) are somewhat open to conjecture. In the past, these animals have been regarded as either troglobiontic or MSS species. Many surface-dwelling species behave at least seasonally as troglophiles (e.g., Eberhard et al. 2014); these species are attracted to the higher moisture conditions underground—especially during drought periods. Some epigean species require the shallow hypogean habitats for estivation or reproduction (including oviparity). The occurrence of subterranean assemblages suggests many of the subterranean habitats have long-term stable environments. In some cases, these habitats were refugia during climatic shifts (e.g., glacial-interglacial periods resulting in surface relatives going extinct with the subterranean population remaining). Examples of presumed subterranean climate relicts (refer to Wynne et al., Chap. 2) include numerous Tertiary relict families, genera, and species (e.g., Maassen 1989; Reischütz et al. 2016; Grego and Szekeres 2017). During the evolution of subterranean air-breathing gastropods, these animals exploited a variety of habitats. Adaptation to various hypogean habitats was driven by historic climatic changes (i.e., increasing regional aridity; Cook and Jones 2012; Morley 2012) and likely reduced exposure to predators and parasites.

A primary example of a nutrient-rich habitat is the rhizosphere (i.e., plant roots, mycorrhizal, and saprophytic fungi) within the deep cave zone. In tropical karst, plant roots penetrate several dozen meters below surface through the limestone and can extend from the cave ceiling to floor (Howarth and Wynne, Chap. 1). In other areas, this habitat occurs within a net-

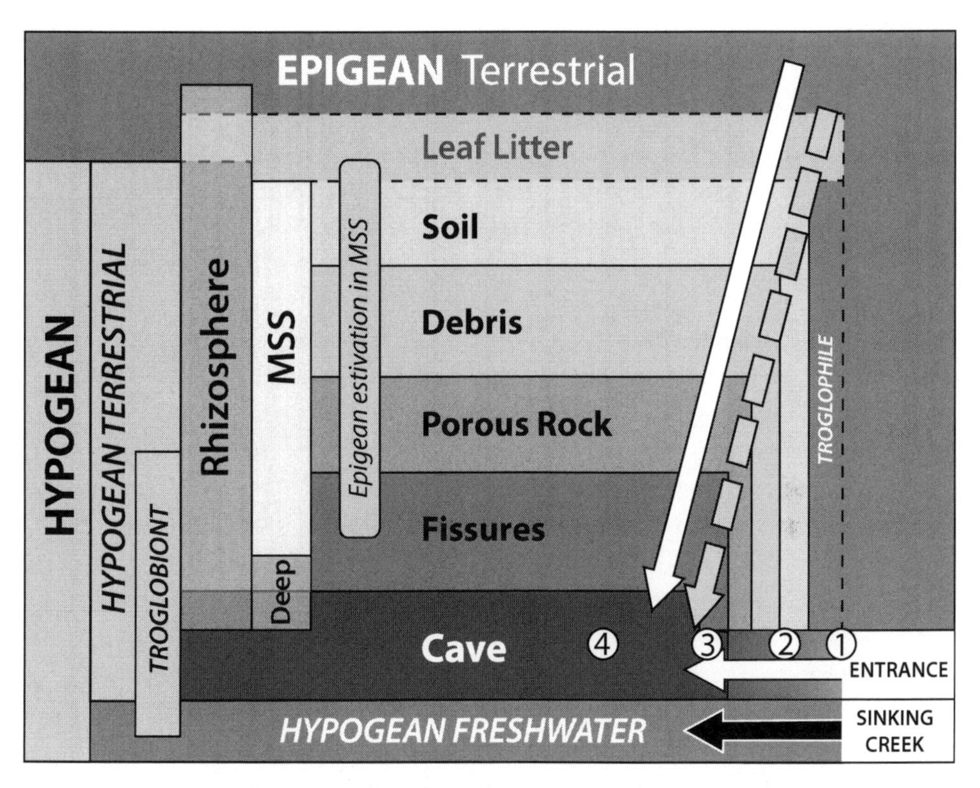

Figure 3.14. Subterranean terrestrial habitats with related evolutionary categories provided. Directions of nutrient pathways are delineated with arrows, while the black dashed line represents interim rainwater percolation. Allochthonous nutrients enter the cave through the entrance, *milieu souterrain superficiel* (MSS) channels, vertical fissures, the rhizosphere (plant roots and decaying organic material), and to a lesser extent, animals. The leaf-litter zone is an intermediate zone between the surface and hypogean habitats. Epigean species may torpor in MSS during periods of drought. Numbers within white circles represent the entrance (1), twilight (2), transition (3), and deep cave (4) zones.

work of small cracks, channels, and crevices resulting from the dissolution of karst. Rhizosphere habitats occur within faults and stone debris (colluvial MSS) or cavernous limestone substrates, such as breccia (macroporous rock MSS); these habitats appear to be especially rich in hypogean terrestrial molluscan diversity. Optimal microclimates occur within fractured rock and stone debris and are saturated by humid subterranean air. The bedrock MSS is frequently a secondary habitat, while the alluvial MSS is unfavorable for troglobiontic gastropods due to the rapidly alternating water conditions and low motility of these taxa. The volcanic MSS is frequently associated with lower pH in acidic rocks and represents the least studied subterranean molluscan habitat.

The narrow specialization induced by the deep cave zone resulted in hypogean terrestrial taxa being extremely vulnerable to stochastic events and environmental change. Similar to hypogean freshwater gastropod species, terrestrial species are not exclusively restricted to small crevices—they also occur in larger cavities, where they can be observed and studied by researchers. Mainly through the passages connected to the rhizosphere, Mollusca may be washed into the cave from colluvial or bedrock MSS during flood events. Hypogean terrestrial species are usually quite minute, ranging from 0.6 to 5 mm in size. Some species with elongate shells (e.g., Clausiliidae) can reach up to 19 mm in length; the largest troglobiontic zonitoid (*Paraegopis oberwimmeri* Klemm, 1965) is up to 40 mm.

As with hypogean freshwater species, nutrient availability is likely the most important factor promoting colonization of specific habitats. Conversely, extreme environmental factors, such as high temperature or low oxygen, seem less important than food availability and have been overcome by adaptation.

Bats also transport large quantities of organic material (i.e., guano) into the cave systems, which can serve as the resource base for many hypogean ecosystems (Ferreira et al. 2007; Howarth and Wynne, Chap. 1). However, guano is high in phosphoric acid and thus has a low pH. Subsequently, this likely limits the ability for gastropods to exploit this nutrient source. Occasionally, species of family Subulinidae are found associated with bat guano (Grego, unpublished data). This group seems to have overcome the acidic condition by developing a thick protein layer (periostracum), which covers the shell to protect the animal (Grego, unpublished data).

In general, cave systems are nutrient limited and subterranean adaptation is usually coupled with a slower, more efficient metabolism (Di Lorenzo et al. 2015) and longevity with longer maturation period (e.g., Voituron et al. 2011; Puljas et al. 2014). Although no data exist for the specific life span of troglobiontic gastropods, it stands to reason these animals adapted similarly to the subterranean environment.

The genus *Zospeum* of the Dinaric and Iberian regions represents a semiaquatic cave-dwelling group. Due to its taxonomy, we consider this genus to be air breathing. However, it is occasionally found in aquatic habitats between the moist layer of the cave walls and directly above this surface. Because of this, *Zospeum* is frequently considered a "freshwater troglobiont." Members of this genus have been documented feeding on chemolithotrophic biofilms in Croatian caves (Lukina Jama) deeper than 1,000 m beneath the surface (Weigand 2013).

Dispersal and Reproductive Biology

While specifics on dispersal and reproductive biology of hypogean terrestrial Gastropoda is deficient, I provide some extrapolations based on closely related leaf-litter-dwelling taxa. For example, the leaf-litter-dwelling *Pontophaedusa funiculum* (Mousson, 1856), which has MSS-dwelling congeners, lays hard-shelled eggs (Páll-Gergely and Németh 2008; Páll-Gergely 2010). This strategy may enable subterranean relatives to overcome variability in moisture within the hypogean environment. It is reasonable to suggest troglobiontic gastropods disperse similarly. Given adequate time, gastropods may disperse to more distant habitats by using mesocaverns and macrocaverns (refer to Howarth and Wynne, Chap. 1).

Some subterranean gastropods have large distributions (e.g., *Spelaeoconcha paganetti* Sturany, 1901 ranges from Brač and Korčula Islands in Croatia to Ilidža near Sarajevo in Bosnia and southward to Trebinje in Hercegovina; Maassen 1989). Additionally, *Pholeoteras eutrix* Sturany, 1904 ranges from Vis Island and Dubrovnik in Croatia 600 km southward through southern Hercegovina into northern Greece (Štamol et al. 1999). Such large distributions suggest that some terrestrial species occasionally overcome classical dispersal barriers (refer to Wynne et al., Chap. 2). Another explanation is that adults and/or eggs are routinely dispersed via surface water streams during flood events. This is evidenced by hypogean terrestrial shells of various species, including two species found within river deposits of nonkarstic areas far from their known karstic distributional range (Grego, unpublished data). Moreover, it is possible these large geographic areas are comprised of clades or taxonomically indistinct, yet genetically distinct, species. Molecular analysis will be required to characterize their distributions more clearly.

Conservation and Future Research

Nonmarine Mollusca are the most sensitive animal group in the world. Their extinctions represent approximately 43% of all extinct animals since 1500 CE (Herbert and Kilburn 2004; Lydeard et al. 2004) and 35.9% in 2021 (IUCN 2021). This high extinction rate emphasizes their value as indicator species of environmental change. Their sensitivity to human activities is further exacerbated by their narrow habitat requirements and small distributional ranges.

Anthropogenic activities, including global climatic change (Mammola et al. 2019), impacts to spring and groundwater habitats (e.g., hydroelectric dam construction projects; Piegay et al. 2009), water diversion (Siebert et al. 2010), drought (Shu et al. 2013), groundwater pollution (Di Lorenzo et al. 2018; Zhao et al., Chap. 7), and deforestation (Trajano 2000; Souza-Silva et al. 2015) can result in habitat loss or degradation. Subsequently, many species have gone, or may go, extinct without being discovered and formally described (Mammola et al. 2019).

For example, human activities are impacting a subterranean biodiversity hotspot, the Dinaric Mountains (Zagmajster et al. 2018), which is tremendously important to global molluscan diversity. This region supports 30% of the hypogean freshwater species and 24% of the terrestrial species globally. In many areas of the Dinaric Alps, hydrology has been altered due to hydroelectric dam projects, which have flooded large karstic areas and inundated caves they support (Milanović 2018). Many spring and cave habitats that once supported unique hypogean fauna are now gone. Subsequently, we have lost the type localities of many stygobiontic species, including *Plagigeyeria tribunicae* Schütt, 1969 and *Paladilhiopsis serbica* Pavlović, 1913 beneath the Bilećko Jezero Dam (Hercegovina) and Jezero Perućac Dam (Serbia/Bosnia) reservoirs, respectively.

Dam construction projects are ongoing in the Dinaric Mountains region (e.g., Sana Spring power plant project); however, today these projects are less frequent. Public opposition to these projects, such as the Ombla River project, offers a ray of hope. This project was suspended due to scientific and public opposition, as the government was not fully considering the environmental impact (Roje-Bonacci and Bonacci 2013; Antonić, Kljaković-Gašpić, Mayer, et al. 2015; Antonić, Kljaković-Gašpić, Hatić, and Peternel 2015; Antonić, Kljaković-Gašpić, Ozimec, et al. 2015; Antonić, Kljaković-Gašpić, Mraković, et al. 2015; Kljaković-Gašpić et al.

2015; Paviša and Sever 2015). Many countries in the western Balkans have gradually adopted European Union environmental rules, such as the Environmental Impact Assessment. Unfortunately, these assessments usually omit the potential impacts of these projects on subterranean fauna.

As most Mollusca are aquatic to semiaquatic, groundwater contamination threatens the persistence of many species globally. During the past several decades, waste production has shifted from degradable organic material to plastics and chemicals. This shift has a dramatic and increasing impact on groundwater quality—especially when the "traditional means" of disposing of waste by rural farmers and small communities (e.g., open-air landfills or placing waste in vertical pits) now includes nonbiodegradable and often toxic refuse.

Because hypogean Mollusca are sensitive to environmental perturbations, and likely contribute to maintaining groundwater quality (Griebler et al. 2014), these organisms may be useful indicator species for groundwater pollution. For example, in 2017, a fish kill event was reported at a sink hole near Nuga Lake in Imotsko Polje in the Bosnian-Croatian region. I later documented a massive stygobiontic Mollusca thanatocoenoses accumulation at the spring of the Tihaljina River in Peć Mlini. This included shells of *Congeria kuščeri*, *Marifugia cavatica* Absolon & Hrabě, 1930, and eight other stygobiontic gastropod species.

Deforestation in karstic areas is another major threat (Whitten 2009; Howarth and Wynne, Chap. 1). The loss of vegetation cover reduces water retention, and in some cases leads to desertification (Trajano 2000). Furthermore, as most hypogean ecosystems are dependent on allochthonous food resources (wood debris, rhizosphere, and plant/tree roots), the removal of surface vegetation will result in a drastic reduction to the complete elimination of a primary nutrient resource (Jiang et al. 2014; Canarini et al. 2019).

Additionally, gravel interstitial habitats are heavily impacted by groundwater overexploitation. These human activities can lead to habitat degradation and/or loss, which can translate into species extirpation and potentially extinction. For example, the springs and alluvial gravel around Viesca, Coahuila, Mexico, dried out 70 to 80 years ago due to intensive groundwater extraction. Today, no live hypogean gastropod specimens have been collected; only empty shells of new gastropod genera and species can be sieved out for study (Czaja et al. 2019).

With respect to the above extensive threats of karstic and groundwater habitats, I believe the documentation of hypogean Mollusca fauna is urgently needed. It is imperative to treat potentially new stygobiont gastropod shells with unique morphology in the same manner as fossil species—thereby, describing the specimens based solely on the availability of shells. Provisional taxonomic positioning within the morphologically closest and geographically most likely genera is a vital exercise. This could be done as a stopgap until molecular analysis and more precise anatomical data become available to accurately establish the species' precise taxonomic position. Although imperfect, even limited information can assist researchers and resource managers in developing conservation strategies that could help safeguard hypogean mollusk habitat and endemic populations.

Conclusions

Our knowledge concerning most hypogean Mollusca, their ecology, zoogeography, physiology, and taxonomy remains insufficient. Most of the hypogean habitats remain inaccessible to humans (Howarth 1983; Ficetola et al. 2019); thus, the majority of the hypogean species remain undescribed (Mammola et al. 2019). Accessing hypogean freshwater habitats (which typically involves only seasonal access) offers limited opportunities for the in situ collection of Mollusca. Subsequently, springs and wells remain the most reliable areas to sample for stygobiontic gastropod species. Obtaining a sufficient number of live individuals for both anatomical comparisons of male and female characters and molecular analyses is often more dependent on luck rather than a systematic approach. Importantly, the ability to collect live individuals at springs is greater during early spring or shortly after seasonal flood events (Grego, unpublished data).

As elucidated above, most hypogean taxa were described based solely on empty shell morphology and frequently just from the type locality (Culver 2012). Single records are clearly inadequate for describing life history, habitat requirements, and distributional range. Importantly, single locality records can hinder researchers and conservation biologists from evaluating these species in terms of their vulnerability to human disturbance.

A persistent challenge is the study of live specimens, which will provide researchers with the ability to examine questions related to the species' molecular taxonomy and ecology. Furthermore, given the potential importance of hypogean Mollusca species as biological indicators for groundwater con-

Table 3.2. Known genera of global hypogean freshwater Gastropoda (with species examples) and their distribution by country or region, including Albania (AL); Australian regions: Lord Howe Island (AU-LH), New South Wales (AU-NSW), Queensland (AU-QLD), Tasmania (AU-TAS), Victoria (AU-VIC), and Western Australia (AU-WA); Austria (AT); Belgium (BE); Bosnia and Hercegovina (BA); Brazil (BR); Bulgaria (BG); Colombia (CO); Croatia (HR); Cyprus (CY); Czech Republic (CZ); Ecuador (EC); Estonia (EE); France (FR); Georgia (GE); Germany (DE); Greece (GR); Hungary (HU); Indonesia (ID); Israel (IL); Iran (IR); Italy (IT); Japan (JP); Kosovo (KS); Laos (LA); Macedonia (MK); Mexico (MX); Montenegro (ME); Morocco (MA); Myanmar (MM); Netherlands (NL); New Zealand (NZ); Oman (OM); Papua New Guinea (PG); Philippines (PH); Poland (PL); Romania (RO); Russia (RU); Serbia (RS); Slovakia (SK); Slovenia (SI); Spain (ES); Sri Lanka (LK); Switzerland (CH); Tajikistan (TJ); Tunisia (TN); Turkey (TR); Turkmenistan (TM); Ukraine (UA); United States of America (USA); and Uzbekistan (UZ). Taxonomy is organized according to morphological homology. An asterisk (*) is used to denote stygophiles; all other species are stygobionts.

Taxonomy	Distribution	Example species
Gastropoda: Heterobranchia		
Acroloxidae		
Acroloxus	BA, GE, SI	*Acroloxus tetensi* (Kuščer, 1932)
Physidae		
Physella	USA	*Physella spelunca* Turner & Clench, 1974
Glacidorbidae		
Glacidorbis	AU-NSW	*Glacidorbis hedleyi* Iredale, 1943
Planorbidae		
Ancylus	BA, HR	**Ancylus fluviatilis* O. F. Müller, 1774
Ferrissia	MX	*Ferissia* sp.
Valvatidae		
Valvata	HR	*Valvata troglobia* Piersanti, 1952
Gastropoda: Neritimorpha		
Neritidae		
Neritilia	PH	*Neritilia cavernicola* Kano & Kase, 2004
Theodoxus	BA, IL, TR	*Theodoxus subterrelictus* Schütt, 1963
Gastropoda: Caenogastropoda		
Emmericiidae		
Emmericia	BA	**Emmericia ventricosa* Brusina, 1870

(*continued*)

Table 3.2. (continued)

Taxonomy	Distribution	Example species
Bythinellidae		
Bythinella	BA, FR, IT, RO, SK, SI	**Bythinella austriaca* (Frauenfeld, 1857)
Cochliopidae		
Andesipyrgus	CO, EC	*Andesipyrgus sketi* Hershler & Velkovrh, 1993
Antrobia	USA	*Antrobia culveri* Hubricht, 1971
Antrobis	USA	*Antrobis breweri* Hershler & Thompson, 1990
Antroselates	USA	*Antroselates spiralis* Hubricht, 1963
Balconorbis	USA	*Balconorbis uvaldensis* Hershler & Longley, 1986
Coahuilix	MX	*Coahuilix hubbsi* Taylor, 1966
Cochliopina	MX	*Cochliopina compacta* (Pilsbry, 1910)
Dasyscias	USA	*Dasyscias franzi* Thompson & Hershler, 1991
Emmericiella	MX	*Emmericiella longa* (Pilsbry, 1909)
Fontigens	USA	*Fontigens cryptica* Hubricht, 1962
Heleobia (Semisalsa)	RO	*Heleobia (Semisalsa) dobrogica* (Grossu & Negrea, 1989)
Cf. *Heleobia* (provisional genus)	OM	Cf. *Heleobia* sp.
Holsingeria	USA	*Holsingeria unthanksensis* Hershler, 1989
Juturnia	MX	*Juturnia coahuilae* (D. W. Taylor, 1966)
Mexicenotica	MX	*Mexicenotica xochi* Grego, Angyal & Bertrán, 2019
Phreatoviesca	MX	*Phreatoviesca spinosa* Czaja & Gladstone, 2021
Pterides	MX	*Pterides rhabdus* Pilsbry, 1909
Pyrgophorus	MX	**Pyrgophorus coronatus* (L. Pfeiffer, 1840)
Stygopyrgus	USA	*Stygopyrgus bartonensis* Hershler & Longley, 1986
Moitessieriidae		
Baldufa	ES	*Baldufa fontinalis* Alba, Tarruella, Prats, Guillén & Corbella, 2010
Bosnidilhia	BA, ME	*Bosnidilhia vreloana* Glöer & Pešić, 2013
Bythiospeum	AT, BE, CH, DE, EE, FR, HU, NL	*Bythiospeum acicula* (Hartmann, 1821)
Clameia	GR	*Clameia brooki* Boeters & Gittenberger, 1990
Corseria	FR	*Corseria corsica* Bernasconi, 1994
Falniowskia	PL	*Falniowskia neglectissima* (Falniowski & Šteffek, 1989)
Henrigirardia	FR	*Henrigirardia wienini* (Girardi, 2001)

Table 3.2. (continued)

Taxonomy	Distribution	Example species
Iglica	AL, AT, BG, BA, GR, HR, IT, KS, MK, MA, SI	*Iglica absoloni* (A. J. Wagner, 1914)
Cf. *Iglica* (provisional genus)	MA	Cf. *Iglica soussensis* Ghamizi & Boulal, 2017
Lanzaia (*Lanzaia*)	BA, HR, KS, ME	*Lanzaia* (*Lanzaia*) *bosnica* (Bole, 1971)
Lanzaia (*Cilgia*)	BA, HR	*Lanzaia* (*Cilgia*) *dalmatica* (Schütt, 1961)
Lanzaia (*Costellina*)	BA, HR	*Lanzaia* (*Costellina*) *turrita* (Kušcer, 1933)
Lanzaia (*Plagigeyeria*)	AL, BA, BG, FR, HR, IT, KS, ME	*Lanzaia* (*Plagigeyeria*) *plagiostoma* (A. J. Wagner, 1914)
Lanzaiopsis	SI	*Lanzaiopsis savinica* Bole, 1989
Moitessieria	ES, FR	*Moitessieria lineolata* Coutagne, 1881
Palacanthilhiopsis	FR	*Palacanthilhiopsis vervierii* Bernasconi, 1988
Paladilhia	FR	*Paladilhia bourguignati* Paladilhe, 1866
Paladilhiopsis	AL, BA, BG, GR, HR, HU, ME, RO, SI, SK, RS, UA	*Paladilhiopsis robiciana* (Clessin, 1882)
Palaospeum	ES	*Palaospeum septentrionalis* (Rolán & Ramos, 1995)
Sardopaladilhia	ES	*Sardopaladilhia plagigeyerica* Manganelli, Bodon, Cianfanelli, Talenti & Giusti, 1998
Saxurinator	BG	*Saxurinator buresi* (Wagner, 1927)
Sorholia	FR	*Sorholia lescherae* (Boeters, 1981)
Spiralix	ES	*Spiralix affinitatis* Boeters, 2003
Tarracospeum	ES	*Tarracospeum raveni* Quiñonero-Salgado, Ruiz-Jarillo, Alonso & Rolán
Trogloiranica	IR	*Trogloiranica tashanica* Fatemi, Malek-Hosseini, Falniowski, Hofman, Kuntner & Grego, 2019
Hydrobiidae		
Agrafia	GR	*Agrafia wiktori* Szarowska & Falniowski, 2011
Alzoniella	AT, CZ, FR, IT, SK	*Alzoniella feneriensis* Giusti & Bodon, 1984
Arganiella	IT	*Arganiella pescei* Giusti & Pezzoli, 1980
Atebbania	MA	*Atebbania bernasconii* Ghamizi, Bodon, Boulal & Giusti, 1998
Avenionia	FR	*Avenionia brevis* (Draparnaud, 1805)
Balkanica	BG	*Balkanica yankovi* Georgiev, 2011
Balkanospeum	BG	*Balkanospeum schniebsae* (Georgiev, 2011)
Belgrandia	HR	*Belgrandia torifera* Schütt, 1961
Belgrandiella	AT, BA, BG, HR, IT, KS, ME, RS, SI	*Belgrandiella cavernica* C. R. Boettger, 1957
Belgrandiellopsis	TN	*Belgrandiellopsis chorfensis* Khalloufi, Béjaoui & Delicado, 2020
Biserta	TN	*Biserta putealis* Khalloufi, Béjaoui & Delicado, 2020

(*continued*)

Table 3.2. (continued)

Taxonomy	Distribution	Example species
Boleana	SI	*Boleana umbilicata* (Kuščer, 1932)
Bracenica	ME	*Bracenica spiridoni* Radoman, 1973
Caucasogeyeria	GE, RU	*Caucasogeyeria gloeri* Grego & Mumladze, 2020
Caucasopsis	GE, RU	*Caucasopsis letsurtsume* Grego & Mumladze, 2020
Corbellaria	ES	*Corbellaria celtiberica* Corbell, Giusti & Boeters, 2012
Dabriana	BA	*Dabriana bosniaca* Radoman, 1974
Dalmatella	HR	*Dalmatella sketi* Velkovrh, 1970
Dalmatinella	HR	*Dalmatinella fluviatilis* Radoman, 1973
Daphniola	CY, GR	*Daphniola exigua* (A. Schmidt, 1856)
Daudebardiella	TR	*Daudebardiella asiana* O. Boettger, 1905
Deganta	ES	*Deganta azarum* (Boeters & Rolán, 1988)
Devetakia	BG	*Devetakia krushunica* Georgiev & Glöer, 2011
Devetakiola	BG	*Devetakiola devetakium* Georgiev & Glöer, 2013
Falsibelgrandiella	TR	*Falsibelgrandiella bunarica* Radoman, 1973
Fissuria	FR	*Fissuria boui* Boeters, 1981
Gocea	MK	*Gocea ohridana* Hadžišce, 1956
Graecoarganiella	GR	*Graecoarganiella parnassiana* Falniowski & Szarowska, 2011
Graziana	SI	**Graziana lacheineri* (Küster, 1853)
Guadiella	ES	*Guadiella ballesterosi* Alba, Tarruella, Prats, Corbella & Guillen, 2009
Hadziella	BA, HR, IT, SI	*Hadziella rudnicae* Bole, 1992
Hauffenia	AT, BA, HR, HU, SI, SK	*Hauffenia edlaueri* (Schütt, 1961)
Hausdorfenia	GE	*Hausdorfenia pseudohauffenia* Grego & Mumladze, 2020
Heideella	MA	*Heideella andreae* Backhuys & Boeters, 1974
Heraultiella	FR	*Heraultiella exilis* (Paladilhe, 1867)
Horatia	ES, FR, HR	*Horatia knorri* Schütt, 1961
Iglicopsis	BA	*Iglicopsis butoti* Falniowski & Hofman, 2021
Imeretiopsis	GE	*Imeretiopsis prometheus* Grego & Palatov, 2020
Insignia	BG	*Insignia macrostoma* Angelov, 1972
Islamia	BA, CY, ES, IT, ME, TR	*Islamia pusilla* (Piersanti, 1952)
Istriana	HR, IT, SI	*Istriana mirnae* Velkovrh, 1971
Iverakia	ME	*Iverakia hausdorfi* Glöer & Pešić, 2014
Kainarella	TM	*Kainarella minima* Starobogatov, 1972
Kartvelobia	GE	*Kartvelobia sinuata* Grego & Mumladze, 2020
Kerkia	BA, HR	*Kerkia kusceri* (Bole, 1961)
Kolevia	BG	*Kolevia bulgarica* Georgiev & Glöer, 2015
Litthabitella	BA, ME, HR	**Litthabitella chilodia* (Westerlund, 1886)
Lyhnidia	AL, MK	*Lyhnidia karamani* Hadžišce, 1956

Table 3.2. (continued)

Taxonomy	Distribution	Example species
Malaprespia	AL	*Malaprespia albanica* Radoman, 1973
Maroccopsis	MA	*Maroccopsis agadirensis* Ghamizi, 1998
Marstoniopsis	BA, FR, HR	*Marstoniopsis croatica* Schütt, 1974
Meyrargueria	FR	*Meyrargueria rasini* Girardi, 2004
Microstygia	BG	*Microstygia deltchevi* Georgiev & Glöer, 2015
Mienisiella	IL	*Mienisiella mienisi* Schütt, 1991
Milesiana	ES	*Milesiana schuelei* Boeters, 1981
Montenegrospeum	BA, HR, ME	*Montenegrospeum bogici* Pešić & Glöer, 2013
Motsametia	GE	*Motsametia borutzkii* Shadin, 1932
Narentiana	BA, HR	*Narentiana albida* Radoman, 1973
Navalis	ES	*Navalis perforatus* Quiñonero-Salgado & Rolán 2017
Neohoratia	ES	*Neohoratia minuta* (Draparnaud, 1805)
Ohridohauffenia	MK	*Ohridohauffenia minuta* (Radoman, 1955)
Ohridohoratia	MK	*Ohridohoratia carinata* (Radoman, 1957)
Ohrigocea	MK	*Ohrigocea miladinovorum* Hadžišce, 1959
Cf. *Orientalina* (provisional genus)	BA	Cf. *Orientalina troglobia* (Bole, 1961)
Palacanthilhiopsis	FR	*Palacanthilhiopsis vervierii* Bernasconi, 1988
Pauluccinella	IT	*Pauluccinella minima* (Paulucci, 1881)
Pezzolia	IT	*Pezzolia radapalladis* Bodon & Giusti, 1986
Phreatica	IT	*Phreatica bolei* Velkovrh, 1970
Plesiella	ES	*Plesiella navarrensis* Boeters, 2003
Pontobelgrandiella	BG	*Pontobelgrandiella nitida* (Angelov, 1972)
Pontohoratia	GE	*Pontohoratia birsteini* Starobogatov, 1962
Pseudamnicola	TJ, UZ	*Pseudamnicola bucharica* Shadin, 1952
Pseudavenionia	IT	*Pseudavenionia pedemontana* Bodon & Giusti, 1982
Pseudocaspia	TM	*Pseudocaspia kainarensis* Starobogatov, 1972
Pseudohoratia	AL, MK	*Pseudohoratia brusinae* (Radoman, 1953)
Pseudoislamia	GR	*Pseudoislamia balcanica* Radoman, 1979
Pyrgula	MK	**Pyrgula annulata* (Linnaeus, 1758)
Rifia	MA	*Rifia yacoubii* Ghamizi, 2020
Sadleriana	HR	*Sadleriana cavernosa* Radoman, 1978
Sarajana	BA	*Sarajana apfelbecki* (Brancsik, 1888)
Schapsugia	RU	*Schapsugia pulcherrima* (Starobogatov, 1962)
Sitnikovia	GE	*Sitnikovia* Chertoprud, Palatov & Vinarski, 2020
Sogdamnicola	TM	*Sogdamnicola pallida* (von Martens, 1874)
Stoyanovia	BG	*Stoyanovia stoyanovi* Georgiev, 2013
Strugia	MK	*Strugia ohridana* Radoman, 1973
Stygobium	ME	*Stygobium hercegnoviensis* Grego & Glöer, 2019
Tachira	RU	*Tachira valvataeformis* (Starobogatov, 1962)
Tadzhikamnicola	TJ	*Tadzhikamnicola likharevi* Izzatulaev, 1973
Tarraconia	ES	*Tarraconia gasulli* (Boeters, 1981)
Terranigra	KS	*Terranigra kosovica* Radoman, 1978

(*continued*)

Table 3.2. (continued)

Taxonomy	Distribution	Example species
Tschernomorica	GE, RU	*Tschernomorica caucasica* (Starobogatov, 1962)
Turkmenamnicola	TM	*Turkmenamnicola lindholmi* Shadin, 1952
Valvatamnicola	UZ	*Valvatamnicola archangelskii* (Shadin, 1952)
Zeteana	ME	*Zeteana ljiljanae* Glöer & Pešić, 2014
Lithoglyphidae		
Phreatoceras	USA	*Phreatoceras taylori* Hershler & Longley, 1986
Phreatodrobia	USA	*Phreatodrobia micra* (Pilsbry & Ferriss, 1906)
Phreatomascogos	MX	*Phreatomascogos gregoi* Czaja, Cardoza-Martinez, Meza-Sanchez, Estrada-Rodriguez & Saenz-Mata, 2019
Amnicolidae		
Amnicola	USA	*Amnicola cora* Hubricht, 1979
Pomatiopsidae		
Akiyoshia	JP	*Akiyoshia kobayashii* Kuroda, 1957
Cochliopopsis	JP	*Cochliopopsis basiangulata* Mori, 1938
Moria	JP	*Moria akiyoshiensis* (Kuroda & Habe, 1958)
Pseudoiglica	LA	*Pseudoiglica olsavskyi* Grego, 2018
Saganoa	JP	*Saganoa akka* Habe, 1965
Spiripockia	BR	*Spiripockia punctata* Simone, 2012
Srilankiella (*nomen nudum*)	LK	*Srilankiella horanae* Bole & Velkovrh (*nomen nudum*)
Thamkhondonia	LA	*Thamkhondonia moureti* Grego, 2018
Tricula	LA	*Tricula valenasi* Grego, 2018
Tateidae		
Austropyrgus	AU-QLD, AU-TAS, AU-VIC	*Austropyrgus lippus* Clark, Miller & Ponder, 2003
Catapyrgus	NZ	*Catapyrgus spelaeus* Climo, 1974
Hadopyrgus	NZ	*Hadopyrgus anops* Climo, 1974
Hemistomia	AU-LH	*Hemistomia minutissima* Ponder, 1982
Kuschelita	NZ	*Kuschelita inflata* Climo, 1974
Leptopyrgus	PG	*Leptopyrgus manneringi* (Climo, 1974)
Nanocochlea	AU-TAS	*Nanocochlea pupoidea* Ponder & Clark, 1993
Opacuincola	NZ	*Opacuincola caeca* Ponder, 1966
Phrantela	AU-TAS	*Phrantela* sp.
Platypyrgus	NZ	*Platypyrgus nelsonensis* Climo, 1977
Potamolithus	BR	*Potamolithus troglobius* Simone & Moracchioli, 1994
Potamopyrgus	IT, NZ	**Potamopyrgus antipodarum* (J. E. Gray, 1843)
Pseudotricula	AU-TAS	*Pseudotricula eberhardi* Ponder, 1992

Table 3.2. (continued)

Taxonomy	Distribution	Example species
Selmistomia	PG	*Selmistomia beroni* Bernasconi, 1995
Westrapyrgus	AU-WA	*Westrapyrgus westralis* Ponder, Clark & Miller, 1999
Assimineidae		
Anaglyphula	ID	*Anaglyphula minutissima* Massen 2000
Cavernacmella	JP	*Cavernacmella kuzuensis* Suzuki, 1937
Pleuroceridae		
Elimia	USA	**Elimia proxima* (Say, 1825)
Pleurocera	USA	**Pleurocera canaliculata* (Say, 1821)
Tornidae		
Teinostoma	MX	*Teinostoma brankovitsi* Rubio, Rolán, Worsaae, Martínez & Gonzalez, 2016
Bivalvia: Heterodonta		
Sphaeriidae		
Neopisidium	JP, LA, MM	*Neopisidium cavernicum* Mori, 1938
Pisidium (Pisidium)	BA, SK	**Pisidium (Pisidium) casertanum* (Poli, 1791)
Pisidium (Euglesa)	GE	*Pisidium (Euglesa) cavaticum* Shadin, 1952
Dreissenidae		
Congeria	HR, BA	*Congeria kusceri* Bole, 1962

tamination, I consider an acceleration of research on their taxonomy and zoogeographic distributions to be the highest of priorities. A firmer comprehension of both aspects will enable researchers and governmental municipalities to both improve our knowledge of these animals and use the information of their distributions to safeguard groundwater quality.

The aim of this chapter was to synthesize available information on subterranean Mollusca, highlight their vulnerability and threats, and identify future research needs related to habitat analysis, assessments, and taxonomic and molecular studies. I hope this work serves to inspire malacologists and speleologists to redouble their efforts in the ecological, ecotoxicological, and taxonomic study of subterranean mollusks. Through these endeavors, we will have the data and information necessary to better manage both these species and the sensitive habitats where they occur.

ACKNOWLEDGMENTS

I would like to express my gratitude to Thomas G. Watters, Zoltán Fehér, Andrzej Falniowski, and Alexander Reschütz for their assistance with proofreading and reviewing this chapter. Anita Eschner, Nesrine Akkari, and Sara Schnedl provided support with microphotography. Photographs were provided by Barna Páll-Gergely and Zoltán Fehér, Hungarian Natural History Museum Budapest, Hungary; Roman Ozimec; Virginie Héros and Manuel Caballer, Muséum National d'Histoire Naturelle Paris (project E-RECOLNAT: ANR-11-INBS-0004); Jochen Gerber, the Field Museum, Chicago, Illinois; Stephanie A. Clark; Winston F. Ponder; and Luiz Ricardo L. Simone, Museum of Zoology, University of São Paulo, Brazil. Shell material used in plates provided by Rajko Slapnik, Dilian Georgiev, Hans-Jürgen Hirschfelder, Antonio Tarruela Ruestestes, Wim Maassen, Peter Subai, Alexander and Peter L. Reischütz, Alain Bertrand, Alexander Czaja, Angyal Dorottya, Zolán Péter Erőss, László Németh, Tamás Németh, Gábor Majoros, and Tamás Déli.

REFERENCES

Angyal, Dorottya, Gergely Balázs, Virág Krízsik, Gábor Herczeg, and Zoltán Fehér. 2018. "Molecular and Morphological Divergence in a Troglobiont *Bythiospeum* Lineage (Gastropoda, Hydrobiidae) within an Isolated Karstic Area in the Mecsek Mts. (Hungary)." *Journal of Zoological Systematics and Evolutionary Research* 18:1–12.

Antonić, Oleg, Fanica Kljaković-Gašpić, Dalibor Hatić, and Hrvoje Peternel. 2015. "Odgovori na Primjedbe s Javne Rasprave i Mišljenja Državnog Zavoda za Zaštitu Prirode Tijekom Postupka glavne ocjene Prihvatljivosti za Ekološku Mrežu za Zahvat Hidroelektrana (HE) Ombla." In *Studija Glavne ocjene Prihvatljivosti Zahvata za Ekološku Mrežu HE OMBLA*. Knjiga 5, Zagreb: OIKON-Geonatura Ltd. Accessed June 15, 2020. https://mzoe.gov.hr/UserDocsImages//NASLOVNE%20FOTOGRAFIJE%20I%20KORI%C5%A0TENI%20LOGOTIPOVI/doc//28_07_2015_studija_glavne_ocjene_prihvatljivosti_zahvata_-_knjiga_5.pdf.

Antonić, Oleg, Fanica Kljaković-Gašpić, Darko Mayer, Jozeip Križan, Nikola Bakšić, and Mirjana Žiljak. 2015. "Hidrogeološka Analiza Šireg Područja Zahvata." In *Studija Glavne ocjene Prihvatljivosti Zahvata za Ekološku Mrežu HE OMBLA*. Knjiga 2, Zagreb: OIKON-Geonatura Ltd. Accessed June 15, 2020. https://mzoe.gov.hr/UserDocsImages//NASLOVNE%20FOTOGRAFIJE%20I%20KORI%C5%A0TENI%20LOGOTIPOVI/doc//studija_glavne_ocjene_prihvatljivosti_zahvata_-_knjiga_2.pdf.

Antonić, Oleg, Fanica Kljaković-Gašpić, Milorad Mraković, Branimir Kutuzovič Hackenberger, and Perica Mustafić. 2015. "Znanstveno—Stručne Podloge za Procjenu Utjecaja HE Ombla na Popovsku Gaovicu (*Delminichthys ghetaldii* Steindachner 1882)." In *Studija Glavne ocjene Prihvatljivosti Zahvata za ekološku mrežu HE OMBLA*. Knjiga 4, Zagreb: OIKON-Geonatura Ltd. Accessed June 15, 2020. https://mzoe.gov.hr/UserDocsImages//

NASLOVNE%20FOTOGRAFIJE%20I%20KORI%C5%A0TENI%20LOGOTIPOVI /doc//studija_glavne_ocjene_prihvatljivosti_zahvata_-_knjiga_4.pdf.

Antonić, Oleg, Fanica Kljaković-Gašpić, Roman Ozimec, Branko Jalžić, Iva Mihoci, Nikola Hanžek, et al. 2015. "Bioraznolikost Špiljskih Objekata na Širem Području Zahvata." In *Studija Glavne ocjene Prihvatljivosti Zahvata za Ekološku Mrežu HE OMBLA. Knjiga 3*, Zagreb: OIKON-Geonatura Ltd. Accessed June 15, 2020. http://www.edubrovnik.org /2015/Ombla/GOPZEM%20HE%20Ombla%20Knjiga%203.pdf.

Arconada, Beatriz, and Marian A. Ramos. 2011. "The Ibero-Balearic Region: One of the Areas of Highest Hydrobiidae (Gastropoda, Prosobranchia, Rissooidea) Diversity in Europe." *Graellsia* 59:91–104.

Arroyo, Gloria, Isabela Arroyo, and Eduardo Arroyo. 1997. "Microbiological Analysis of Maltravieso Cave (Caceres), Spain." *International Biodeterioration & Biodegradation* 40:131–139.

Badri, Dayakar V., and Jorge M. Vivanco. 2009. "Regulation and Function of Root Exudates." *Plant, Cell & Environment* 32:666–681.

Balázs, Gergely, and Brian Lewarne. 2017. "Observations on the Olm *Proteus anguinus* Population of the Vrelo Vruljak System (Eastern Herzegovina, Bosnia and Herzegovina)." *Natura Sloveniae* 19:39–41.

Beran, Luboš, Marco Bodon, and Simone Cianfanelli. 2014. "Revision of '*Hauffenia jadertina*' Kuscer 1933, and Description of a New Species from Pag Island, Croatia (Gastropoda: Hydrobiidae)." *Journal of Conchology* 41:585–601.

Beran, Luboš, Artur Osikowski, Sebastian Hofman, and Andrzej Falniowski. 2016. "*Islamia zermanica* (Radoman, 1973) (Caenogastropoda: Hydrobiidae): Morphological and Molecular Distinctness." *Folia Malacologica* 24:25–30.

Bernasconi, Reno. 1995. "Two New Cave Prosobranch Snails from Papua New Guinea: *Selmistomia beroni* n. gen. n. sp. (Caenogastropoda: Hydrobiidae) and *Georissa papuana* n. sp. (Archaeogastropoda: Hydrocenidae)." *Revue Suisse de Zoologie* 102:373–386.

Bilandžija, Helena, Brian Morton, Martina Podnar, and Helena Ćetković. 2013. "Evolutionary History of Relict *Congeria* (Bivalvia: Dreissenidae): Unearthing the Subterranean Biodiversity of the Dinaric Karst." *Frontiers in Zoology* 10:1–17.

Bodon, Marco, and Simone Cianfanelli. 2012. "Il Genere *Islamia* Radoman, 1973, Nell'Italia Centro–Settentrionale (Gastropoda: Hydrobiidae)." *Bolletino Malacologico* 48:1–37.

Bodon, Marco, Simone Cianfanelli, and Alessandro Montanari. 2009. "Mollusks of the Frasassi Karstic Complex and Adjacent Sulfidic Springs." *The Frasassi Stygobionts and Their Sulfidic Environment* 10:9–11.

Bole, Jože, and France Velkovrh. 1986. "Mollusca from Continental Subterranean Aquatic Habitats." In *Stygofauna Mundi*, edited by Lazare Botosaneanu, 177–208. Leiden: Brill and Backhuys Publishers.

Canarini Alberto, Christina Kaiser, Andrew Merchant, Andreas Richter, and Wolfgang Wanek. 2019. "Root Exudation of Primary Metabolites: Mechanisms and Their Roles in Plant Responses to Environmental Stimuli." *Frontiers in Plant Science* 10:1–17.

Chertoprud, Elizaveta M., Dimitri M. Palatov, and Maxim V. Vinarski. 2020. "Reveling the Stygobiont and Crenobiont Mollusca Biodiversity Hotspot of Caucasus: Part II. *Sitnikovia* gen. nov., a New Genus of Stygobiont Microsnails (Gastropoda: Hydrobiidae) from Georgia." *Zoosystematica Rossica* 29:258–266.

Clark, Stephanie A., Alison C. Miller, and Winston F. Ponder. 2003. "Revision of the Snail Genus *Austropyrgus* (Gastropoda: Hydrobiidae): A Morphostatic Radiation of Freshwater Gastropods in Southeastern Australia." *Records of the Australian Museum* 28:1–109.

Climo, Frank M. 1974. "Description and Affinities of the Subterranean Molluscan Fauna of New Zealand." *New Zealand Journal of Zoology* 1:247–284.

Cook, Charlotte G., and Richard Jones. 2012. "Paleoclimate Dynamics in Continental Southeast Asia over the Last ~30,000 Cal Yrs BP." *Palaeogeography Palaeoclimatology Palaeoecology* 339:1–11.

Coughlan, Neil E., Andrew L. Stevens, Thomas C. Kelly, Jaimie T. A. Dick, and Marcel A. K. Jansen. 2017. "Zoochorous Dispersal of Freshwater Bivalves: An Overlooked Vector in Biological Invasions?" *Knowledge Management of Aquatic Ecosystems* 418:1–8.

Culver, David C. 2012. "Mollusks." In *Encyclopedia of Caves*, 2nd ed., edited by William B. White and David C. Culver, 382–386. New York: Elsevier, Academic Press.

Culver, David C., and Tanja Pipan. 2014. *Shallow Subterranean Habitats: Ecology, Evolution, and Conservation*. Oxford: Oxford University Press.

Czaja, Alexander, Gabriel F. Cardoza-Martínez, Iris G. Meza-Sánchez, José L. Estrada-Rodríguez, Jorge Saenz-Mata, Jorge L. Becerra-López, et al. 2019. "New Genus, Two New Species and New Records of Subterranean Freshwater Snails (Caenogastropoda; Cochliopidae and Lithoglyphidae) from Coahuila and Durango, Northern Mexico." *Subterranean Biology* 29:89–102.

Czaja, Alexander, José Luis Estrada-Rodriguez, Ulises Romero-Mendez, Verónica Ávila-Rodriguez, Iris Gabriela Meza-Sánchez, and Alan P. Covich. 2017. "New Species and Records of Phreatic Snail (Caenogastropoda: Cochliopidae) from the Holocene of Coahuila, Mexico." *Archiv für Molluskenkunde* 146:227–232.

Darwall, William, Savrina Carrizo, Catherine Numa, Violeta Barrios, Jörg Freyhof, and Kevin Smith. 2014. *Freshwater Key Biodiversity Areas in the Mediterranean Basin Hotspot: Informing Species Conservation and Development Planning in Freshwater Ecosystems*. Cambridge, UK: IUCN.

Davis, George M., Cui-E. Chen, Chun Wu, Tie-Fu Kuang, and Xin-Guo Xing. 1992. "The Pomatiopsidae of Hunan, China (Gastropoda: Rissoacea)." *Malacologia* 34:143–342.

de la Torre, M. Angeles, and Gonzalo Gomez-Alarcon. 1994. "Manganese and Iron Oxidation by Fungi Isolated from Building Stone." *Microbial Ecology* 27:177–118.

Delicado, Diana. 2018. "A Rare Case of Stygophily in the Hydrobiidae (Gastropoda: Sadleriana)." *Journal of Molluscan Studies* 84:480–485.

Deming, Jody W., and John A. Baross. 1993. "Deep-Sea Smokers. Windows to a Subsurface Biosphere?" *Geochimica et Cosmoschimica Acta* 57:3219–3230.

Dillon, Jesse G., Susan Fishbain, Scott R. Miller, Brad M. Bebout, Kirsten S. Habicht, Samuel M. Webb, et al. 2007. "High Rates of Sulfate Reduction in a Low-Sulfate Hot Spring Microbial Mat Are Driven by a Low Level of Diversity of Sulfate-Respiring Microorganisms." *Applied and Environmental Microbiology* 73:5218–5226.

Di Lorenzo, Tiziana, Marco Cifoni, Barbara Fiasca, Alessia Di Cioccio, and Diana M. P. Galassi. 2018. "Ecological Risk Assessment of Pesticide Mixtures in the Alluvial Aquifers of Central Italy: Toward More Realistic Scenarios for Risk Mitigation." *Science of the Total Environment* 644:161–172.

Plate 1.1. Examples of the five evolutionary categories of subterranean-dwelling organisms. [A] *Troglobiont*: the highly specialized dragon millipede, *Hylomus yuani* Liu & Wynne, 2019 from the South China Karst, Guangxi, China. Courtesy of J. Judson Wynne. [B] *Stygobiont*: the Dougherty Plain cave crayfish, *Cambarus cryptodytes* Hobbs, 1941, from Georgia, USA. Courtesy of Danté B. Fenolio. [C] *Obligate troglophile*: *Pratherodesmus voylesi* Shear, 2009 from north rim Grand Canyon, Arizona. Courtesy of J. Judson Wynne. [D] *Troglophile*: mating pair of cave crickets, Rhaphidophoridae n. gen. n. sp. from Grand Canyon Parashant National Monument, Arizona (Wynne and Voyles 2014). Courtesy of Kyle Voyles. [E] *Trogloxene*: *Peridroma* sp. from the island of Hawaiʻi. Courtesy of Fred D. Stone. Prior to the arrival of the introduced rat to the Hawaiian Islands, these moths reportedly would darken the sky during their evening emergence (Howarth et al. 2020). Because accidentals are random and reflective of the epigean community where the cave is located, we felt including an example would not be useful.

1
2
3
1 mm
4
5
6
7
8
9
10
11
12
13
14
15
16
17
18
19
5 mm
1 mm
3 mm

Plate 3.1. (*opposite*) Hypogean freshwater limpets (Gastropoda), neritids (Gastropoda), Bivalvia, and hypogean terrestrial gastropod species of families Zonitidae and Carychiidae. [1] *Ferissia* sp., Cenote Xoch, Yucatán, Mexico; [2] *Acroloxus tetensi* (Kuščer, 1932), Škocjan Cave, Slovenia; [3] *Acroloxus* sp., Tihajlina, Grude District, Bosnia; [4] *Ferissia* cf. *californica* (Rowell, 1863), Sakvarilje Cave, Georgia; [5] *Pisidium* (*Euglesa*) *cavaticum* Shadin, 1952, Letsurtsume Cave, Samegrelo, Caucasus, Georgia; [6] *Pisidium* (*Euglesa*) *ljovushkini* Starobogatov, 1962, Iazoni Cave, Samegrelo, Georgia; [7] *Neopisidium* sp., Na Li Cave, Khammouane, Laos; [8] *Pisidium* sp., Cenote Xoch, Yucatán, Mexico; [9] *Neopisidium* sp., entrance of Tham Nam Don Cave, Khammouane, Laos; [10] *Congeria mulaomerovici* Bilandžija, Morton, Podnar, Ćetković, 2013, Suvaja Cave, Lušći Palanka, Bosnia; [11] *Troglaegopis mosorensis* (Kuščer, 1933), Provalja Cave, Hercegovina; [12] *Balcanodiscus carinatus* P. L. Reischütz, 1983, Thasos, Greece; [13] *Balcanodiscus mirus* A. Reischütz, N. Steiner-Reischütz, & P. L. Reischütz, 2016, Komito, Evia, Greece; [14] *Aegopis spelaeus* A. J. Wagner, 1914, Popovo Polje, Hercegovina; [15] *Meledella werneri* Sturany, 1908, Sobra, Mljet Island, Croatia; [16] *Balcanodiscus stummerorum* A. Reischütz, P. L. Reischütz & W. Fischer, 2008, Ipapanitstou Christou Monastery, Greece; [17] *Zospeum vasconicum* Prieto, De Winter, Weigand, Gómez, Jochum, 2015, Cueva de Otxas, Pais Vasco, Spain; [18] *Zospeum isselianum* Pollonera, 1887, Vedro Polje, Bihač, Bosnia; [19] *Theodoxus subterrelictus* Schütt, 1963, Gorica, Trebinje, Hercegovina. Images courtesy of Jozef Grego.

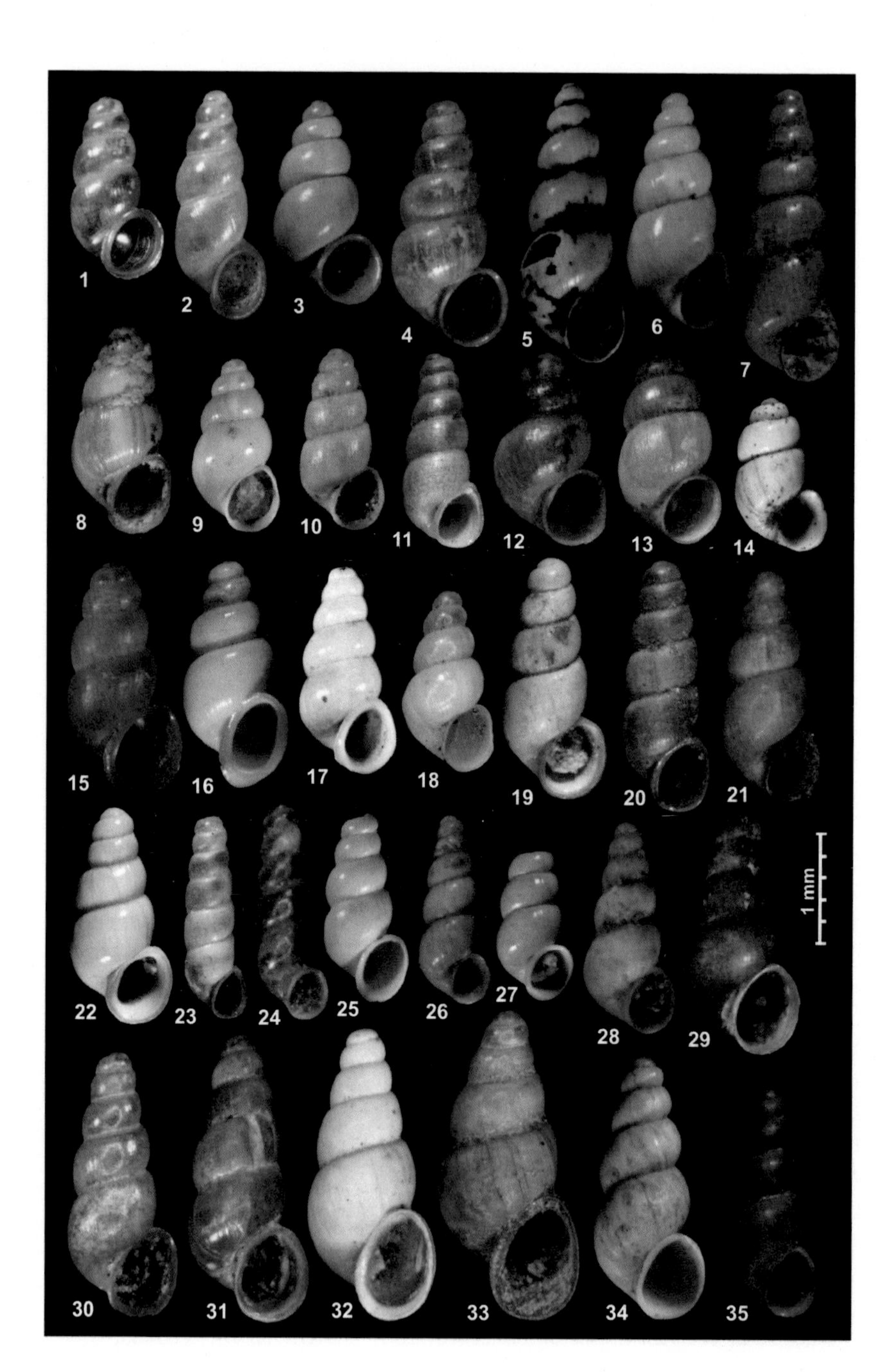
1
2
3
4
5
6
7
8
9
10
11
12
13
14
15
16
17
18
19
20
21
22
23
24
25
26
27
28
29
1 mm
30
31
32
33
34
35

Plate 3.2. (*opposite*) Examples of European hypogean freshwater gastropod species of the genera *Paladilhiopsis, Bythiospeum, Iglica, Stoyanovia, Plesiella, Alzoniella, Saxurinator, Narentiana*, and *Vinodolia*. [1] *Saxurinator schlickumi* Schütt, 1960, Pejë, Rugovska Klisura, Kosovo; [2] *Paladilhiopsis gittenbergeri* (A. Reischütz & P. L. Reischütz, 2006), Vau i Dejës, Albania; [3] *Stoyanovia jazzi* Georgiev & Glöer, 2013, Vodnata Cave, Musina, Bulgaria; [4] *Paladilhiopsis maroskoi* (Glöer & Grego, 2015), Donja Pecka, Bosnia; [5] *Iglica cf. absoloni* (A. J. Wagner, 1914), Izvor Peranovac below Vjetrenica Cave, Hercegovina; [6] *Pleisella guipuzcoa* Boeters & Bertand, 2003, Manantial Zazpi, Spain; [7] *Iglica forumjuliana* Pollonera, 1877, Papariano, Italy; [8] *Bythiospeum bourguignati* (Paladilhe, 1866), Laure Hérault, France; [9] *Pleisella guipuzcoa* Boeters & Bertand, 2003, Pais Asco, Spain; [10] *Iglica concii* (Allegretti, 1844), Volbarno springs, Nalmase, Brescia, Lombardy, Italy; [11] *Moitessieria fontsaintei* Bertrand, 2001, Foint Santé, Ustou, France; [12] *Alzoniella junqua* Boeters, 2000, Hérault, France; [13] *Alzoniella slovenica* (Ložek & Brtek, 1964), Hrušovo, Drienčanský Kras, Slovakia; [14] *Alzoniella slovenica bojnicensis* (Ložek & Brtek, 1964), Bojnice, Slovakia; [15] *Paladilhiopsis thessalica* Schütt, 1970, Agia Paraskevi, Greece; [16] *Paladilhiopsis hrustovoensis* (Glöer & Grego, 2015), Hrustovo, Bosnia; [17] *Paladilhiopsis szarowskae* (Glöer, Grego, Erőss, Fehér, 2015), Krumë, Albania; [18] *Paladilhiopsis wohlberedti* Grego, Glöer, Fehér, Erőss, 2017, Tamarë, Albania; [19] *Paladilhiopsis petroedei* (Glöer & Grego, 2015), Izvor Plive, Bosnia; [20] *Paladilhiopsis kanalitensis* (A. Reischütz, N. Reischütz, & P. L. Reischütz, 2016), Dukat Fshat, Albania; [21] *Paladilhiopsis falniowskii* Grego, Glöer, Fehér & Erőss, 2017, Krumë, Albania; [22] *Spiralix valenciana castellonica* Boeters, 2003, Castellón, Spain; [23] *Iglica illyrica* Schütt, 1975, Deçan, Kosovo; [24] *Iglica sp.*, Deçan, Kosovo; [25] *Paladilhiopsis prekalensis* Grego, Glöer, Fehér, Erőss, 2017, Shpellë e Zhylit, Prekal, Albania; [26] *Iglica bagliviaeformis* Schütt, 1970, Spring Ombla, Komolac, Croatia; [27] *Paladilhiopsis szekeresi* Grego, Glöer, Fehér, Erőss, 2017, Tamarë, Albania; [28] *Iglica absoloni* (A. J. Wagner, 1914), Spring Ombla, Komolac, Croatia; [29] *Montenegrospeum sketi* Grego & Glöer, 2018, Izvor Grab, Grabske Mlinice, Croatia; [30] *Paladilhiopsis hungaricum* (Pintér, 1968), Nagy Víztó Cave, Mecsek, Hungary; [31] *Paladilhiopsis oshanovae* (Soós, 1927), Jónás Island, Szigetköz, Hungary; [32] *Bythiospeum sandbergeri* (Flach, 1886), Kelheim, Bavaria, Germany; [33] *Bythiospeum senefelderi* (Geyer, 1907), Solnhofen, Germany; [34] *Vinodolia matjasici* (Bole, 1961), Izvor Vitoja, Drume, Montenegro; [35] *Narentiana* cf. *albida* Radoman, 1973, Izvor Buna, Blagaj, Hercegovina. Images courtesy of Jozef Grego.

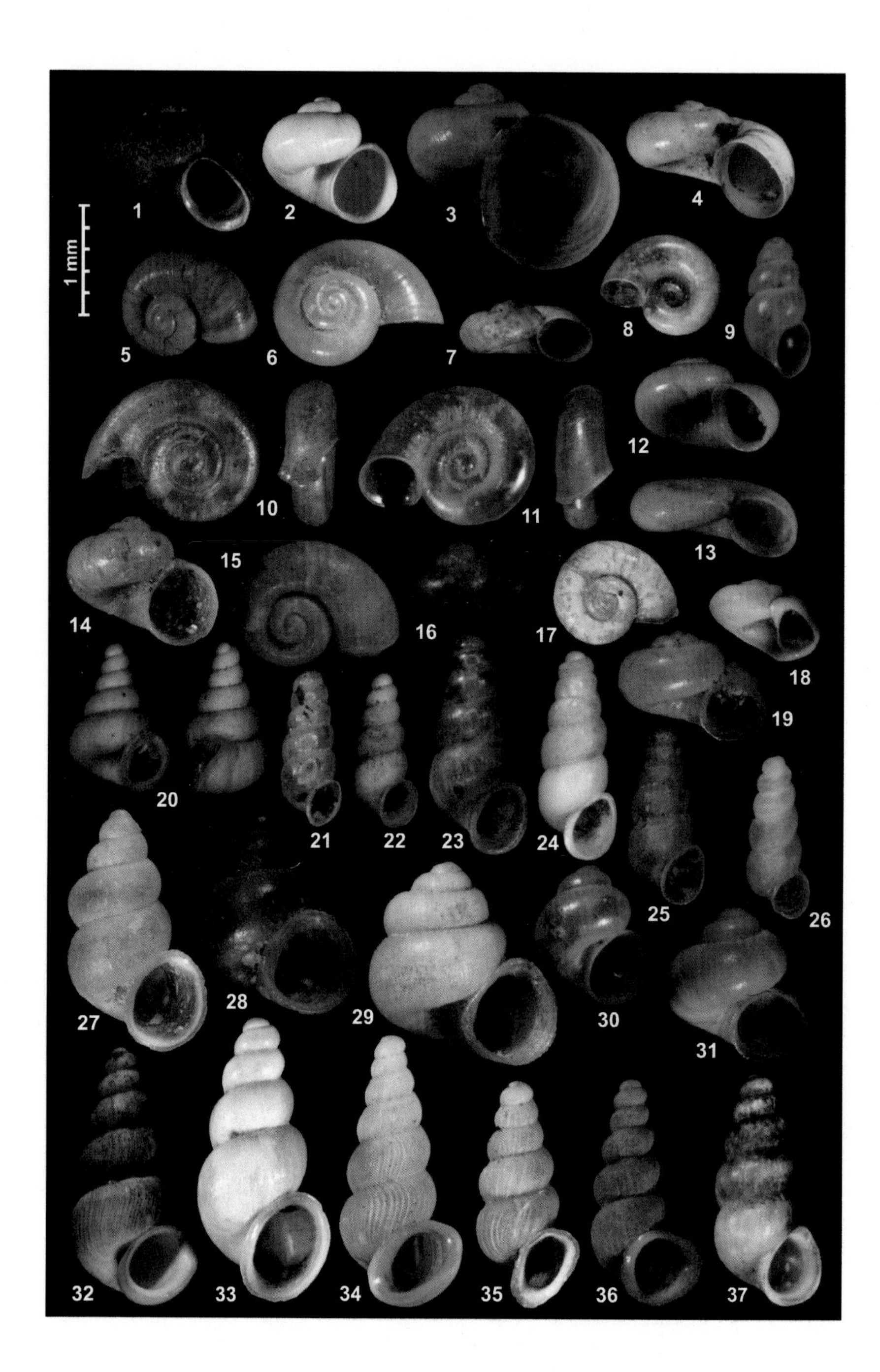
1 mm
1
2
3
4
5
6
7
8
9
10
11
12
13
14
15
16
17
18
19
20
21
22
23
24
25
26
27
28
29
30
31
32
33
34
35
36
37

Plate 3.3. (*opposite*) Examples of European hypogean freshwater gastropod species of the genera *Lanzaia, Saxurinator, Dabriana, Islamia, Hauffenia, Bracenica, Daphniola, Hadziella, Corbellaria, Milesiana, Heraultiella, Henrigirardia, Navalis, Sardopaladilhia,* and *Moitessiera.* [1] *Islamia montenegrina* Glöer, Grego, Erőss, Fehér, 2015, Vitoja Spring, Montenegro; [2] *Islamia pusilla* Piersanti, 1952, Giuliano, Italy; [3] *Dabriana bosniaca* Radoman, 1974, Dabarska Pećina Cave, Bosnia; [4] *Islamia germaini* Boetters & Falkner, 2003, Source de la Grozonne, France; [5] *Hauffenia wienerwaldenis* Haase, 1992, Kaumberg, Austria; [6] *Hauffenia tellinii* Polonera, 1899, Udine, Italy; [7] *Bracenica vitojaensis* Glöer, Grego, Erőss, Fehér, 2015, Vitoja Spring, Montenegro; [8] *Corbellaria celtiberica* Corbell, Guisti & Boeters, 2012, Soria, Spain; [9] *Spiralix asturica* Quiñonero-Salgado, Ruiz-Cobo & Rolán, 2017, Priorio, Asturia, Spain; [10] *Hadziella ephippiostoma* Kuščer, 1932, Timavo Spring, Trieste, Italy; [11] *Hadziella anti* Schütt, 1960, Udine, Italy; [12] *Hauffenia michleri* Kuščer, 1932, Vrchnika Spring, Slovenia; [13] *Kerkia* sp., Žabljak Spring, Livno, Bosnia; [14] *Islamia moquiniana* Dupuy, 1851, Lagorce Ardèche, France; [15] *Hauffenia kisdalmae* Erőss & Petró, 2008, Kismaros, Börzsöny, Hungary; [16] *Milesiana schuelei* Boeters, 1981, Genal River, Málaga, Spain; [17] *Heraultiella exilis* (Paladilhe, 1867), Hérault, France; [18] *Navalis perforatus* Quiñonero-Salgado & Rolán, 2017, Fuente del Hambe, Segorbe, Spain; [19] *Islamia henrici giennensis* Arconada & Ramos, 2006, Acequia, Granada, Spain; [20] *Henrigirardia wienini* Giardi, 2001, Source de la Rome, Hérault, France; [21] *Moitessieria barrinae* Alba, Corbella, Prats, Tarruella & Guilien, 2007, Horta de St. Joan, Tarragona, Spain; [22] *Moitessieria rhodani* Bouguignat, 1883, St. Julien de Roisters, France; [23] *Saxurinator buresi* (A. J. Wagner, 1928), Temnata Dupka Cave, Lakatnik, Bulgaria; [24] *Moitessieria nezi,* Boeters & Beltrand 2001, Quell du Nez, Rébenacq, France; [25] *Moitessieria aurea* Tarruella, Corbella, Prats, Guillén & Alba, 2012, Callers, Lleida, Spain; [26] *Moitessieria locardi* Coutagne, 1885, Source Lauret, St. Jean de Gard, France; [27] *Sardopaladilhia buccina* Rolán & Ortí, 2003, Castellón, Spain; [28] *Sardopaladilhia subdistorta* Rolán & Ortí, 2003, Sueras, Castellón, Spain; [29] *Sardopaladilhia distorta* Rolan & Martinez-Ortí, 2003, Piscina de Sueras, Onda, Spain; [30] *Islamia lagari* Altimra, 1960, Font de los Dons, Barcelona, Spain; [31] *Daphniola exigua* A. Schmidt, 1856, Tembi Valley, Agia Paraskevi, Greece; [32] *Paladilhiopsis montenegrinus* (Schütt, 1959), Vruljak 1 Cave, Gorica, Hercegovina; [33] *Paladilhiopsis plivensis* Glöer & Grego, 2015, Draganić, Izvor Plive, Bosnia; [34] *Lanzaia vjetrenicae* (Kuščer, 1933), Vrelo Tučevac, Trebinje, Hercegovina; [35] *Lanzaia pesici* (Glöer, Grego, Erőss, Fehér, 2015), Vitoja Spring, Montenegro; [36] *Lanzaia bosnica* (Bole, 1970), Dabarska Pećina Cave, Bosnia; [37] *Lanzaia matejkoi* Grego & Glöer, 2019, Nemila Spring, Herceg Novi, Montenegro. Images courtesy of Jozef Grego.

Plate 3.4. Hypogean terrestrial gastropod species of the Dinaric Karst, eastern Europe. [1] *Aegopis spelaeus* A. J. Wagner, 1914, Bjelušica Cave, Popovo Polje, Hercegovina; [2] *Meledella werneri* Sturany, 1908, Otaševica Cave, Mljet Island, Croatia; [3] *Virpazaria pageti kleteckii* Štamol & Subai, 2012, Vilina Spilja Cave, Croatia; [4] *Pholeoteras euthrix* Sturany, 1904, Vilina Peć Cave, Ombla, Croatia; [5] *Spelaeoconcha paganetti polymorpha* A. J. Wagner, 1914, Baba Peć Cave, Biokovo, Croatia; [6] *Zospeum* sp., Vrelo Cave, Croatia; [7] *Zospeum kusceri* A. J. Wagner, 1912, Lokvarka, Croatia. Images courtesy of Roman Ozimec.

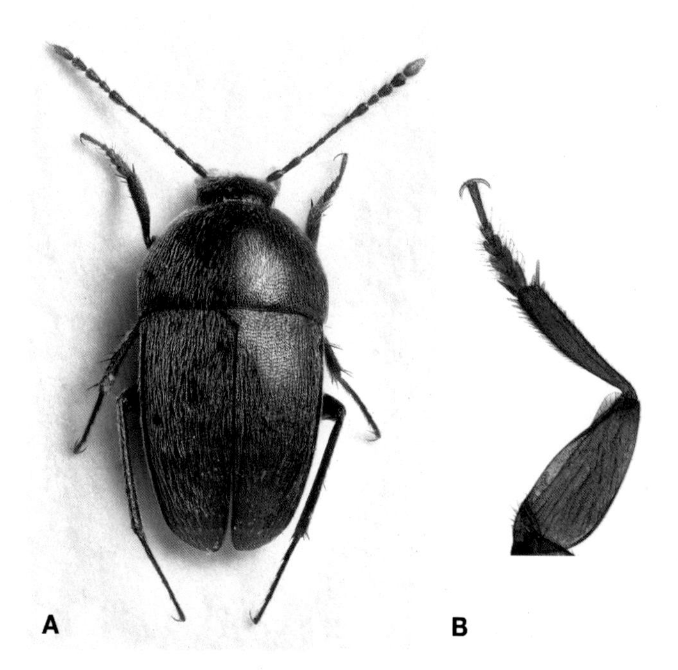

Plate 4.1. [A] The reddish-brown habitus, covered with golden pubescence, is common to many Cholevini species. *Bathysciola* (s. str.) *fabiolae* Latella, Sbordoni & Allegrucci, 2017. Image courtesy of Marcel Clemens. [B] Right foreleg of a male, *Bathysciola* (s. str.) *octaviani* Latella, Sbordoni & Allegrucci, 2017. The pentamerous and dilated tarsomeres, spines, and spurs on the tibia are visible. Image courtesy of Leonardo Latella.

Plate 4.2. *Leptodirus hohenwarthi reticulatus* (Muller, 1904) (*left*) and *Bathysciola* (s. str.) *valeriae* Latella, Sbordoni, Allegrucci, 2017 (*right*) for comparison. In the former (leptodiroid type), the head and legs are extremely long, the prothorax is cylindrical, and the antennae are highly elongated. Scale bar: 1 mm. Images courtesy of Leonardo Latella.

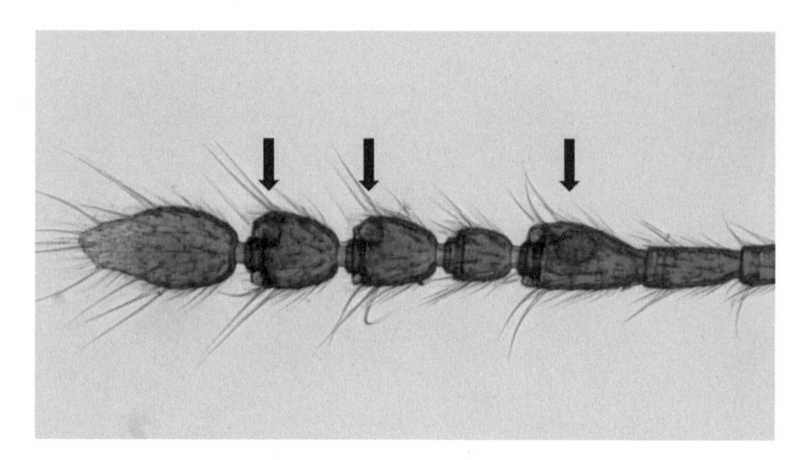

Plate 4.3. Right antenna of *Bathysciola octaviani* Latella, Sbordoni & Allegrucci 2017; red arrows indicate the ovate Hamann's organs (i.e., sensory structures) present on the seventh, ninth, and tenth antennomeres. Image courtesy of Leonardo Latella.

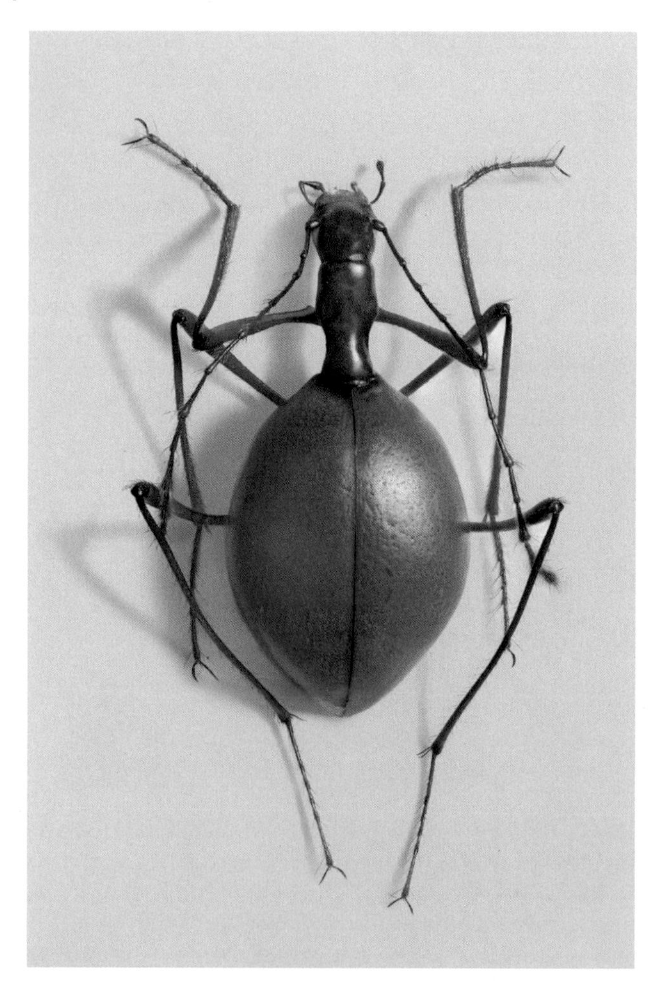

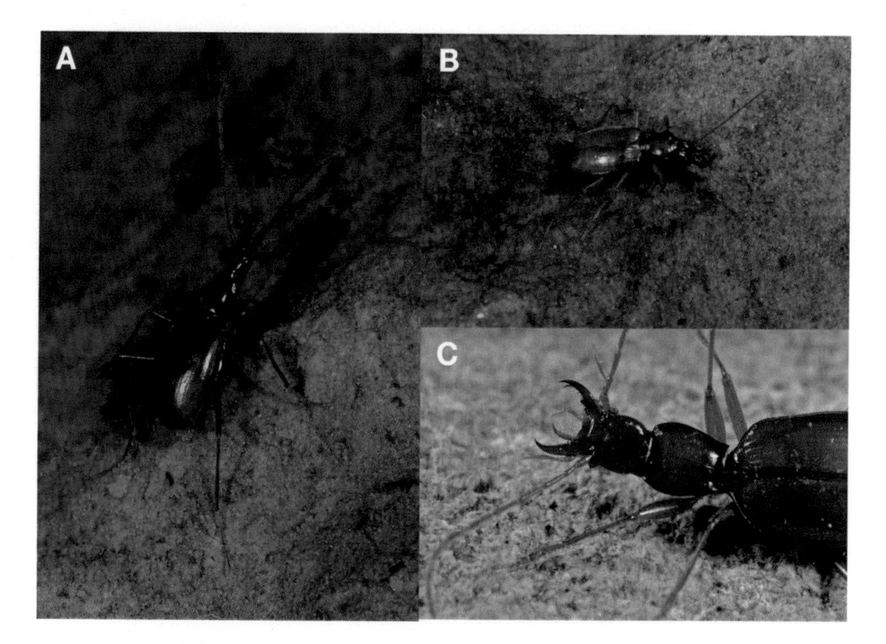

Plate 5.1. Three genera of cave trechine beetles from the Interior Low Plateau, southeastern United States. [A] *Neaphaenops tellkampfii* from Breckinridge Co., Kentucky; [B] *Pseudanophthalmus barberi* from Breckinridge Co., Kentucky; and [C] *Darlingtonea kentuckensis* from Rockcastle Co., Kentucky. Images courtesy of Matthew L. Niemiller.

Plate 4.4. (*opposite*) The European *Leptodirus hohenwarthi reticulatus* illustrates many of the morphological adaptations of subterranean beetles, such as absence of eyes, elongated legs and antennae, and false physogastry. Image courtesy of Marcel Clemens.

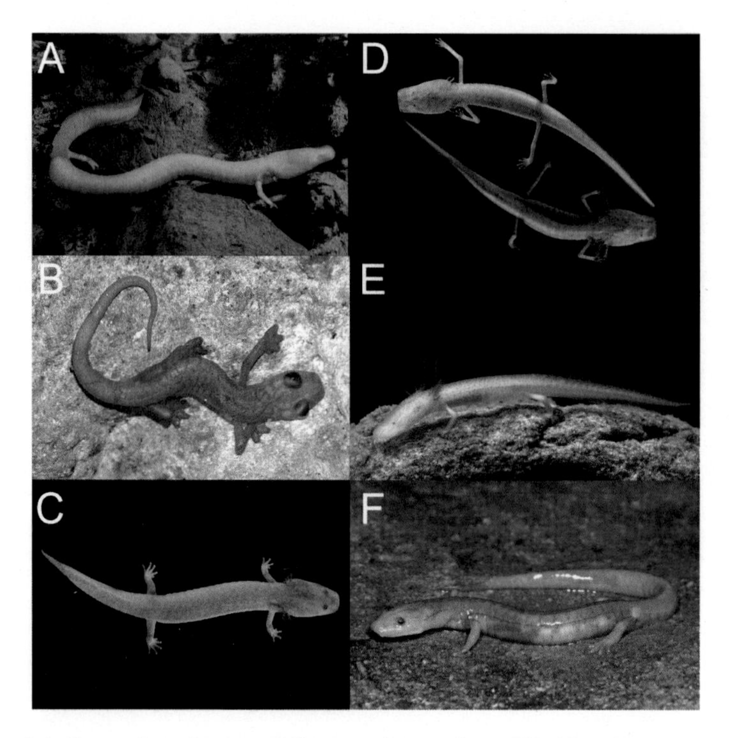

Plate 6.1. Examples of cave-obligate salamanders. [A] Olm (*Proteus anguinus*) from the region of Postojna, Slovenia; [B] bigfoot splayfoot salamander (*Chiropterotriton magnipes*) from the region of Querétaro, Mexico (courtesy of Jerry Fant); [C] Tennessee cave salamander (*Gyrinophilus palleucus*) from Jackson County, Alabama; [D] Texas blind salamander (*Eurycea rathbuni*) from Hayes County, Texas; [E] Georgia blind salamander (*Eurycea wallacei*) from Washington County, Florida; and [F] grotto salamander (*Eurycea spelaea*) from Delaware County, Oklahoma. Images courtesy of Danté B. Fenolio unless otherwise noted.

Plate 6.2. (*opposite*) Mitochondrial DNA phylogenetic tree of *Blepsimolge* (modified from Bendik et al. 2013). Major habitat (hypogean and epigean) was reconstructed and shows subterranean colonization events. Subterranean populations are not necessarily stygobionts. Phenotypes observed among hypogean populations span a continuum from epigean to highly stygobiontic. Species assignments are based on existing literature and determinations were made by the author (AGG). Tip labels are habitat (*left*) and species assignment (*right*). An array of cave-dwelling forms exists within *Paedomolge*, exemplified by *Eurycea sosorum* from Taylor Spring, Hays County, Texas (*upper image*) and *E. neotenes* from Hector Hole, Bexar County, Texas (*lower image*). Images courtesy of Nathan F. Bendik.

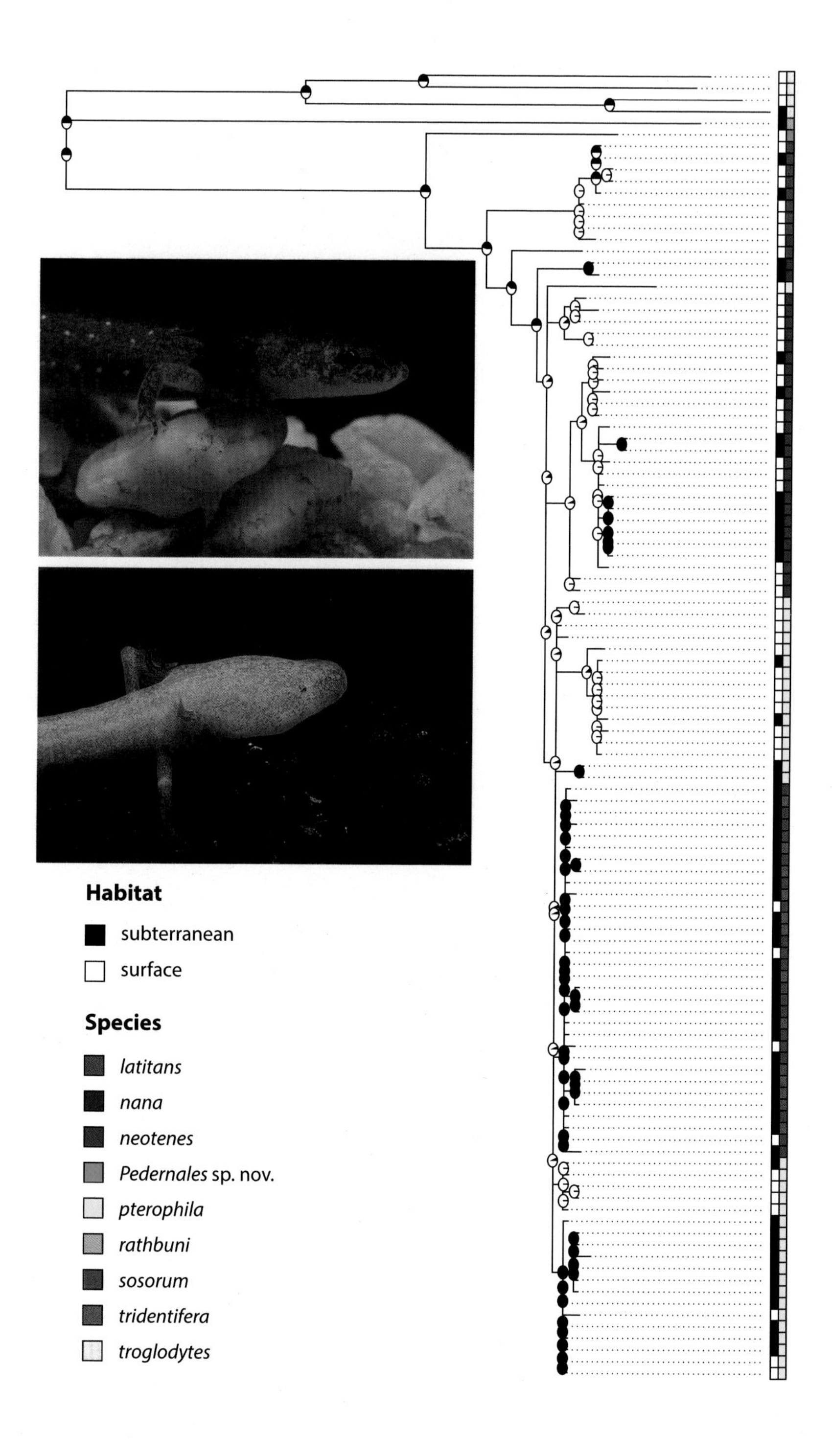
Habitat
subterranean
surface
Species
latitans
nana
neotenes
Pedernales sp. nov.
pterophila
rathbuni
sosorum
tridentifera
troglodytes

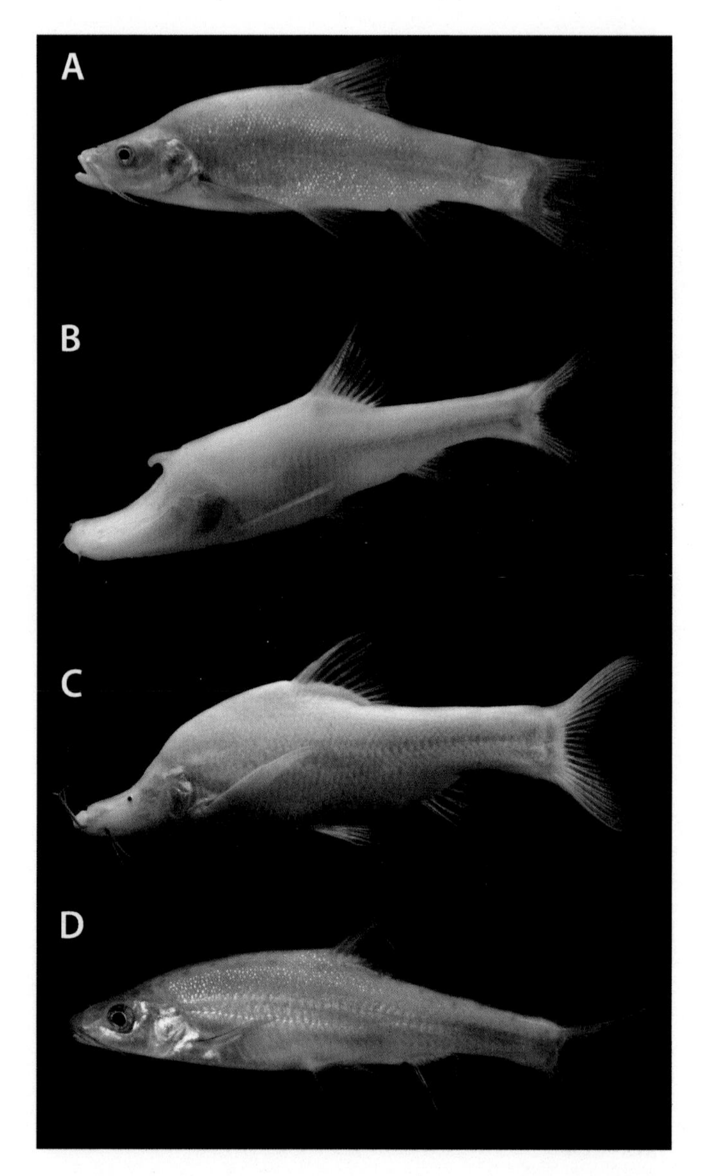

Plate 7.1. Morphological variation within the genus *Sinocyclocheilus*, South China Karst. [A] *Sinocyclocheilus guilinensis* Ji, 1982 (clade "*jii*") from Guangxi Zhuang Autonomous Region; [B] *Sinocyclocheilus aquihornes* Li & Yang, 2007 (clade "*angularis*") from Yunnan Province; [C] *Sinocyclocheilus microphthalmus* Li, 1989 (clade "*cyphotergous*") from Guangxi Zhuang Autonomous Region; and [D] *Sinocyclocheilus angustiporus* Zheng & Xie, 1985 (clade "*tingi*") from Guizhou and Yunnan Provinces. Images courtesy of Danté B. Fenolio and Yahui Zhao.

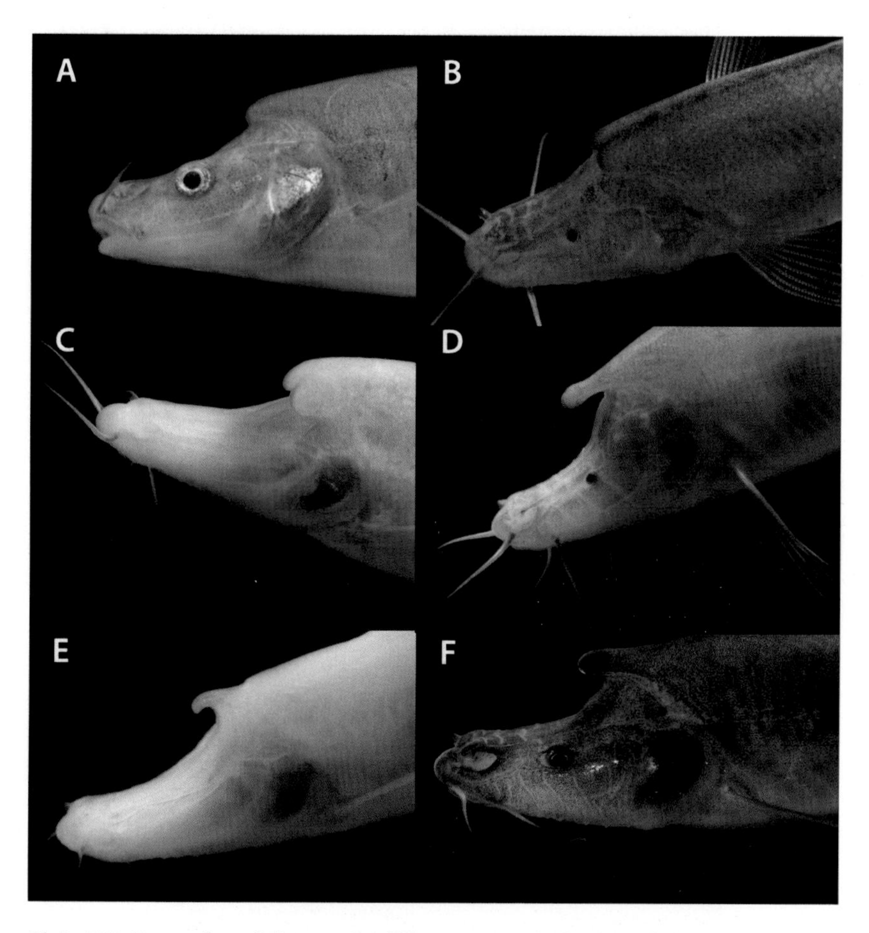

Plate 7.2. Examples of the vastly different morphologies of nape horn shape of the genus *Sinocyclocheilus*, South China Karst. [A] *S. angularis* Zhang & Wang, 1990 from Guizhou Province; [B] *S. bicornatus* Wang & Liao, 1997 from Guizhou Province; [C] *S. furcodorsalis* Chen, Yang & Lan, 1997 from Guangxi Zhuang Autonomous Region; [D] *S. rhinocerous* Li & Tao, 1994 from Yunnan Province; [E] *S. aquihornes* Li & Yang, 2007 from Yunnan Province; and [F] *S. tileihornes* Mao, Lu & Li, 2003 from Yunnan Province. Images courtesy of Danté B. Fenolio and Yahui Zhao.

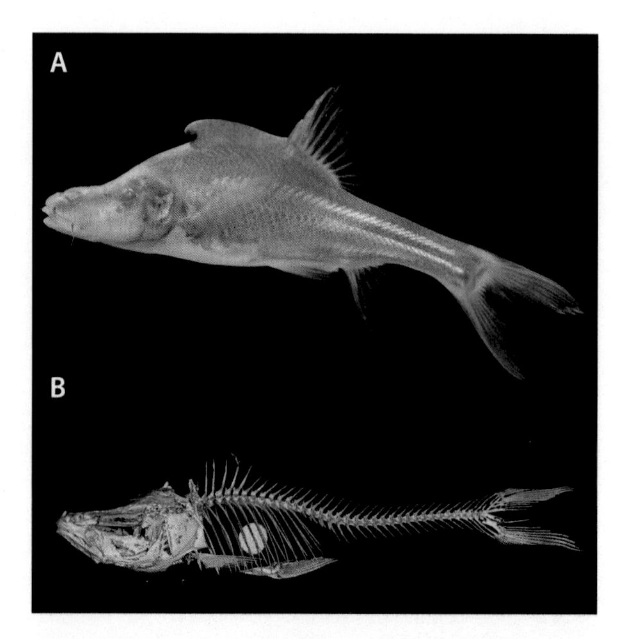

Plate 7.3. [A] *Sinocyclocheilus cyphotergous* (Dai, 1988), and [B] a micro-computed tomography scan of the same species, Guizhou Province, China. Both images illustrate the horn-like humpback on the dorsum is composed of adipose tissue. Images courtesy of Danté B. Fenolio and Yahui Zhao.

Plate 7.4. Representatives of the family Nemacheilidae, including [A] *Troglonectes macrolepis* (Huang, Chen & Yang, 2009) from Guangxi Zhuang Autonomous Region, and [B] *Triplophysa rosa* Chen & Yang, 2005 from Chongqing Municipality, China. Images courtesy of Danté B. Fenolio and Yahui Zhao.

Di Lorenzo, Tiziana, Walter D. Di Marzio, Daniele Spigoli, Mariella Baratti, Giuseppe Messana, Stefano Cannicci, et al. 2015. "Metabolic Rates of a Hypogean and an Epigean Species of Copepod in an Alluvial Aquifer." *Freshwater Biology* 60:426–435.

Dourson, Daniel C., Ronald S. Caldwell, and Judy A. Dourson. 2018. *Land Snails of Belize, Central America*. Stanton: Goatslug Publications.

Dudich, Endre. 1930. "Az Aggteleki-barlang Állatvilágának Élelemforássai." *Állattani Közlemények* 27:62–85.

Dudich, Endre. 1932. "A Barlangok Biológiai Osztályozása." *Barlangvilág* 2:1–9.

Eberhard, Stefan M., Graeme B. Smith, Michael M. Gibian, Helen M. Smith, and Michael R. Gray. 2014. "Invertebrate Cave Fauna of Jenolan." *Proceedings of the Linnean Society of New South Wales* 136:35–67.

Eme, David, Maja Zagmajster, Teo Delić, Cene Fišer, Jean-François Flot, Lara Konecny-Dupré, et al. 2018. "Do Cryptic Species Matter in Macroecology? Sequencing European Groundwater Crustaceans Yields Small Ranges but Does Not Challenge Biodiversity Determinants." *Ecography* 41:424–436.

Engel, Annette Summers. 2007. "Observations on the Biodiversity of Sulfidic Karst Habitats." *Journal of Cave and Karst Studies* 69:187–206.

Engel, Annette Summers. 2012. "Chemoautotrophy." In *Encyclopedia of Caves*, 2nd ed., edited by William B. White and David C. Culver, 90–102. New York: Elsevier, Academic Press.

Engel, Annette Summers, Megan L. Porter, Brian. K. Kinkle, and Thomas C. Kane. 2001. "Ecological Assessment and Geological Significance of Microbial Communities from Cesspool Cave, Virginia." *Geomicrobiology Journal* 18:259–274.

Erőss, Zoltán Péter, and Ede Petró. 2008. "A New Species of the Valvatiform Hydrobiid Genus *Hauffenia* from Hungary (Mollusca: Caenogastropoda: Hydrobiidae)." *Acta Zoologica Academiae Scientiarum Hungaricae* 54:73–81.

Falniowski, Andrzej, and Serban Sarbu. 2015. "Two New Truncatelloidea Species from Melissotrypa Cave in Greece (Caenogastropoda)." *ZooKeys* 530:1–14.

Falniowski, Andrzej, Magdalena Szarowska, Ioan Sîrbu, Allexandra Maria Hillebrand-Voiculescu, and Mihail Baciu. 2008. "*Heleobia dobrogica* (Grossu and Negrea, 1989) (Gastropoda: Rissooidea: Cochliopidae) and the Estimated Time of Its Isolation in a Continental Analogue of Hydrothermal Vents." *Molluscan Research* 28:165–170.

Fatemi, Yaser, Mohammad Javad Malek-Hosseini, Andrzej Falniowski, Sebastian Hofman, Matjaž Kuntner, and Jozef Grego. 2019. "Description of a New Genus and Species as the First Stygobiont Gastropod Species from Caves in Iran." *Journal of Cave and Karst Studies* 81:233–243.

Fehér, Zoltán, Tamás Deli, Zoltán P. Eerőss, and Romilda Lika. 2019. "Taxonomic Revision of the Subterranean Genus *Virpazaria* Gittenberger, 1969 (Gastropoda, Spelaeodiscidae)." *European Journal of Taxonomy* 558:1–25.

Ferreira, Rodrigo Lopes, Xavier Prous, and Rogério Parentoni Martins. 2007. "Structure of Bat Guano Communities in a Dry Brazilian Cave." *Tropical Zoology* 20:55–74.

Ficetola, Gentile Francesco, Claudia Canedoli, and Fabio Stoch. 2019. "The Racovitzan Impediment and the Hidden Biodiversity of Unexplored Environments." *Conservation Biology* 33:214–216.

Fišer, Cene, Maja Zagmajster, and Valerija Zakšek. 2013. "Coevolution of Life History Traits and Morphology in Female Subterranean Amphipods." *Oikos* 122:770–778.

Gargani, Julien, and Christophe Rigollet. 2007. "Mediterranean Sea Level Variations during the Messinian Salinity Crisis." *Geophysical Research Letters* 34:L10405.

Ghamizi, Mohamed, and Mokhtar Boulal. 2017. "New Stygobiont Snail from Groundwater of Morocco (Gastropoda: Moitessieriidae)." *Ecologica Montenegrina* 10:11–13.

Ghamizi, Mohamed, Marco Bodon, Mokhtar Boulal, and Folco Giusti. 1998. "A New Genus from Subterranean Waters of the Tiznit Plain, Southern Morocco (Gastropoda: Prosobranchia: Hydrobiidae)." *Haliotis* 26:1–8.

Gladstone, Nicholas S., Evin T. Carter, Michael L. McKinney, and Matthew L. Niemiller. 2018. "Status and Distribution of the Cave-Obligate Land Snails in the Appalachians and Interior Low Plateau of the Eastern United States America." *American Malacological Bulletin* 36:62–78.

Gladstone, Nicholas S., Matthew L. Niemiller, Evelyn B. Pieper, Katherine E. Dooley, and Michael L. McKinney. 2019. "Morphometrics and Phylogeography of the Cave-Obligate Land Snail *Helicodiscus barri* (Gastropoda, Stylommatophora, Helicodiscidae)." *Subterranean Biology* 30:1–32.

Gladstone Nicholas S., Kathryn E. Perez, Evelyn B. Pieper, Evin T. Carter, Katherine E. Dooley, Nathaniel F. Shoobs, et al. 2019. "A New Species of Stygobitic Snail in the Genus *Antrorbis* Hershler & Thompson, 1990 (Gastropoda, Cochliopidae) from the Appalachian Valley and Ridge of Eastern Tennessee, USA." *ZooKeys* 898:103–120.

Glavaš, Olga Jovanović, Branko Jalžić, and Helena Bilandžija. 2017. "Population Density, Habitat Dynamic and Aerial Survival of Relict Cave Bivalves from Genus *Congeria* in the Dinaric Karst." *International Journal of Speleology* 46:13–22.

Glöer, Peter, Jozef Grego, Zoltán Peter Erőss, and Zoltán Fehér. 2015. "New Records of Subterranean and Spring Molluscs (Gastropoda: Hydrobiidae) from Montenegro and Albania with the Description of Five New Species." *Ecologica Montenegrina* 4:70–82.

Grego, Jozef. 2018. "First Record of Subterranean Rissoidean Gastropod Assemblages in Southeast Asia (Mollusca, Gastropoda, Pomatiopsidae)." *Subterranean Biology* 25:9–34.

Grego, Jozef. 2020. "Revision of the Stygobiont Gastropod Genera *Plagigeyeria* Tomlin, 1930 and *Travunijana* Grego & Glöer, 1919 (Mollusca; Gastropoda; Hydrobiidae) in Hercegovina and Adjacent Regions." *European Journal of Taxonomy* 691:1–56.

Grego, Jozef, and Jozef Šteffek. 2007. "Biodiversity of Macaya N. P. and Biosphere Reserve in Condition of Global Changes." In *Proceedings of UNESCO–MAB Conference Stara Lesna, Slovakia,* edited by Julius Oslanyi, 85–112. Bratislava: Slovak Academy of Sciences.

Grego, Jozef, and Jozef Šteffek. 2010. "Ulitníky Podzemného druhu *Hauffenia* sp. Vo Vyvieračkách Plešiveckej Planiny." In *Plešivecká Planina*, edited by Jaroslav Stankovič, Václav Cílek, and Radka Schmelzová, 150–152. Liptovský Mikuláš: Slovenská Speleologická Spoločnosť.

Grego, Jozef, and Miklós Szekeres. 2017. "*Cotyorica nemethi* n. gen. n. sp., a Remarkable Tertiary Relict of the Subfamily Phaedusinae (Gastropoda: Pulmonata: Clausiliidae) from Northern Turkey." *Ecologica Montenegrina* 10:26–30.

Grego, Jozef, Dorottya Angyal, and Luis Arturo Liévano Beltrán. 2019. "First Record of Subterranean Freshwater Gastropods (Mollusca, Gastropoda, Cochliopidae) from the Cenotes of Yucatán State." *Subterranean Biology* 29:79–88.

Grego, Jozef, Peter Glöer, Zoltán Peter Erőss, and Zoltán Fehér. 2017. "Six New Subterranean Freshwater Gastropod Species from Northern Albania and Some New Records from Albania and Kosovo (Mollusca, Gastropoda, Moitesieriidae and Hydrobiidae)." *Subterranean Biology* 23:85–107.

Grego, Jozef, Peter Glöer, Andrzej Falniowski, Sebastian Hofman, and Artur Osikowski. 2019. "New Subterranean Freshwater Gastropod Species from Montenegro (Mollusca, Gastropoda, Moitessieriidae, and Hydrobiidae)." *Ecologica Montenegrina* 20:71–90.

Grego, Jozef, Peter Glöer, Aleksandra Rysiewska, Sebastian Hofman, and Andrzej Falniowski. 2018. "A New Montenegrospeum Species from South Croatia (Mollusca: Gastropoda: Hydrobiidae)." *Folia Malacologica* 26: 25–34.

Grego, Jozef, Levan Mumladze, Andrzej Falniowski, Artur Osikowski, Aleksandra Rysiewska, Dimitry M. Palatov, et al. 2020. "Revealing the Stygobiotic and Crenobiotic Molluscan Biodiversity Hotspot in Caucasus: Part I. The Phylogeny of Stygobiotic Sadlerianinae Szarowska, 2006 (Mollusca, Gastropoda, Hydrobiidae) from Georgia with Descriptions of Five New Genera and Twenty-One New Species." *ZooKeys* 955:1–77.

Griebler, Christian, Florian Malard, and Tristan Lefébure. 2014. "Current Developments in Groundwater Ecology: From Biodiversity to Ecosystem Function and Services." *Current Opinion in Biotechnology* 27:159–167.

Grossu, Alexandru, and Alexandrina Negrea. 1989. "*Paladilhia (Paladilhiopsis) dobrogica* n. sp. une Nouvelle Espece de la Familie Moitessieriidae (Gasteropoda, Prosobranchia)." *Miscellanea Speologica Romana* 1:33–37.

Habe, Tadashige. 1965. "Descriptions of One New Species and One New Subspecies of Freshwater Gastropods from Japan." *Venus* 23:20–209.

Harzhauser, Mathias, and Oleg Mandic. 2004. "The Muddy Bottom of Lake Pannon—a Challenge for Dreissenid Settlement (Late Miocene; Bivalvia)." *Palaeogeography, Palaeoclimatology, Palaeoecology* 204:331–352.

Herbert, Dai, and Dick Kilburn. 2004. "Conservation." In *Field Guide to the Snails and Slugs of Eastern South Africa,* edited by Natal Museum, 44–47. Pietermaritzburg: Natal Museum.

Hershler, Robert. 1985. "Systematic Revision of the Hydrobiidae (Gastropoda: Rissoacea) of the Cuatro Cienegas Basin, Coahuila, Mexico." *Malacologia* 26:31–123.

Hershler, Robert, and Winston F. Ponder. 1998. "A Review of Morphological Characters of Hydrobioid Snails." *Smithsonian Contributions to Zoology* 600:1–55.

Hershler, Robert, and France Velkovrh. 1993. "A New Genus of Hydrobiid Snails (Prosobranchia: Rissooidea) from Northern South America." *Proceedings of the Biological Society of Washington* 106:182–189.

Hofman, Sebastian, Aleksandra Rysiewska, Artur Osikowski, Jozef Grego, Boris Sket, Simona Prevorčnik, et al. 2018. "Phylogenetic Relationships of the Balkan Moitessieriidae (Caenogastropoda: Truncatelloidea)." *Zootaxa* 4486:311–339.

Holmes, Andrew James, Niina A. Tujula, Marita Holey, Annalisa Contos, Julia M. Hames, Peter Rogers, et al. 2001. "Phylogenetic Structure of Unusual Aquatic Microbial Formation in Nullarbor Caves, Australia." *Environmental Microbiology* 3:256–264.

Howarth, Francis G. 1980. "The Zoogeography of Specialized Cave Animals: A Bioclimatic Model." *Evolution* 34:394–406.

Howarth, Francis G. 1983. "Ecology of Cave Arthropods." *Annual Review of Entomology* 28:365–389.

Inkhavilay, Khamla, Chirasak Sutcharit, Piyoros Tongkerd, and Somsak Panha. 2016. "New Species of Micro Snails from Laos (Pulmonata: Vertiginidae and Diapheridae)." *Journal of Conchology* 42:213–232.

[IUCN] International Union for Conservation of Nature. 2021. "Red List of Threatened Species." Accessed January 14, 2021. http://www.iucnredlist.org.

Jasinska, Edyta J., Brenton Knott, and Arthur J. McComb. 1996. "Root Mats in Ground Water: A Fauna-Rich Cave Habitat." *Journal of the North American Benthological Society* 15:508–519.

Jaume, Damia, Geoffrey A. Boxshal, and William F. Humphreys. 2001. "New Stygobiont Copepods (Calanoida; Misophrioida) from Bundera Sinkhole, an Anchialine Cenote in North-Western Australia." *Zoological Journal of the Linnean Society* 133:1–24.

Jiang, Zhongcheng C., Yanqing Q. Lian, and Xiaoqun Q. Qin. 2014. "Rocky Desertification in Southwest China: Impacts, Causes, and Restoration." *Earth Science Reviews* 132:1–12.

Jochum, Adrienne, Anton J. de Winter, Alexander M. Weigand, Benjamín Gómez, and Carlos Prieto. 2015. "Two New Species of *Zospeum* Bourguignat, 1856 from the Basque-Cantabrian Mountains, Northern Spain (Eupulmonata, Ellobioidea, Carychiidae)." *ZooKeys* 483:81–96.

Juberthie, Christian, Bernard Delay, and Michael Bouillon. 1980. "Extension du Milieu Souterrain en Zone Non Calcaire: Description d'un Nouveau Milieu et de son Peuplement par les Coléoptères Troglobies." *Mémoirs de Biospéologie* 7:19–52.

Kano, Yasunori, and Tomoki Kase. 2004. "Genetic Exchange between Anchialine Cave Populations by Means of Larval Dispersal: The Case of a New Gastropod Species *Neritilia cavernicola*." *Zoologica scripta* 33:423–437.

Kappes, Heike, and Peter Haase. 2012. "Slow, but Steady: Dispersal of Freshwater Molluscs." *Aquatic Science* 74:1–14.

Kelley, Joanna L., Lenin Arias-Rodriguez, Dorrelyn Patacsil Martin, Muh-Ching Yee, Carlos D. Bustamante, and Michael Tobler. 2016. "Mechanisms Underlying Adaptation to Life in Hydrogen Sulfide-Rich Environments." *Molecular Biology and Evolution* 33:1419–1434.

Kljaković-Gašpić, Franica, Oleg Antonić, Božica Šorgić, Tomi Haramina, Ines Horvat, Nikolina Bakšič, et al. 2015. In *Studija Glavne ocjene Prihvatljivosti Zahvata za Ekološku Mrežu HE OMBLA*, Knjiga 1, Zagreb: OIKON-Geonatura Ltd. Accessed July 20, 2020. https://mzoe.gov.hr/UserDocsImages//NASLOVNE%20FOTOGRAFIJE%20I%20KORI%C5%A0TENI%20LOGOTIPOVI/doc//28_07_2015_studija_glavne_ocjene_prihvatljivosti_zahvata_-_knjiga_1.pdf.

Kováč, Ľubomír, Dana Elhottová, Andrej Mock, Alena Nováková, Václav Krištůfek, Alica Chroňáková, et al. 2014. "Mollusca." In *The Cave Biota of Slovakia*, edited by Ľubomír Kováč. Banská Bystrica: State Nature Conservancy of the Slovak Republic.

Kuroda, Tokubei. 1963. *A Catalogue of the Non-marine Molluscs of Japan, including the Okinawa and Ogasawara Islands*. Tokyo: Malacological Society of Japan.

Kuroda, Tokubei, and Tadashige Habe. 1957. "Troglobiontic Aquatic Snails from Japan." *Venus* 19:183–196.

Latella, Leonardo, Mauro Cobolli, and Mauro Rampini. 1999. "La Fauna delle Grotte nei Gessi dell´Alto Crotonese (Calabria)." *Thalassa Salentina* 23:103–113.

Liew, Thor-Seng, Jaap Jan Vermeulen, Mohammad Effendi bin Marzuki, and Menno Schilthuizen. 2014. "A Cybertaxonomic Revision of the Micro-Landsnail Genus *Plectostoma* Adam (Mollusca, Caenogastropoda, Diplommatinidae), from Peninsular Malaysia, Sumatra and Indochina." *ZooKeys* 393:1–107.

Ložek, Vojen, and Ján Brtek. 1964. "Neue *Belgrandiella* aus den Westkarpaten." *Archive für Molluskenkunde* 93:201–207.

Lydeard, Charles, Robert H. Cowie, Winston F. Ponder, Arthur E. Bogan, Philipe Bouchet, Stephanie A. Clark, et al. 2004. "The Global Decline of Nonmarine Mollusks." *BioScience* 54:321–330.

Maassen, Wim J. M. 1989. "Eine Neue Unterart von *Spelaeoconcha paganettii* Sturany, 1901 (Gastropoda Pulmonata: Enidae) aus Jugoslawien." *Basteria* 53:93–96.

Maassen, Wim J. M. 2008. "Remarks on a Small Collection of Terrestrial Molluscs from North–West Laos, with Descriptions of Three New Species (Mollusca: Pulmonata: Streptaxidae, Vertiginidae)." *Basteria* 72:233–240.

Macalady, Jennifer L., Daniel S. Jones, and Ezra H. Lyon. 2007. "Extremely Acidic, Pendulous Cave Wall Biofilms from the Frasassi Cave System, Italy." *Environmental Microbiology* 9:1402–1414.

Mammola, Stefano, Pedro Cardoso, David C. Culver, Louis Deharveng, Rodrigo L. Ferreira, Cene Fišer, et al. 2019. "Scientists' Warning on the Conservation of Subterranean Ecosystems." *BioScience* 69:641–650.

Mammola, Stefano, Pier Mauro Giachino, Elena Piano, Alexandra Jones, Marcel Barberis, Giovanni Badino, et al. 2016. "Ecology and Sampling Techniques of an Understudied Subterranean Habitat: The *Milieu Souterrain Superficiel* (MSS)." *The Science of Nature* 103:1–24.

Milanović, Petar. 2018. *Engineering Karstology of Dams and Reservoirs*. New York: CRC Press.

Monteiro, Mervin J., Milton A. Typas, Burton F. Moffett, and Brian W. Bainbridge. 1982. "Isolation and Characterization of a High Molecular Weight Plasmid from the Obligate Methanol-Utilising Bacterium *Methylomonas (Methanomonas) methylovora*." *FEMS Microbiology Letters* 15:235–237.

Morley, Robert J. 2012. "A Review of the Cenozoic Paleoclimate History of Southeast Asia." In *Biotic Evolution and Environmental Change in Southeast Asia (Systematics Association Special Volume Series, p. I)*, edited by David J. Gower, Kenneth Johnson, James Richardson, Brian Rosen, Lukas Rüber, and Suzanne Williams, 79–114. Cambridge: Cambridge University Press.

Morton, Brian, and Sanja Puljas. 2013. "Life-History Strategy, with Ctenidial and Pallial Larval Brooding of the Troglodytic 'Living Fossil' *Congeria kusceri* (Bivalvia: Dreissenidae) from the Subterranean Dinaric Alpine Karst of Croatia." *Biological Journal of the Linnean Society* 108:294–314.

Mousavi-Sabet, Hamed, Saber Vatandoust, Yaser Fatemi, and Soheil Eagderi. 2016. "Tashan Cave a New Cave Fish Locality for Iran; and *Garra tashanensis*, a New Blind Species from the Tigris River Drainage." *Fish Taxa* 1:133–148.

Mouthon, Jacques, and Martin Daufresne. 2008. "Population Dynamics and Life Cycle of *Pisidium amnicum* (Müller) (Bivalvia: Sphaeriidae) and *Valvata piscinalis* (Müller)

(Gastropoda: Prosobranchia) in the Saône River, A Nine-Year Study." *International Journal of Limnology: Annales de Limnologie* 44:241–251.

Murphy, Lisa N., Daniel Kirk-Davidoff, Natalie Mahowald, and Bette L. Otto-Bliesner. 2009. "A Numerical Study of the Climate Response to Lowered Mediterranean Sea Level during the Messinian Salinity Crisis." *Palaeogeography, Palaeoclimatology, Palaeoecology* 279:41–59.

Muyzer, Gerard, and Alfons J. M. Stams. 2008. "The Ecology and Biotechnology of Sulphate-Reducing Bacteria." *Nature Reviews Microbiology* 6:441–454.

Nordsieck, Hartmut. 2007. "Pre-Pleistocene Clausiliidae of Central and Western Europe." In *Worldwide Door Snails*, 134–139. Hackenheim, Germany: Conch Books.

Odabashi, Deniz Anil, and Naime Arlarslan. 2015. "Description of a New Subterranean Nerite: *Theodoxus gloeri* n. sp. with Some Data on the Freshwater Gastropod Fauna of Balıkdamı Wetland (Sakarya River, Turkey)." *Ecologica Montenegrina* 2:327–333.

Odhner, Nils Hjalmar. 1921. "On Some Species of *Pisidium* in the Swedish State Museum." *Journal of Conchology* 16:218–223.

Orghidan, Traian. 1955. "Un Nou Domeniu de Viata Acvatica Subterana 'Biotopul Hiporeic.'" *Buletin Stiintific Sectia de Biologie si Stiinte Agricole si Sectia de Geologie si Geografie* 7:657–676.

Oromí, Pedro, Anna L. Medina, and Marisa L. Tejedor. 1986. "On the Existence of a Superficial Underground Compartment in the Canary Islands." *Actas del IX Congreso Internacional de Espeleología, Barcelona* 2:147–151.

Ortuño, Vicente M., José D. Gilgado, Alberto Jiménez-Valverde, Alberto Sendra, and Gonzalo Pérez-Suárez. 2013. "The 'Alluvial Mesovoid Shallow Substratum,' a New Subterranean Habitat." *PLOS ONE* 8:e76311.

Osikowski, Artur. 2018. "Isolation as a Phylogeny-Shaping Factor: Historical Geology and Cave Habitats in the Mediterranean Truncatelloidea Gray, 1840 (Caenogastropoda)." *Folia Malacologica* 25:213–229.

Osikowski, Artur, Sebastian Hofman, Dilian Georgiev, Aleksandra Rysiewska, and Andrzej Falniowski. 2017. "Unique, Ancient Stygobiont Clade of Hydrobiidae (Truncatelloidea) in Bulgaria: The Origin of Cave Fauna." *Folia Biologica (Kraków)* 65:79–93.

Páll-Gergely, Barna. 2010. "Additional Information on the Reproductive Biology and Development of the Clausilial Apparatus in *Pontophaedusa funiculum* (Mousson, 1856) (Gastropoda, Pulmonata, Clausiliidae, Phaedusinae)." *Malacologica Bohemoslovaca* 9:1–4.

Páll-Gergely, Barna. 2014. "Description of the Second *Laotia* Saurin, 1953; a Genus New to the Fauna of Vietnam (Gastropoda: Cyclophoroidea)." *Folia Malacologica* 22:289–292.

Páll-Gergely, Barna, and László Németh. 2008. "Observations on the Breeding Habits, Shell Development, Decollation, and Reproductive Anatomy of *Pontophaedusa funiculum* (Mousson 1856) (Gastropoda, Pulmonata, Clausiliidae, Phaedusinae)." *Malacologica Bohemoslovaca* 7:11–14.

Páll-Gergely, Barna, Tamás Deli, Zoltán Péter Erőss, Peter L. Reischütz, Alexander Reischütz, and Zoltán Fehér. 2018. "Revision of the Subterranean Genus *Spelaeodiscus* Brusina, 1886 (Gastropoda, Pulmonata, Spelaeodiscidae)." *ZooKeys* 769:13–48.

Páll-Gergely, Barna, Zoltán Fehér, András Hunyadi, and Takahiro Asami. 2015. "Revision of the Genus *Pseudopomatias* and Its Relatives (Gastropoda: Cyclophoroidea: Pupinidae)." *Zootaxa* 3937:1–49.

Páll-Gergely, Barna, Jozef Grego, Jaap J. Vermeulen, Alexander Reischütz, and Adrienne Jochum. 2019. "New *Tonkinospira* Jochum, Slapnik and Páll-Gergely, 2014 Species from Southeast Asia (Gastropoda: Pulmonata: Hypselostomatidae)." *Raffles Bulletin of Zoology* 67:517–535.

Páll-Gergely, Barna, András Hunyadi, Đỗ Đuc Sân, Fred Naggs, and Takahiro Asami. 2017. "Revision of the Alycaeidae of China, Laos and Vietnam (Gastropoda: Cyclophoroidea) I: The Genera *Dicharax* and *Metalycaeus*." *Zootaxa* 4331:1–124.

Páll-Gergely, Barna, Igor V. Muratov, and Takahiro Asami. 2016. "The Family Plectopylidae (Gastropoda, Pulmonata) in Laos with the Description of Two New Genera and a New Species." *ZooKeys* 592:1–26.

Páll-Gergely, Barna, Alexander Reischütz, Wim J. M. Maassen, Jozef Grego, and András Hunyadi. 2020. "New Taxa of Diapheridae Panha & Naggs in Sutcharit et al., 2010 from Laos and Thailand (Gastropoda: Eupulmonata: Stylommatophora)." *Raffles Bulletin of Zoology* 68:1–13.

Palmer, Arthur N. 2007. *Cave Geology*. Dayton, OH: Cave Books.

Paviša, Tomislav and Zvonimir Sever. 2015. "Hidroelektrana Ombla—Projekt za Energetsko Iskorištenje Podzemnih Voda u Kršu." *Građevinar* 67:471–483.

Pešić, Vlado, and Peter Glöer. 2013. "A New Freshwater Snail Genus (Hydrobiidae, Gastropoda) from Montenegro, with a Discussion on Gastropod Diversity and Endemism in Skadar Lake." *ZooKeys* 281:69–90.

Peterson, Dawn E., Kenneth L. Finger, Sandra Iepure, Sandro Mariani, Alessandro Montanari, and Tadeusz Namiotko. 2013. "Ostracod Assemblages in the Frasassi Caves and Adjacent Sulfidic Spring and Sentino River in the Northeastern Apennines of Italy." *Journal of Cave and Karst Studies* 75:11–27.

Pettinelli, Roberto, and Maria Clara Bicchierai. 2009. "Life Cycle of *Pisidium henslowanum* (Sheppard, 1823) (Bivalvia, Veneroida, Sphaeriidae) from Piediluco Lake (Umbria, Italy)." *Fundamental and Applied Limnology—Archiv für Hydrobiologie* 175:79–92.

Piegay Hervé, Adrien Alber, Louise Slater, and Laurent Bourdin. 2009. "Census and Typology of Braided Rivers in the French Alps." *Aquatic Sciences* 71:371–388.

Pintér, László Ernest. 1968. "Zur Kenntnis der Hydrobiiden des Mecsek—Gebirges (Ungarn) (Gastropoda: Prosobranchia)." *Acta Zoologica Academiae Scientiarium Hungaricae* 14:441–445.

Ponder, Winston F. 1992. "A New Genus and Species of Aquatic Cave-Living Snail from Tasmania (Mollusca: Gastropoda: Hydrobiidae)." *Papers and Proceedings of the Royal Society of Tasmania* 126:23–28.

Ponder, Winston F., Stephanie A. Clark, Stefan Eberhard, and Joshua B. Studdert. 2005. "A Radiation of Hydrobiid Snails in the Caves and Streams at Precipitous Bluff, Southwest Tasmania, Australia (Mollusca: Caenogastropoda: Rissooidea: Hydrobiidae)." *Zootaxa* 1074:1–66.

Por, Francis D. 1963. "The Relict Aquatic Fauna of the Jordan Rift Valley: New Contributions and Review." *Israel Journal of Zoology* 12:47–58.

Por, Francis D. 2007. "Ophel: A Groundwater Biome Based on Chemoautotrophic Resources. The Global Significance of the Ayalon Cave Finds, Israel." *Hydrobiologia* 592:1–10.

Por, Francis D. 2011. "Groundwater Life: Some New Biospeleological Views Resulting from the Ophel Paradigm." *Travaux l'Institut de Speologie Emile Racovitza* 50:61–76.

Por, Francis D. 2014. "Sulfide Shrimp? Observations on the Concealed Life History of the Thermosbaenacea (Crustacea)." *Subterranean Biology* 14:63–77.

Porter, Megan L., Sarah Russell, Anette Summers Engel, and Libby A. Stern. 2002. "Population Studies of the Endemic Snail *Physa spelunca* (Gastropoda: Physidae) from Lower Kane Cave, Wyoming (Abstract)." *Journal of Cave and Karst Studies* 64:181.

Prevorčnik Simona, Sebastian Hofman, Teo Delić, Aleksandra Rysiewska, Artur Osikowski, and Andrzej Falniowski. 2019. "*Lanzaiopsis* Bole, 1989 (Caenogastropoda: Truncatelloidea): Its Phylogenetic and Zoogeographic Relationships." *Folia Malacologica* 27:193–201.

Prié, Vincent. 2019. "Molluscs." In *Encyclopedia of Caves*, 3rd ed., edited by William B. White, David Culver, and Tanja Pipan, 725–731. Amsterdam: Elsevier, Academic Press.

Puljas, Sanja, Melita Peharda, Brian Morton, Nives Štambuk Giljanović, and Ivana Jurić. 2014. "Growth and Longevity of the 'Living Fossil' *Congeria kusceri* (Bivalvia: Dreissenidae) from the Subterranean Dinaric Karst of Croatia." *Malacologia* 57:353–364.

Racovitza, Gheorghe. 1983. "Sur les Relations Dynamiques entre le Milieu Souterrain Superficiel et le Milieu Cavernicole." *Mémoires de Biospéologie* 10:85–90.

Rada, Biljana, and Tonći Rada. 2012. "New Data about the Reproductive Cycle of *Congeria kusceri* Bole, 1962 (Bivalves: Dreissenidae) from the Pit 'Jama u Predolcu' (Croatia)." *Italian Journal of Zoology* 79:105–110.

Raschmanová, Natália, Vladimír Šustr, Ľubomír Kováč, Andrea Parimuchová, and Miroslav Devetter. 2018. "Testing the Climatic Variability Hypothesis in Edaphic and Subterranean Collembola (Hexapoda)." *Journal of Thermal Biology* 78:391–400.

Reischütz, Alexander, and Peter L. Reischütz. 2009. "Ein Beitrag zur Kenntnis der Molluskenfauna von Montenegro, nebst Beschreibung zweier neuer Arten der Gattung *Virpazaria* Gittenberger 1969." *Nachrichtenblatt der Ersten Vorarlberger Malakologischen Gesellschaft* 16:51–60.

Reischütz, Alexander, Peter L. Reischütz, and Miklós Szekeres. 2016. "The Clausiliidae Subfamily Phaedusinae (Gastropoda, Pulmonata) in the Balkans." *Nachrichtenblatt der Ersten Vorarlberger Malakologischen Gesellschaft* 23:93–117.

Rendoš, Michal, Andrej Mock, and Tomáš Jászay. 2012. "Spatial and Temporal Dynamics of Invertebrates Dwelling Karstic Mesovoid Shallow Substratum of Sivec National Nature Reserve (Slovakia), with Emphasis on Coleoptera." *Biologia* 67:1143–1151.

Richling, Ira, Yaron Malkowsky, Jaqueline Kuhn, Hans-Jörg Niederhöfer, and Hans D. Boeters. 2016. "A Vanishing Hotspot—Impact of Molecular Insights on the Diversity of Central European *Bythiospeum* Bourguignat, 1882 (Mollusca: Gastropoda: Truncatelloidea)." *Organisms, Diversity and Evolution* 17:67–85.

Riesch, Rüdiger, Martin Plath, and Ingo Schlupp. 2010. "Toxic Hydrogen Sulfide and Dark Caves: Life-History Adaptations in a Livebearing Fish (*Poecilia mexicana*, Poeciliidae)." *Ecology* 91:1494–1505.

Roach, Katherine A., Michael Tobler, and Kirk O. Winemiller. 2011. "Hydrogen Sulfide, Bacteria, and Fish: A Unique, Subterranean Food Chain." *Ecology* 92:2056–2062.

Roje-Bonacci, Tanja, and Ognjen Bonacci. 2013. "Interactive Comment on the Possible Negative Consequences of Underground Dam and Reservoir Construction and Operation in Coastal Karst Areas: An Example of the HEPP Ombla near Dubrovnik (Croatia)." *Natural Hazards and Earth System Sciences Discussion* 1:C248–C254.

Rubio, Federico, Emilio Rolán, Katrine Worsaae, Alejandro Martínez, and Brett C. Gonzalez. 2016. "Description of the First Anchialine Gastropod from a Yucatán Cenote, *Teinostoma brankovitsi* n. sp. (Caenogastropoda: Tornidae), including an Emended Generic Diagnosis." *Journal of Molluscan Studies* 82:169–177.

Ruffo, Sandro. 1957. "Le Attuali Conoscenze sulla Fauna Cavernicola della Regione Pugliese." *Memorie di Biogeografica Adriatica* 3:1–143.

Rysiewska, Aleksandra, Simona Prevorčnik, Artur Osikowski, Sebastian Hofman, Luboš Beran, and Andrzej Falniowski. 2017. "Phylogenetic Relationships in *Kerkia* and Introgression between *Hauffenia* and *Kerkia* (Caenogastropoda: Hydrobiidae)." *Journal of Zoological Systematics and Evolutionary Research* 55:106–117.

Sands, Artur F., Sergej V. Sereda, Björn Stelbrink, Thomas A. Neubauer, Sergei Lazarev, Thomas Wilke, et al. 2019. "Contributions of Biogeographical Functions to the Species Accumulation May Change Over Time in Refugial Regions." *Journal of Biogeography* 46:1274–1286.

Sarbu, Serban. M. 2000. "Movile Cave: A Chemoautotrophically Based Groundwater Ecosystem." In *Ecosystems of the World: Subterranean Ecosystems*, edited by Horst Wilkens, David C. Culver, and William F. Humphreys, 319–343. Amsterdam: Elsevier.

Sarbu, Serban M., Thomas C. Kane, and Brian K. Kinkle. 1996. "A Chemoautotrophically Based Cave Ecosystem." *Science* 272:1953–1955.

Schabereiter-Gurtner, Claudia, Cesareo Saiz-Jimenez, Guadalupe Piñar, Werner Lubitz, and Sabine Rölleke. 2002. "Phylogenetic 16S rRNA Analysis Reveals the Presence of Complex and Partly Unknown Bacterial Communities in Tito Bustillo Cave, Spain, on Its Paleolithic Paintings." *Environmental Microbiology* 4:392–400.

Schmid-Araya, Jenny M. 2000. "Invertebrate Recolonization Patterns in the Hyporheic Zone of a Gravel Stream." *Limnology and Oceanography* 45:1000–1005.

Schütt, Hartwig. 1972. "Ikonographische Darstellung der Unterirdisch lebenden Molluskengattung *Plagigeyeria* Tomlin (Prosobranchia: Hydrobiidae)." *Archiv für Molluskenkunde* 102:113–123.

Schwartz, Megane, and Ronald V. Dimock. 2001. "Ultrastructural Evidence for Nutritional Exchange between Brooding Unionid Mussels and their Glochidia Larvae." *Invertebrate Biology* 120:227–236.

Shu, Shu-Sen, Wan-Sheng Jiang, Tony Whitten, Jun-Xing Yang, and Xiao-Yong Chen. 2013. "Drought and China's Cave Species." *Science* 340: 272.

Siebert, Stefan, Jacob Burke, Jean Marc Faures, Karen Frenken, Petra Döll, and Felix T. Portman. 2010. "Groundwater Use for Irrigation: A Global Inventory." *Hydrology and Earth System Sciences* 14:1863–1880.

Simone, Ricardo Luiz. 2012. "A New Genus and Species of Cavernicolous Pomatiopsidae (Mollusca, Caenogastropoda) in Bahia, Brazil." *Papéis Avulsos de Zoologia* 52: 515–524.

Simone, Ricardo Luiz. 2018. "*Lavajatus moroi*, New Cavernicolous Subulininae from Ceará, Brazil (Gastropoda, Eupulmonata, Achatinidae)." *Spixiana* 41:173–187.

Sket, Boris. 2012. "Dinaric Karst, Diversity" In *Encyclopedia of Caves*, 2nd ed., edited by William B. White and David C. Culver, 158–165. New York: Elsevier, Academic Press.

Slack-Smith, Shirley M. 1993. "The Non-marine Molluscs of the Cape Range Peninsula, Western Australia." *Records of the Western Australian Museum* 45:87–108.

Soós, Lajos. 1940. "Adatok az Északkeleti Kárpátok Mollusca-faunájának Ismeretéhez." *Állattani Közlemények* 37:140–154.

Souza-Silva, Marconi, Rogério Parentoni Martins, and Rodrigo Lopes Ferreira. 2015. "Cave Conservation Priority Index to Adopt a Rapid Protection Strategy: A Case Study in Brazilian Atlantic Rain Forest." *Environmental Management* 55:279–295.

Štamol, Vesna, Branko Jalžić, and Eduard Kletečki. 1999. "A Contribution to Knowledge about the Distribution of the Troglobiontic Snail *Pholeoteras eutrix* Sturany, 1904 (Mollusca, Gastropoda)." *Natura Croatica* 8:407–419.

Starobogatov, Yaroslav Igorevich. 1962. "Contribution to Molluscs from Subterranean Waters of Caucasus." *Bulletin of Moscow Society of Naturalists—Biological Series* 67:42–54.

Šteffek, Jozef, and Jozef Grego. 2005. "Preliminary Research Report on the Genus cf. *Hauffenia* (Mollusca: Gastropoda: Hydrobiidae) Distribution Research in Slovakia." In *Molluscs, Quaternary, Faunal Changes and Environmental Dynamics, A Symposium on Occasion of 80th Birthday of Vojen Ložek*, edited by Ivan Horáček, 32–33. Prague: Charles University.

Stevanović, Zoran, Želimir Pekaš, Boban Jolović, Arben Pambuku, and Dragan Radojević. 2014. "Classical Dinaric Karst Aquifer—an Overview of Its Past and Future." In *Proceedings Conference Karst without Boundaries*, edited by Neno Kukurić, Zoran Stevanović, and Neven Krešić, 1–6. Trebinje: Diktas.

Szarowska, Magdalena. 2006. "Molecular Phylogeny, Systematics and Morphological Character Evolution in the Balkan Rissooidea (Caenogastropoda)." *Folia Malacologica* 14:99–168.

Tobler, Michael T., Katherine A. Roach, Kirk O. Winemiller, Reid L. Morehouse, and Martin Plath. 2013. "Population Structure, Habitat Use, and Diet of Giant Waterbugs in a Sulfidic Cave." *The Southwestern Naturalist* 58:420–426.

Trajano, Eleonora. 2000. "Cave Faunas in the Atlantic Tropical Rain Forest: Composition, Ecology, and Conservation." *Biotropica* 32:882–893.

Vaxevanopoulos, Markos. 2009 "Identifying Hypogenic Features in Greek Caves." In *15th International Congress of Speleology Proceedings, Earth Sciences*, edited by William B. White, 1713–1716. Kerrville: International Union of Speleology—Greyhound Press.

Vinarski, Maxim V., and Yuri I. Kantor. 2016. *Analytical Catalogue of Fresh and Brackish Water Molluscs of Russia and Adjacent Countries*. Moscow: Tovarishchestvo Nauchnyckh Izdanii KMK.

Vinarski, Maxim V., Dimitri M. Palatov, and Peter Glöer. 2014. "Revision of '*Horatia*' Snails (Mollusca: Gastropoda: Hydrobiidae *Sensu Lato*) from South Caucasus with Description of Two New Genera." *Journal of Natural History* 48:2237–2253.

Voituron, Yann, Michelle de Fraipont, Julien Issartel, Olivier Guillaume, and Jean Clobert. 2011. "Extreme Lifespan of the Human Fish (*Proteus anguinus*): A Challenge for Ageing Mechanisms." *Biology Letters* 7:105–107.

Weck, Robert G., and Steven J. Taylor. 2016. "Life History Studies of a Cave-Dwelling Population of *Physa* Snails (Gastropoda: Basommatophora: Physidae) from Southwestern Illinois." *Speleobiology Notes* 8:1–9.

Weigand, Alexander M. 2013. "New *Zospeum* Species (Gastropoda, Ellobioidea, Carychiidae) from 980 m Depth in the Lukina Jama–Trojama Cave System (Velebit Mts., Croatia)." *Subterranean Biology* 11:45–53.
Wethington, Amy, and Robert Guralnick. 2004. "Are Populations of Physids from Different Hot Springs Distinctive Lineages?" *American Malacological Bulletin* 19:135–144.
Whitten, Tony. 2009. "Applying Ecology for Cave Management in China and Neighbouring Countries." *Journal of Applied Ecology* 46:520–523.
Wilke, Thomas, George M. Davis, Andrzej Falniowski, Folco Giusti, Marco Bodon, and Magdalena Szarowska. 2009. "Molecular Systematics of Hydrobiidae (Mollusca: Gastropoda: Rissooidea): Testing Monophyly and Phylogenetic Relationships." *Proceedings of the Academy of Natural Sciences of Philadelphia* 151:1–21.
Yildirim, Mehmet Zeki, Duygu Ceren Çağlan Kaya, Mustafa Emre Gürlek, and Seval Bahadir Koca. 2017. "A New Species of *Islamia* (Caenogastropoda: Hydrobiidae) from Lakes Region of Turkey." *Ecologica Montenegrina* 11:9–13.
Zagmajster, Maja, David C. Culver, Marry C. Christman, and Boris Sket. 2010. "Evaluating the Sampling Bias in Pattern of Subterranean Species Richness: Combining Approaches." *Biodiversity and Conservation* 19:3035–3048.
Zagmajster, Maja, Florian Malard, David Eme, and David C. Culver. 2018. "Subterranean Biodiversity Patterns from Global to Regional Scales." In *Cave Ecology*, edited by Oana Moldovan, Ľubomír Kováč, and Stuart Halse, 195–227. Switzerland: Springer, Nature.
Zupan-Hajna, Nadja. 2015. "What is Karst?" In *Life and Water on Karst: Monitoring of Transboundary Water Resources of Northern Istria*, edited by Nadja Zupan-Hajna, Nataša Ravbar, Josip Rubinić, and Metka Petrič, 6–15. Postojna: Collegium Graphicum.

4

The Subterranean Cholevinae of Italy

Leonardo Latella

Introduction

Cholevinae, a subfamily of Leiodidae (order Coleoptera) of small to medium sized beetles (typically between 1 and 10 mm), is widespread on all continents globally, except Antarctica. Dating back to at least the Jurassic, cholevine beetles once inhabited the forests that covered the southern part of Pangea (Giachino and Vailati 1993). When plate tectonics subdivided Pangea around 175 million years ago (mya), the Cholevinae dispersed with the continents, which partially explains the present distribution of the seven tribes (Perreau 1989; Trewick 2017). In fact, Ptomaphagini, Anemadini, and Cholevini are distributed almost globally with the exception of the poles, while Eucatopini occurs only in South America, with Leptodirini primarily distributed across Europe, the Middle East, and Central Asia. Additionally, Oriticatopini is exclusive to sub-Saharan Africa, while Sciaphyni is presently known only from Japan and Siberia. Worldwide, Cholevinae has colonized both edaphic and hypogean habitats.

Currently, 335 genera and nearly 2,050 species of Cholevinae have been described globally (Antunes-Carvalho et al. 2019). With 62 genera and about 310 known species and subspecies of Cholevinae, Italy is the second-most diverse region of the world (the Balkan Peninsula is the richest). Four tribes (Ptomaphagini, Anemadini, Cholevini, and Leptodirini) are widespread throughout the country—with the highest number of species concentrated in the central-east part of the Alps.

The tribe Leptodirini constitutes the largest group in Italy. Rich in endemic species of subterranean and in some cases endogean habits, representatives are distributed throughout the peninsula and islands. Currently, at least 46 genera and 202 species of Leptodirini have been described from Italy. However, several groups are under revision, including the polyphyletic genus *Bathysciola* Jeannel, 1910 (Giachino and Vailati 2019; Giachino, pers. com. 2020). The greatest diversity, which is also defined by many short-range endemic species, occurs in the pre-Alpine area of Northern Italy (which is part of the "Alpine" biogeographic province). Overall, this province supports 30 genera, 113 endemic species, and no fewer than 97 troglobionts (Table 4.1). Fewer species (primarily belonging to the genera *Bathysciola* and *Parabathyscia*) are known from peninsular Italy. Of which, many are epigean with widespread distributions. Additional studies will contribute to the already high number of known taxa within this tribe, especially research focusing on the relatively unstudied mountains of the Apennine biogeographic province (Zoia and Latella 2006; Latella et al. 2017).

Depigmented and normally anophthalmic, many Italian species of Leptodirini live in the humic or deeper strata of soil and subterranean habitats, where they show both morphological and physiological adaptation to the hypogean environment (Zoia and Latella 2006). Immediate ancestors of the subterranean Leptodirini (in pre-Alpine and Apennine temperate regions) are forest soil- and leaf-litter-dwelling beetles that colonized subterranean environments due to the climatic oscillations from Miocene epoch through the Quaternary period. Their isolation within subterranean environments, coupled with their low dispersal capability, resulted in a high rate of speciation and endemism (Barr 1968; Culver 1982; Sbordoni 1982; Humphreys 2000; Latella et al. 2017). The other three Cholevinae tribes consist primarily of widespread surface-dwelling species. Although only a few species have been documented using caves, these taxa are normally winged and often good dispersers (Zoia and Latella 2006; Latella 2015a).

In this chapter, I used Italy's six biogeographic provinces (*sensu* Ruffo and Vigna Taglianti 2002; Fig. 4.1) to frame cholevine richness, troglomorphy, and natural history, as well as future research and conservation needs.

Taxonomy and Systematics

Taxonomic History

The first species in family Cholevinae, *Silpha subvillosa*, was described in 1777 by German zoologist J. A. E. Goeze. While it was originally placed in the family Silphidae, it was later moved to Cholevinae, and the name was

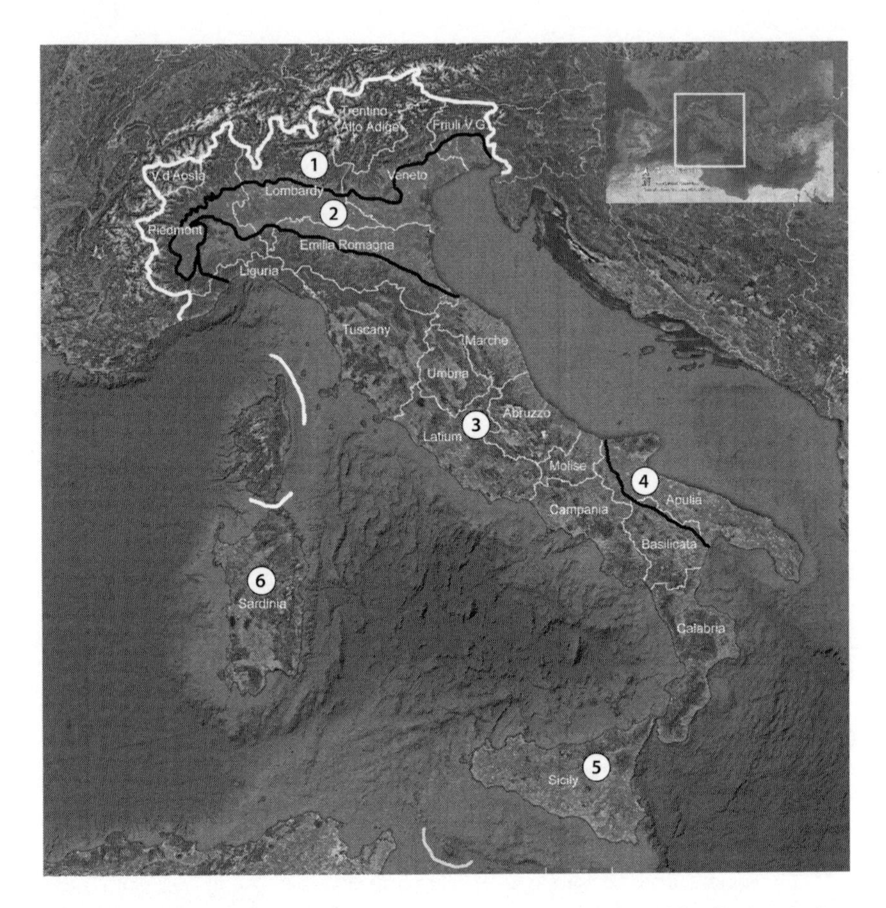

Figure 4.1. Map of Italy, northernmost border at top left (*white line*), administrative regions (*white lines with names*), and the six biogeographic provinces (*black lines*), which include (1) Alpine, (2) Po, (3) Apennine, (4) Apulian, (5) Sicilian, and (6) Sardinian Provinces. Image from Google Earth, version 7.3.4 (January 2022); *Italy and surrounding environs*.

changed to *Ptomaphagus subvillosus* (Goeze, 1777). Twenty years later, Latreille (1796), Paykull (1798), and Hellwig (1795) created, independently of one another, the genera *Choleva*, *Catops*, and *Ptomaphagus*, respectively. These genera were used to group the different species they described. Throughout the nineteenth century, Cholevinae remained part of the family Silphidae. In 1837, Kirby presented the family name, Cholevidae (Newton 1998). However, the classification of Cholevinae has been subjected to several changes and modifications over time.

Table 4.1. Number of cholevine endemic genera, endemic species, and troglobionts for each Italian biogeographic province

Biogeographic province	Endemic genera	Endemic species	Troglobionts
Alpine province	30	113	97
Po province	—	—	—
Apennine province	—	61	16
Apulian province	—	—	—
Sicilian province	—	3	—
Sardinian province	3	21	15
Total	33	198	128

Jeannel made significant advances in Cholevinae taxonomy. He established a framework for interpreting the phylogenetic relationships between the different groups. In the first half of the twentieth century, he revised this group, placing them in a nearly equivalent family, Catopidae (Jeannel 1936). He also produced monographs on Catopidae, where he reported the description and distribution of all known species within this family (Jeannel 1922; 1936). However, his most comprehensive contribution was the revision of cave-dwelling Leptodirini (= Bathysciini) (Jeannel, 1911, 1914, 1924).

Cholevinae has since undergone further revision. In compliance with the International Code of Zoological Nomenclature, the name was later changed from Catopidae to Cholevidae (Zwick 1979). The subfamily name was also changed from Bathysciinae to Leptodirinae (Silfverberg 1990), based on priority of first usage.

Recent discoveries of new taxa have highlighted additional characters and taxonomic relationships, as well as facilitated the formulation of new hypotheses about the phylogenetic relationships within the group. Several authors (Crowson 1955; Perreau 1989, 2000, 2004, 2015; Newton 1998; Fresneda et al. 2011) considered Cholevinae a subfamily of Leiodidae. Additional subfamilies within Leiodidae are Coloninae, Platypsyllinae, Camiarinae, Catopocerinae, and Leiodinae. With the exception of Camiarinae (known only in south-temperate countries) and Catopocerinae (which is known only in North America and Russia), the remaining subfamilies occur in Italy.

The systematics of Italian Cholevinae has had recent advancements. Many researchers have examined and revised both the systematic position (describing and reorganizing new genera and species) and phylogeny (i.e., the evolutionary affinities between the different recently studied taxa) of genera within these tribes. These include *Parabathyscia* (Zoia 1986), the phyletic series of both *Dellabeffaella* and *Boldoria* (Vailati 1988), a systematic review of the tribe Anemadini (Giachino and Vailati 1993), and taxonomical analysis of several genera of Leptodirini from the Venetian Prealps (Giachino and Vailati 2005). An update of the endemic genus *Ovobathysciola* from Sardinia led to the description of five new species and the revision of phylogenetic affinities (Casale 2014). Additionally, Vailati (2017a) completed a critical review of the systematics of the genera *Boldoria* and *Pseudoboldoria*, which were subsequently split into seven different genera. The taxonomy of the genus *Bathysciola* from Northern Italy has also been revised, where seven species were split into as many new genera (Giachino and Vailati 2019). Finally, Latella et al. (2017) compared molecular and morphological characters in 11 pre-Apennine species from the genus *Bathysciola* of Central Italy, and a morphological review of the Bathysciotina subtribe from Northern Italy was recently completed by Perreau (2019).

Morphology

The morphology of cholevine beetles varies with the degree of adaptation to the subterranean environment. Surface-dwelling and/or troglophilic species range in color from reddish brown to brown and black. Hypogean species are typically less pigmented. Their lighter colors of reddish brown or yellowish are due to a thinning of chitin—the polysaccharide of which the exoskeleton is composed (Rabasović et al. 2015; Noh et al. 2016). Furthermore, in both epigean and hypogean species, these beetles are often covered with a golden pubescence, which is arranged lengthwise along the body (Plate 4.1A).

The head is almost always retractable, while the mouthparts are hypognathous (or downward facing), and degree of eye development is dependent upon the degree of troglomorphy. However, some subterranean-specialized species of Italian Leptodirini have mouthparts in a prognathic position (e.g., the highly subterranean-adapted *Leptodirus hohenwarthi reticulatus* Müller, 1905). In the Italian subterranean-adapted Leptodirini, eyes are often absent, with some exceptions, including *Baudiola tarsalis*

(Kiesenwetter, 1861) in the Piedmont, *Reitteriola pumilio* (Reitter, 1884) from the Western Alps and northern Apennines, and *Phaneropella lesinae* (Reitter, 1881) in the Apulia region (Casale et al. 1991; Giachino and Vailati 2019). An evident occipital keel is present on the back of the head of most surface-dwelling species and is either reduced or absent in subterranean-adapted species. Antennae are composed of 11 articles of variable length, with the eighth antennomere always smaller than the previous and the next two articles.

Regarding the body, the thorax has a pronotum that is generally broad and transverse with edges regularly curved and convex, while in hypogean species it can become more elongate and narrower. The prosternum is always well developed and procoxal cavities are closed behind. Elytra are variable in shape, and either completely, or nearly, cover the abdomen. Some species have a sutural stria essentially parallel to the junction of the elytra. Hind wings are generally well developed in the epigean species but absent in endogean and hypogean species. The mesosternum may have a median keel between the two coxae; the mesosternum is either separated or confluent. Metacoxal cavities are elongated laterally and coalescent with one another; although in some Leptodirini, these characters are separated by more or less well-developed apophyses. Legs are often short and strong—except in some species of subterranean Leptodirini and Cholevini, which have more elongate and thin legs. Also, legs have spines and spurs, and their arrangements vary across taxonomic groups. In the majority of Cholevini, protarsi are pentamerous (i.e., each protarsus composed by five tarsomeres), and in males, the tarsomeres are dilated (Plate 4.1B). In many Italian Leptodirini genera of the Central and Western Alps and Prealps, males have protarsi with only four tarsomeres and often tarsomeres are not dilated. Female Leptodirini always have four tarsomeres. The meso- and meta-legs always have five tarsomeres. Finally, the abdomen has six visible sternites in Italian Cholevinae, with the exception of Anemadini, which has seven sternites.

Cholevinae larvae are campodeiform (i.e., elongated shape with legs ambulatory and body clearly divided into head, thorax, and abdomen) with the head and some thoracic segments sclerotized. The head is prognathous with well-developed mandibles. Maxillary palps are formed by three articles with the first larger than the others. Labial palps consist of three articles and antennae. Development and presence of the eyes is different in each taxonomic group. The thorax is formed by three sclerotized segments. Tergites have some setae and are typically long; their arrangement is often diagnostic in identifying across species. Legs are well developed and similar to one another. The abdomen is composed of nine well-differentiated segments with two biarticulated urogomphi on the ninth, and the tenth segment is a pygopod. Also, the chaetotaxy of the abdominal tergites is often useful in species identification.

Pupae are exarate (i.e., their appendages are free yet close to the body), with the tegument slightly sclerotized and whitish. Wing outlines are clearly striated, and the pterothecae (protective covering that outlines the wing) of the hind wing are present in all Italian tribes with the exception of Leptodirini.

Cholevinae Biology

Life History

Our knowledge of life history requirements of cholevine beetles is based primarily upon several captive-breeding studies, and to a limited extent, on direct field observations. As would be expected, the number of developmental stages varies due to degree of specialization to the cave environment. Few studies have documented cholevine larval development; however, the most common larval cycle consists of three instars (Lucarelli et al. 1980; Lucarelli and Sbordoni 1986; Latella, unpublished data). After a period of roaming freely (from 1 to 40 days, depending on the species), the larva builds a protective case in which it spends a variable period (from 2 to 15 days, depending on the species). The larva remains in its case through pupation. In some troglobiontic species, the number of larval stages was decreased (Deleurance-Glaçon 1961; Sbordoni 1980). Additionally, studies on species within the genera *Pseudoboldoria* and *Leptodirus* Schmidt, 1832 in Northern Italy and *Bathysciola* in Central Italy confirmed that, in subterranean species, the larval stages are reduced to one or two instars (Deleurance-Glaçon 1961; Rossi 1976; Zoia 1996). For one troglobiont (*Bathysciola sisernica* Cerutti & Patrizi, 1952), there are two to three instar stages (Lucarelli and Sbordoni 1986). Interestingly, instar period for epigean and troglophilic species (e.g., *Bathysciola sarteanensis* (Bargagli, 1870) and *B. derosasi* Jeannel, 1914; Lucarelli et al. 1980; Latella, unpublished data) is always three stages. Comparatively, North American cave *Ptomaphagus* have similar life histories but are less modified (Peck 1986).

Larvae morphology varies. Deleurance-Glaçon (1963) described differences in the morphology of epigean/troglophilic species and troglobionts. In general, the former have asymmetrical mandibles, equipped with prostheca (a sclerite articulated to the base of the mandible) and a retinaculum (a tooth on the masticatory margin of the mandible), and biarticulated cerci, while the latter have symmetrical mandibles, without prostheca or a retinaculum, and rudimentary cerci. Importantly, Sbordoni et al. (1982) observed variations in cave-dwelling larvae, which had intermediate characters between these two endmember types.

In various studies, adult cholevine lifespan ranged from three to more than five years in captivity. *Bathysciola derosasi* Jeannel, 1914 (Lucarelli et al. 1980) and *B. sisernica* Cerruti & Patrizi, 1952 (Lucarelli and Sbordoni 1986) both lived at least three years. *Neobathyscia mancinii* Jeannel, 1924 lived for about four years (Latella, unpublished data), while the longest-lived species, *Leptodirus hohenwarti* Schmidt, 1832, lived for over five years (Sbordoni et al. 1981).

Foraging Guilds and Behavior

Both larval and adult stage members of Cholevinae are mainly scavengers, feeding on decaying animal (zoosaprophagous) or plant matter (phytosaprophagous; Jeannel 1936). Initially, Jeannel (1936) suggested foraging guilds may have contributed in unique ways to cholevine evolution in the subterranean environment: phytosaprophagous beetles associated with leaf litter became specialized to life in humid environments until ultimately colonizing subterranean environments, while the zoosaprophagous species dwelled in the dens of mammals and other vertebrates, and in some cases, evolved to colonize hypogean ecosystems. However, current data does not support this idea. In many cases, the distinction between phyto- and zoosaprophagous is unclear. For example, Italian species of the tribe Ptomaphagini, *Ptomaphagus pius* Seidlitz, 1887, *P. varicornis* (Rosenhauer, 1847), and *P. sericatus* (Chaudoir, 1845), are found in moss and leaf litter, as well as on mammal carcasses. Species of the Anemadini tribe (e.g., *Nemadus colonoides* (Kraatz, 1851)) occur in decomposing vegetation and soil, and in ant nests of the genus *Lasius* Fabricius, 1804 (Jeannel 1936; Giachino and Vailati 1993). Additionally, many Italian species belonging to the genera *Catops* and *Sciodrepoides* are attracted to carrion and the dung of nonherbivorous mammals—provided relative humidity is adequate to prevent desiccation (Jeannel 1936; Zoia and Latella 2006). Finally, while many Leptodirini species depend on mold and fungi on both carcasses and feces (Zoia and Latella 2006), large quantities of nutrients (e.g., guano accumulations and large vertebrate carcasses) are reportedly unsuitable for Cholevinae (Zoia 1989).

To date, only two studies have examined the gut microbiota of Italian subterranean-adapted Leptodirini. Latella et al. (2008) revealed a complex microbial community plays a substantial biological role in the host organism and contributes to host fitness and adaptation within *Neobathyscia mancinii* Jeannel, 1924 and *N. pasai* Ruffo, 1950. Both animals are short-range endemics occurring in a few caves of the Venetian Prealps, with their subterranean distribution dependent on food resource availability (Latella et al. 2017). Interestingly, Paoletti et al. (2013) reported that *Cansiliella servadeii* Paoletti, 1980, an endemic troglobiont of northeastern Italy, may consume the microorganisms on moonmilk formations—moonmilk is a soft whitish formation of carbonate minerals (primarily calcite), which occurs in limestone caves (Cacchio et al. 2014).

Colonization and Adaptation

Subterranean Colonization

Colonization of subterranean environments is followed by speciation, but this is not a direct, compulsory consequence (Wynne et al., Chap. 2). For speciation to occur in temperate regions, the karstic massif must be isolated, with interruption of gene flow toward the surface or adjacent karstic massifs (Holsinger 2000). The success of epigean species colonizing the subterranean environment and becoming troglophilic may ultimately lead to that population evolving toward troglomorphism (Christiansen 2012). However, in some cases (e.g., Orthoptera: Rhaphidophoridae), populations may remain troglophiles and may establish a sort of adaptive equilibrium (Latella and Sbordoni 2002). In temperate regions, immediate ancestors of the hypogean organisms (*sensu* Giachino and Vailati 2010) were most likely soil- and leaf-litter-dwelling species believed to have become isolated during a Pleistocene interglacial period (Sbordoni 1982; Barr and Holsinger 1985; Holsinger 2000).

Concerning the current distribution of Italian Cholevinae, particularly Leptodirini, these populations have been strongly influenced by paleogeographic and paleoclimatic events. In some cases, radiation events occurred long before the Pleistocene, dating back to the Miocene (Caccone and Sbordoni 2001; Zoia and Latella 2006: Latella et al. 2017). For example, the Alpine chain region was affected by paleogeographic events during the Quaternary period (van Husen 2000). These events strongly influenced the current distribution of Cholevinae (Giachino and Vailati 2005). Classical refugial areas for subterranean-adapted endemic Leptodirini include the Alpine chain, specifically pre-Alpine areas (Sbordoni 1982; Latella and Sbordoni 2002; Fig. 4.2).

In central Italy and Sardinia, current distributions and phylogenetic relationships depict a similar story. A recent molecular study on selected species of the genus *Bathysciola* in the Apennines Mountains and foothills (pre-Apennines) suggested that successional geologic events influenced the current distribution and colonization of subterranean environments in peninsular Italy since the Miocene (Latella et al. 2017). Estimated

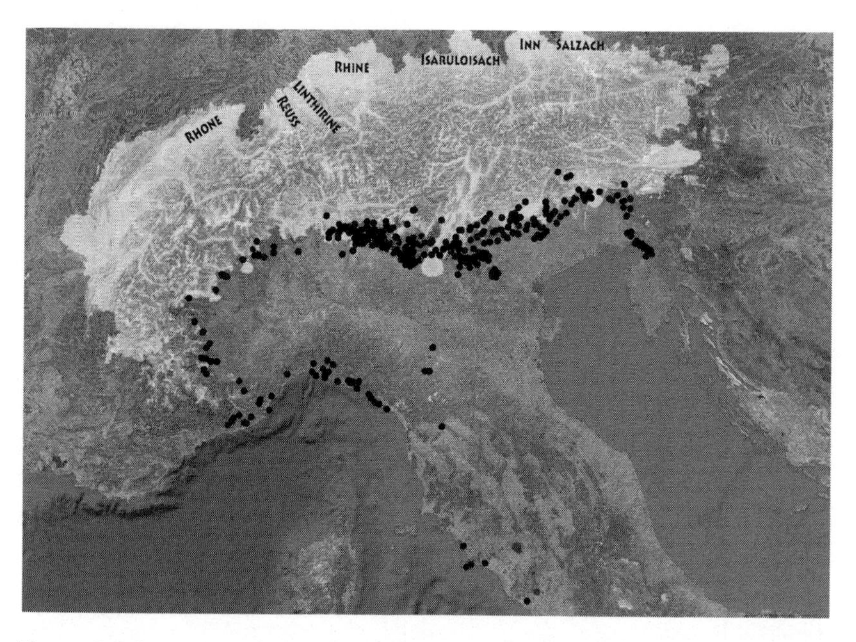

Figure 4.2. Ice extent during the Last Glacial Maximum (*white mass at top of figure*) over the Alpine Mountain chain. Dots denote the present distribution of troglobiontic Leptodirini in Northern Italy along the southern margin of the glacial ice. Image from Google Earth, version 7.3.2 (June 2019); *Northern Italy and surrounding environs.*

divergence time of the pre-Apennine-region species from the Apennine lineages ranges between 21.39 and 11.9 mya (Latella et al. 2017). These dates are consistent with the rise of the Apennine chain in northern Tuscany (Cipollari and Cosentino 1995). Additionally, the rise of this mountain range may have isolated ancestors of the *B. sarteanensis* group from other lineages in the Apennine region from northern Tuscany through southern Basilicata, and some coastal areas of central Italy (Latella et al. 2017).

During the late Miocene (11.27 to 6.41 mya), the central portion of the Apennine paleogeographic domain was geologically active and the tectonically driven sedimentary basins were formed (Cipollari and Cosentino 1995). Such events could have been responsible for the isolation and ultimate vicariance events of the two main pre-Apennine *Bathysciola* lineages. Colonization and subsequent isolation of the species currently populating different pre-Apennine groups probably occurred between the Pliocene and Pleistocene following the repeated isolation of these areas due to marine transgressions (Latella et al. 2017).

In Sardinia, genera of Leptodirini were also influenced by paleogeographic events (Casale et al. 2009). The movement and counterclockwise rotation of the Corso-Sardinian microplate, which started in the southern Iberic peninsula (Boccaletti et al. 1990), linked most troglobiontic species of the genera *Speonomus* (*Batinoscelis*), *Patriziella*, and *Ovobathysciola* to the congeneric taxa of *Speonomus* (or closely related genera—e.g., the phyletic lineage of *Anillochlamys*) and many present-day species in the Baetic-Pyrenean area (Caccone and Sbordoni 2001). Moreover, during the Messinian age (late Miocene), the salinity crisis of the Mediterranean produced land bridges that existed until 3.5 mya (Hsu et al. 1977). This phenomenon may also explain the affinity of *Bathysciola damry* (Abeille, 1881) with the species of the same genus living in the Apennine. Numerous regressions and transgressions of the Mediterranean Sea (some fluctuating by as much as 120 m above and below present-day sea level) occurred between the Pliocene and Quaternary periods and were driven by glacial and interglacial events. These fluctuations may have created land bridges between Sardinia and Corsica, the Tuscan Archipelago, and some parts of the Italian peninsula that permitted Leptodirini to migrate throughout these areas (Casale et al. 2009).

Adaptation

With at least 850 known species globally and about 320 in Italy, Leptodirini represents one of the most extraordinary examples of adaptive radiation in the subterranean environment. These highly specialized species occur in caves, *milieu souterrain superficiel* (MSS; Juberthie et al. 1980), and artificial cavities (e.g., mines and basements of ancient buildings) in the Italian mainland and islands. Diversity in this tribe exhibits morphological, physiological, and ecological adaptations for life underground (Delay 1978; Racovitza 1980; Sbordoni 1980; Latella and Stoch 2001). The most remarkable morphological adaptations consist of elongated and slender legs and antennae, reduced or absent eyes, elongation of body shape, and depigmentation.

One of the most interesting morphological characteristics of hypogean Leptodirini is the progressive elongation of the head. In the least specialized forms, the head is more or less round, but it is lengthened in the more subterranean-specialized forms. For some species (e.g., *Leptodirus hohenwarthi*, *Astagobius angustatus* (Schmidt, 1852), and the various species of the genus *Anthroherpon*), it has become up to three times longer than wide (Jeannel 1924; Latella 2011). Also, head elongation coincides with a backward directional shift of the antennae in the most highly troglobiontic species (Plate 4.2). Furthermore, mouth parts transition from a hypognathic to a prognathic position (Sbordoni et al. 1982; Latella 2015b).

In several Italian genera (e.g., some species of the genera *Ovobathysciola*, *Bathysciola*, and *Parabathyscia*), the antennae are elongate, while the body maintains a bathyscioid shape. Elongated antennae favor the development of the numerous sensory organs, including tactoreceptors, chemoreceptors, and hygroreceptors (Culver 1982; Latella 2011). The presence of internal sensory structures on the seventh, ninth, and tenth antennomeres has been documented in leptodirine antennae (Accordi and Sbordoni 1978). These structures, called Hamann's organs, seem to serve as hygroreceptors, representing progressively increasing degrees of complexity from epigean species to troglobionts (Lucarelli and Sbordoni 1978; Plate 4.3).

The prothorax shape differs between epigean to highly subterranean-specialized forms. In endogean and many hypogean species, especially in peninsular Italy, the prothorax is largely indistinguishable from the abdomen, giving the animal an ovoid appearance (i.e., the bathyscioid type).

The intermediate shape (pholeuonoid type) occurs in middling species, while the most specialized troglobiontic Leptodirini have a narrow cylindrical prothorax constricted at the base (leptodiroid type; Sbordoni et al. 1982).

A peculiar adaptation typical of some extremely subterranean-specialized Leptodirini is the development of enlarged globular-shaped elytra that covers the abdomen. This character has a spherical appearance, called "false physogastry." The best example of this character is exhibited in *Leptodirus hohenwarthi reticulatus* (Latella 2011). Adaptation may be related to the high hygrophilia in this group—as there is also a reduction of spiracles and tracheae (Sbordoni 1980). Respiration seems to occur mainly in the subelytral cavity, whereby the animal stores moisture under the elytra. This enables it to continue to breathe when the subterranean environment becomes less hydric (Accordi et al. 1980; Plate 4.4).

Another classic adaptation of highly subterranean-specialized animals is elongate antennae and long, slender legs. Long antennae allow increased sensory perception of the surrounding environment and longer legs facilitate energy-efficient movement (Latella 2011). In less subterranean-specialized taxa, legs are short and retractable under the abdomen, which is due to the flat femora that have a concavity in which the tibia can be inserted (Sbordoni et al. 1982). Conversely, troglobionts have long, cylindrical femora that far exceed the margin of the body (Sbordoni 1980).

Other adaptations of note involve myrmecophilous species. While not cave-dwelling per se, these species are indeed hypogean species. Several cholevine genera (e.g., *Attumbra*, *Philomessor*, *Catopsimorphus*, and *Parabathyscia*) complete their entire life cycle within the host ants' nests and exhibit a shorter, more convex body shape, shorter tarsi, and shorter, more compact antennae (Latella 2015a). Additionally, the species of the genus *Attumbra*, which live in the nests of *Messor*, are hairless (Latella 2015a). Overall, antennomeres are usually more flattened and narrower, body plan is "limuloid" (i.e., broadly oval, attenuated posteriorly), and legs are short and thickened (Kistner 1979; Parker 2016). While a large number of arthropods have evolved some capacity to target ant colonies (i.e., living as social parasites with varying degrees of intimacy with their hosts; Parker 2016), Cholevinae are not strongly socially integrated into host colonies. However, some species inhabit ants' nests almost exclusively (Jeannel 1936; Peck and Cook 2007).

Genetic Variability

The amount of genetic variability expressed by troglobiontic populations is a controversial issue (Sbordoni et al. 2012). After initial isolation underground, heterozygosity in subterranean species increases with evolutionary time, so that genetic variability could be greater in a small population of an old troglobiont than in a larger population of epigean relatives (Sbordoni 1982; Holsinger 2000). Conversely, Culver (1982) noted no differences in heterozygosity both among subterranean populations and between hypogean and epigean species. Studies conducted on Italian Leptodirini confirmed an increase of heterozygosity in cave populations and the role of bottlenecks in enhancing genetic differentiation in isolated subterranean populations (Sbordoni 1982; Sbordoni et al. 2012; Latella et al. 2017). The bottleneck effect was studied in 1956, when a population of *Bathysciola derosasi* (Jeannel, 1914) was artificially introduced to a cave in central Italy where no other *Bathysciola* were known to occur (Patrizi 1956).

Thirty years hence, a genetic study was conducted. Sbordoni (1982) and De Matthaeis et al. (1983) reported a genetic distance of 0.012 (on a set of 13 loci) from the original population. Additionally, a high level of heterozygosity was confirmed by Sbordoni et al. (1981) in a study on the troglobiontic *Leptodirus hohenwarthi, Orostygia doderoi* J. Müller, 1919, and *O. meggiolaroi* Agazzi, 1968. Sbordoni et al. (2012) interpreted this variability as adaptation to a discontiguous environment where different cave microhabitats and heterogeneity of resources favor a multiple-niche polymorphism strategy.

Habitat Selection

Hypogean habitats are interconnected, ranging in size from the small pore spaces of the MSS to macrocaverns (i.e., caves). As caves are the only environment accessible to humans, they have mistakenly been considered the primary habitat for troglobionts. Subterranean-adapted animals primarily occur in the network of small cracks and fissures (i.e., mesocaverns) in the host bedrock (Howarth 1983; Latella and Sbordoni 2002; Giachino and Vailati 2010; Latella 2015b; Howarth and Wynne, Chapt. 1) and the MSS. The latter is the hypogean environment constituted by a network of interstices generally formed by fragmentation of the bedrock and the accumulation of debris (Juberthie 1983; Giachino and Vailati 2010; Nae and

Băncilă 2017). Other subterranean environments include animal burrows, which range up to several meters deep and are excavated in different environments (Nevo 1979).

All four Italian tribes of Cholevini have species related to the abovementioned subterranean habitats, although with different levels of adaptation. Ptomaphagini, Anemadini, and Cholevini use subterranean habitats, either as troglophiles or, in some cases, as seasonal inhabitants, while Leptodirini includes highly troglobiontic forms. For Ptomaphagini, there are a few examples of subterranean dwellers in Italy. These include *Ptomaphagus sericatus* (Chaudoir, 1845), which is found frequently in cave entrances in Northern Italy (Vailati 1986); also, stable populations are known from caves in the province of Verona (Latella, unpublished data). *Ptomaphagus clavalis* Reitter, 1884 from Sardinia is considered troglophilic (Casale et al. 2009) and *P. varicornis* (Rosenhauer, 1847) can be found in burrows of small mammals (Jeannel 1936). Anemadini (e.g., *Anemadus acicularis* (Kraatz, 1852) and *Speonemadus orchesioides* (Fairmaire, 1879)) are frequently found in caves in central Italy and Sicily, respectively (Sbordoni et al. 1982); however, they are troglophilic.

Regarding Cholevini, many species occur in caves permanently or seasonally. *Catops speluncarum* Reitter, 1884 occurs in caves in Sardinia (Casale at al. 2009), while *Choleva cisteloides* (Flolich, 1799) is distributed throughout the Italian territory and is also cave-dwelling (Zoia and Latella 2006). *Choleva sturmi* Brisout, 1863 occurs in natural caves and artificial cavities from northern to southern Italy (Vailati 1986; Latella 1995). *Choleva sturmi* uses caves seasonally in Lombardy (Northern Italy; Vailati 1986); it is commonly found from fall through spring and is largely absent during the summer. Additionally, variations in seasonal use across species has also been documented. Specifically, Vailati (1988) reported seasonal variations of some pre-Alpine species across three habitat types: fissure networks, caves, and the MSS. In Lombardy, the author also demonstrated that some species of the genera *Cacciamallia* and *Ragazzonia* inhabited caves during the winter and the fissure network and MSS in autumn and spring.

In addition to seasonal variations, geology also influences the occurrence of Leptodirini in caves, MSS, fissures, and deep soil. In a recent study of Leptodirini across different habitats in Veneto, we documented seasonal differences in their distributions by geologic substrate (Latella, unpublished data). For example, in Lessini Mountains (Veneto, Verona province), species of the genera *Neobathyscia* and *Bathysciola* occurred seasonally in caves (season varies according to genus and species), MSS, and deep soil in an area characterized by highly fractured Jurassic limestone. In dolomitic limestone strata containing alteration material of fine-grain size (reducing the availability of micro- and mesocaverns and MSS), Leptodirini occurred in caves throughout the year (Latella, unpublished data).

Regarding habitat selection within caves, Latella et al. (2008) and Bernabò et al. (2011) examined the distribution of two stenothermal cold-dwelling Leptodirini, *Neobathyscia mancinii* and *N. pasai*. They found that both species appear to have different environmental thresholds to high temperature and survived short-term heat shock by developing a heat shock response (HSR) based on the synthesis of heat shock proteins (HSPs). Under environmentally stressful conditions, HSPs act as molecular chaperones to stabilize actively denaturing proteins, refold proteins that have already denatured, and direct irreversibly denatured proteins to the proteolytic machinery of the cell (Feder and Hofmann 1999). Also, under nonstressful conditions, HSPs facilitate the correct folding of proteins during translation and their transport across membranes (Lindquist and Craig 1988; Walsh et al. 1997; Hartl and Hayer-Hartl 2002). Among HSPs, the 70-kilodalton family, consisting of inducible (HSP70 coded by *hsp70* gene) and constitutive (heat shock cognate, HSC70 coded by *hsc70* gene) forms, is most studied in relation to thermal stress and has been found in all organisms investigated to date (Chen et al. 2018). In both *Neobathyscia mancinii* and *N. pasai*, *hsc70* mRNA levels were constant with increasing temperature, whereas a significant increase of the inducible member (*hsp70*) mRNA was observed in *N. pasai* (Bernabò et al. 2011). This difference might be due to their in-cave distribution: *N. pasai* colonizes the cave entrance where temperature is more variable than in the deeper reaches of the cave, where *N. mancinii* occurs (Lencioni et al. 2010). For the first time, these results highlighted the occurrence of an HSR in subterranean arthropods and suggested the correlation between the intensity of this response and the adaptation to the subterranean environment.

Concerning commensal species occurring within animal nests and burrows, species within the genera *Choleva* and *Catops* may occur exclusively within these habitats (Jeannel 1936; Sokolowsky 1942). For example, *Catops joffrei* Sainte-Claire Deville, 1927 in the Piedmont region of northwest Italy occurs in the burrows of marmots (*Marmota marmota* (Linnaeus, 1758)), while *Catops dorni* Reitter, 1913 uses mole (*Talpa europaea* Linnaeus, 1758) burrows throughout the peninsula and Sicily (Zoia and Latella 2006). Additionally, several Italian species of the Cholevinae genera *Attumbra, Philomessor, Catopsimorphus*, and *Parabathyscia* occur within the nests of *Messor, Aphaenogaster*, and *Lasius* ant species (Zoia and Latella 2006).

Diversity and Distribution

As would be expected, our knowledge of Italian Cholevinae diversity has continued to improve over time. Jeannel developed two global lists of Leptodirini (Jeannel 1924) and Cholevini (Jeannel 1924), which also included both species and collection localities. Through these efforts, he enumerated over 800 species and subspecies globally. Later, Sbordoni et al. (1982) published an overview of the cave-dwelling Italian species, where they reported distributional information on 161 species and provided several taxonomic revisions. The first actual checklist of all Italian Cholevinae, which was part of the project "*Checklist delle specie della fauna italiana*," enumerated 46 genera and 238 species (Angelini et al. 1995). About 10 years later, as part of the project "Checklist and Distribution of Italian Fauna (CKmap)," Zoia and Latella (2006) expounded upon these previous efforts and reported 47 genera and 251 species. Currently, an update to the checklist of Italian Cholevinae is underway by Pier Mauro Giachino and the author, which includes 63 genera and 285 species. Of these, 202 species belong to the Leptodirini tribe (71% of the known cholevine species). Regarding species endemic to Italy, there are at least 196 Cholevinae species with 183 species (or ~91%) occurring within the tribe Leptodirini.

To reiterate, the tribe Leptodirini boasts the richest diversity in subterranean species. Species of this tribe occur across all Italian administrative regions—although with varying degrees of troglomorphy. About three-quarters of the 202 species of Leptodirini are troglobionts. Of these, although some lack morphological adaptations to the subterranean environment, all are known only to hypogean systems. Including the 62.5% of known Italian subterranean species, the Prealps regions of Lombardy and Veneto represent the most diverse areas with 25 genera of Leptodirini—of which, 22 genera and at least 90 species are troglobionts (Sbordoni 1982; Holsinger 2000; Fig. 4.3).

With approximately 180 endemic species, the Alpine regions also supports a high diversity of troglobiontic cholevine beetles (Giachino and

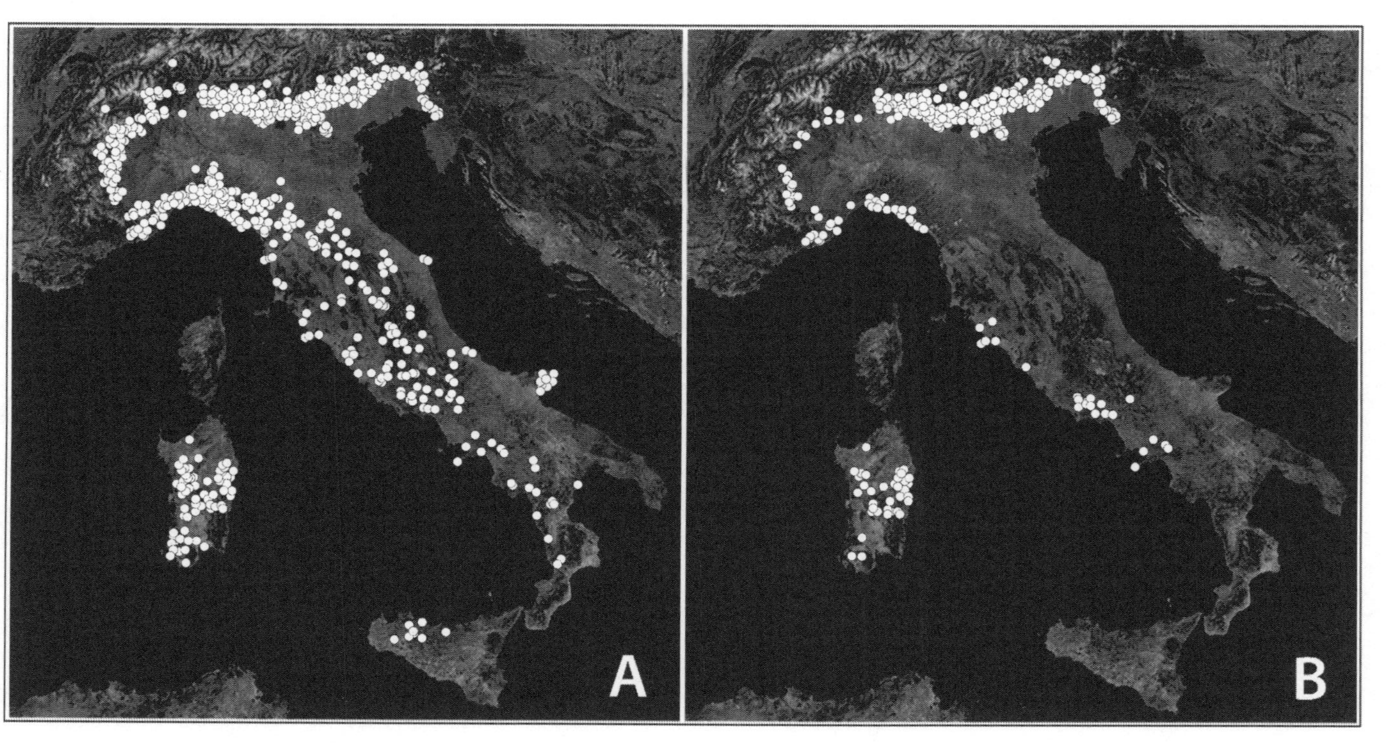

Figure 4.3. The Leptodirini tribe in Italy with [A] the distribution of all known species, and [B] subterranean-dwelling species only. Image from Google Earth, version 7.3.2 (June 2019); *Italy and surrounding environs*.

Vailati 2005; Zoia and Latella 2006; Giachino et al. 2011). More than half (or 97 species) are identified as highly specialized to the subterranean environment (see Table 4.1). Other regions of endemic diversity, listed in rank order, are Apennines (61 species), Sardinia (21 species), and Sicily (3 species; Zoia and Latella 2006; Giachino and Vailati 2007; Casale et al. 2009; Giachino et al. 2011).

Regional lists have been compiled for the following areas. In the Central Eastern Alps, 27 genera and 131 species are listed, which include about 100 subterranean-dwelling species (Giachino and Vailati 2005). Latella (2015b) identified nine genera and 21 species in the Maritime Alps; of these, only five were considered subterranean dwelling. Fourteen genera with 78 species have been confirmed in the Central and Northern Apennine regions (Giachino and Vailati 2007), while Sicily supports at least 12 genera and 35 species (Giachino et al. 2011) and Sardinia contains 11 genera and at least 26 species (Casale et al. 2009).

Along the Italian-Slovenian border, the karst area of Friuli Venezia Giulia supports five highly troglobiontic species. Additional subterranean-dwelling Leptodirini occurring in this area include *Bathysciotes khevenhulleri* (Miller, 1851), *Sphaerobathyscia hoffmanni* (Motschulsky, 1856), and two species of the genus *Pretneria*.

In the Western Alps (Piemonte and Liguria), five genera, including 20 troglobiontic species, have been cataloged. Lower diversity of troglobionts in this region is believed to be related to geology—as this region is composed primary of granitic-crystalline rocks, and karst is largely absent. The genus *Parabathyscia*, which occurs in both the Maritime and Ligurian Alps, is known to occur only within subterranean environments, while in the Apennines conspecifics are always endogean (Zoia 1986; Zoia and Latella 2006).

For the Apennines (primarily in the coastal mountainous region of pre-Apennines), all known subterranean-dwelling species occur within the polyphyletic genus *Bathysciola* (Latella et al. 2017). At least 15 subterranean species are distributed in caves and MSS in the Argentario and Uccellina Mountains (Tuscany) and the foothills of Latium and Campania, while the troglobiontic *Bathysciola raveli* (Dodero, 1904) is known from only one cave on the island of Capri (Campania; Zoia and Latella 2006). On Sardinia, three endemic subterranean genera and one subgenus, *Ovobathysciola*, *Patriziella*, *Sardostygia*, and *Speonomus* (*Batinoscelis*), consist of 15 species (Casale et al. 2009).

Future Research and Conservation

Due to the high rate of endemism of Italian Cholevinae, especially within the largely troglobiontic tribe of Leptodirini, this group represents an important component of Italy's biological diversity. Most species of subterranean Leptodirini are considered short-range endemics and are typically known from only a few caves or a specific karst area. Subsequently, they are sensitive to human activities, including the destruction of karst due to mining or quarrying activities, urbanization and deforestation, the pollution of subterranean waters, and overextraction of aquifers (Casale 2000; Faille 2019; Howarth and Wynne, Chap. 1).

At present, we know little regarding the ecology and population dynamics of many of the Leptodirini species. Modern studies on diet (Paoletti et al. 2013; Latella et. al 2017) and habitat selection (Latella et al. 2008) have begun to demonstrate connections between habitat use, climatic factors, and nutrient resources. However, this information is lacking for most species. Except for two studies (Sbordoni 1982; Latella et al. 2017), we still know too little concerning the demography, specific distributions, and population dynamics for most species. Even the reactions to environmental disturbances and the factors influencing seasonal movements between and within underground habitats are still little studied (Vailati 1988; Lencioni et al. 2010; Mammola et al. 2015, 2016).

Importantly, ecological data, in addition to ongoing research on the bioaccumulation of certain contaminants (such as pesticides and microplastics), will ultimately be required to identify whether Leptodirini species are viable as biological-indicator species. To date, Latella (unpublished data) has identified the presence of microplastics in the gut of some Leptodirini species. These preliminary data provide provisional support for the sensitivity of underground environments to external pollutants and the importance of troglobionts as potential indicators of environmental change in subterranean habitats. While these results are indeed compelling, additional studies will be required to better measure and ultimately quantify the impacts of anthropogenic contamination and the responses of Leptodirini species to subterranean contamination.

Similar to most countries globally (Mammola et al. 2019), Italy lacks the environmental legislation needed to protect subterranean fauna and ecosystems. However, some regions and protected areas have regulations in place concerning the collection of species occurring within caves and other subterranean habitats. For example, the Friuli Venezia Giulia, Marche, and Abruzzo regions have passed laws to protect cave-dwelling fauna and require research permits for specimen collection. Presently, only one Italian Cholevinae species, *Leptodirus hohenwarthi*, is included in the European Habitats Directive (Trizzino et al. 2013)—the primary legislative agreement on Europe's nature conservation policy (refer to Council of the European Communities 1992 for more information). However, many other troglobiontic Cholevinae species are short-range endemics with highly specific habitat requirements; thus, these species are expected to be highly vulnerable to environmental and climatic changes. As the European Habitats Directives species list is in need of revision due to potential biases related to the listing of arthropod species (refer to Cardoso 2012), more troglobiontic Italian cholevine species will likely be added once they have been properly assessed.

As also reported by Ober et al. (Chap. 5) for cave trechine beetles (family Carabidae) of eastern North America, none of the Italian Cholevinae species have been evaluated using the International Union for Conservation of Nature's criteria in the *IUCN Red List of Threatened Species*. Minimally, Red List assessments for short-range endemic troglobiontic species should be initiated as a first step to both elevate the global importance of this family and to provide some of the information necessary to develop evidence-based cave conservation policy in Italy.

ACKNOWLEDGMENTS

The author would like to thank Stewart Peck, Valeria Lencioni, and the editor for providing useful comments and advice leading to the improvement of this chapter. Roberta Salmaso, Andrea Latella, and Marcel Clemens provided images for some of the figures.

REFERENCES

Accordi, Fiorenza, and Valerio Sbordoni. 1978. "The Fine Structure of Hamann's Organ in *Leptodirus hohenwarti*, a Highly Specialized Cave Bathysciinae (Coleoptera, Catopidae)." *International Journal of Speleology* 9:153–165.

Accordi, Fiorenza, Mauro Rampini, and Valerio Sbordoni. 1980. "Ultrastruttura della Cuticola e Falsa Fisogastria nei Batiscini Cavernicoli." In *Atti XII Congresso Naz. Entomologia*, 79–86. Roma: Mario Sticca.

Angelini, Fernando, Paolo Audisio, Giorgio Castellini, Roberto Poggi, Dante Vailati, Adriano Zanetti, et al. 1995. "Coleoptera Polyphaga II (Staphylinoidea escl. Staphylinidae)." In *Checklist delle specie della fauna italiana*, edited by Alessandro Minelli, Sandro Ruffo, and Sandro La Posta, 47, 1–39. Bologna: Calderini ed.

Antunes-Carvalho, Caio, Ignacio Ribera, Rolf G. Beutel, and Pedro Gnaspini. 2019. "Morphology-Based Phylogenetic Reconstruction of Cholevinae (Coleoptera: Leiodidae): A New View on Higher-Level Relationships." *Cladistics* 35:1–41.

Barr, Thomas C., Jr., 1968. "Cave Ecology and the Evolution of Troglobites." *Evolutionary Biology* 2:35–102.

Barr, Thomas C., Jr., and John R. Holsinger. 1985. "Speciation in Cave Faunas." *Annual Review of Ecology and Systematics* 16:313–337.

Bernabò, Paola, Leonardo Latella, Olivier Jousson, and Valeria Lencioni. 2011. "Cold Stenothermal Cave-Dwelling Beetles Do Have an HSP70 Heat Shock Response." *Journal of Thermal Biology* 36:206–208.

Boccaletti, Mario, Neri Ciaranfi, and Domenico Casentino. 1990. "Palinspastic Restoration and Paleogeographic Reconstruction of the Peri-Tyrrhenian Area during the Neogene." *Palaeogeography, Palaeoclimatology, Palaeoecology* 77:41–50.

Bognolo, Mario, and Dante Vailati 2010. "Revision of the Genus *Aphaobius* Abelle de Perrin, 1878 (Coleoptera, Cholevidae, Leptodirinae)." *Scopolia* 68:1–75.

Cacchio, Paola, Gianluca Ferrini, Claudia Ercole, Maddalena Del Gallo, and Aldo Lepidi. 2014. "Biogenicity and Characterization of Moonmilk in the Grotta Nera (Majella National Park, Abruzzi, Central Italy)." *Journal of Cave and Karst Studies* 76:88–103.

Caccone, Adalgisa, and Valerio Sbordoni. 2001. "Molecular Biogeography of Cave Life: A Study Using Mitochondrial DNA from Bathysciine Beetles." *Evolution* 55:122–130.

Cardoso, Pedro. 2012. "Habitats Directive Species Lists: Urgent Need of Revision." *Insect Conservation and Diversity* 5:169–174.

Casale, Achille. 2000. "Impatto Antropico e Biomonitoraggio in Ecosistemi Sotterranei." *Atti e Memorie dell'Ente Fauna Siciliana* 6:61–76.

Casale, Achille. 2014. "Il genere *Ovobathysciola* Jeannel, 1924, Endemico di Sardegna, con Descrizione di Cinque Nuove Specie (Coleoptera, Cholevidae, Leptodirini)." *Annali del Museo Civico di Storia Naturale "G. doria"* 106:223–289.

Casale, Achille, Pier Mauro Giachino, Dante Vailati, and Mauro Rampini. 1991. "Note Sulla Linea Filetica di *Phaneropella* Jeannel, 1910 con Descrizione di Tre Nuovi Sottogeneri e di Una Nuova Specie di Turchia (Coleoptera, Cholevidae, Bathysciinae)." *Natura Bresciana: Annuario del Museo Civico di Storia Naturale di Brescia* 26:197–222.

Casale, Achille, Giuseppe Grafitti, and Leonardo Latella. 2009. "The Cholevidae of Sardinia (Coleoptera)." *Zootaxa* 2318:290–316.

Chen, Bing, Martin E. Feder, and Le Kang. 2018. "Evolution of Heat-Shock Protein Expression Underlying Adaptive Responses to Environmental Stress." *Molecular Ecology* 27:3040–3054.

Christiansen, Kenneth. 2012. "Morphological Adaptations." In *Encyclopedia of Caves*, 2nd ed., edited by David C. Culver and William B. White, 517–528. Amsterdam: Elsevier, Academic Press.

Cipollari, Paola, and Domenico Cosentino. 1995. "Il Sistema Tirreno-Appennino: Segmentazione Litosferica e Propagazione del Fronte Appenninico." *Studi Geologici Camerti* 2:125–134.

Council of the European Communities. 1992. “Council Directive 92/43/EEC of 21 May 1992 on the Conservation of Natural Habitats and of Wild Fauna and Flora.” *Official Journal of the European Communities* 35:7–50.

Crowson, Roy A. 1955. *The Natural Classification of the Families of Coleoptera*. London: N. Lloyd Press.

Culver, David C. 1982. *Cave Life: Evolution and Ecology*. Cambridge, MA: Harvard University Press.

Delay, Bernard. 1978. “Milieu Souterrain et Écophysiologique de la Reproduction et du Développement des Coléoptères Bathysciinae Hypogés.” *Mémoire de Biospéologie* 5:1–349.

Deleurance-Glaçon, Sylvie 1961. “Morphologie des Larves de *Royerella tarissani* Bedel et *Leptodirus hohenwarti* Schm.” *Annales de Spéléologie* 16:193–198.

Deleurance-Glaçon, Sylvie 1963. “Recherches sur les Coléoptères Troglobies de la Sous-famille des Bathysciinae.” *Annales des Sciences Naturelles, Zoologie* 12:1–172.

De Matthaeis, Elvira, Valerio Sbordoni, Marco Mattoccia, Giuliana Allegrucci, Adalgisa Caccone, Donatella Cesaroni, et al. 1983. “Struttura Genetica di Popolazioni Cavernicole di *Bathysciola derosasi* (Col. Cat.): Bottleneck e Divergenza Genetica.” *Atti XII Congresso Nazionale di Entomologia, Roma* 1983:253–254.

Faille, Arnaud, 2019. “Beetles.” In *Encyclopedia of Caves*, 3rd ed., edited by William B. White, David C. Culver, and Tanja Pipan, 102–108. Amsterdam: Elsevier, Academic Press.

Feder, Martin E., and Gretchen E. Hofmann. 1999. “Heat-Shock Proteins, Molecular Chaperones, and the Stress Response: Evolutionary and Ecological Physiology.” *Annual Review of Physiology* 61:243–282.

Fresneda, Javier, Vasily V. Grebennikov, and Ignacio Ribera. 2011. “The Phylogenetic and Geographic Limits of Leptodirini (Insecta: Coleoptera: Leiodidae: Cholevinae), with a Description of *Sciaphyes shestakovi* sp. n. from the Russian Far East.” *Arthropod Systematics & Phylogeny* 69:99–123.

Giachino, Pier Mauro. 1990. “Note Sulle *Bathysciola* di Sardegna. Sistematica, Corologia e Zoogeografia delle Specie Affini a *B. damryi* (Abeille, 1881) (Col., Cholevidae, Bathysciinae).” *Annali Museo Civico di Storia Naturale Giacomo Doria* 88:301–329.

Giachino, Pier Mauro, and Dante Vailati. 1993. “Revisione degli Anemadinae Hatch, 1928 (Coleoptera Cholevidae).” *Monografie di Natura Bresciana* 18:1–314.

Giachino, Pier Mauro, and Dante Vailati. 2005. “I Cholevidae Alpi e Prealpi Italiane. Inventario, Analisi Faunistica e Origine del Popolamento nel Settore Compreso Fra i Corsi del Fiume Ticino e Tagliamento (Coleoptera).” *Biogegraphia, Lavori della Società Italiana di Biogeografia* 26:229–378.

Giachino, Pier Mauro, and Dante Vailati. 2007. “I Coleotteri Colevidi dell’Appennino Settentrionale e Centrale: Inventario, Analisi Faunistica e Origine del Popolamento (Coleoptera Cholevidae).” *Biogegraphia, Lavori della Società Italiana di Biogeografia* 28:365–420.

Giachino, Pier Mauro, and Dante Vailati. 2010. *The Subterranean Environment. Hypogean Life, Concepts and Collecting Techniques*. Verona: WBA Handbooks 3.

Giachino, Pier Mauro, and Dante Vailati. 2019. “Il Genere *Bathysciola* Jeannel, 1910. Revisione della Sezione III (*pars*) (*sensu* Jeannel, 1924) con Descrizione di Sette Nuovi Generi (Coleoptera: Leiodidae: Cholevinae: Leptodirini).” *Memorie della Società Entomologica Italiana* 95:83–109.

Giachino, Pier Mauro, Dante Vailati, and Cosimo Baviera. 2011. "I Coleotteri Colevidi della Sicilia: Inventario, Analisi Faunistica e Origine del Popolamento (Coleoptera, Cholevidae)." *Biogeographia, Lavori della Società Italiana di Biogeografia* 30:467–484.

Hartl, F. Ulrich, and Manajit Hayer-Hartl. 2002. "Molecular Chaperones in the Cytosol: From Nascent Chain to Folded Protein." *Science* 295:1852–1858.

Hellwig, Johann Christian Ludwig. 1795. "Favna Etrvsca Sistens Insecta qvae in Provinciis Florentina et Pisana Praesertim Collegit Petrvs Rossivs." In *Regio Pisano Aethenaeo Publ. prof. et Soc. Ital.* Tomus primus XXVIII+ 457, IX pl. Helmstadii: C. G. Fleckeisen.

Holsinger, John R. 2000. "Ecological Derivation, Colonization, and Speciation." In *Ecosystems of the World: Subterranean Ecosystems*, edited by Horst Wilkens, David C. Culver, and William F. Humphreys, 399–415. Amsterdam: Elsevier.

Howarth, Francis G. 1983. "Ecology of Cave Arthropods." *Annual Review of Entomology* 28:365–389.

Hsu, Kenneth J., Lucien Montadert, Daniel Bernouilli, Maria Bianca Cita, Albert Erickson, Robert E. Garrison, et al. 1977. "History of the Mediterranean Salinity Crisis." *Nature* 267:399–403.

Humphreys, William F. 2000. "Relict Fauna and Their Derivation." In *Ecosystems of the World: Subterranean Ecosystems*, edited by Horst Wilkens, David C. Culver, and William F. Humphreys, 417–432. Amsterdam: Elsevier.

Jeannel, René. 1911. "Révision des Bathysciinae (Coleoptéres, Silphides). Morphologie, Distribution Géographique, Systématique." *Archives de Zoologie Expérimentale et Génèrale* 47:1–641.

Jeannel, René. 1914. "Sur la Systématique des Bathysciinae (Colèoptères, Silphides). Les Séries Phylétiques de Cavernicoles." *Archives de Zoologie Expérimentale et Génèrale* 54:57–78.

Jeannel, René. 1922. "La Variation des Piéces Copulatrices chez les Coléoptéres." *Comptes-Rendus de l'Académie des Sciences Naturelles* 174:324–327.

Jeannel, René. 1924. "Monographie des Bathysciinae." *Archives de Zoologie Expérimentale et Générale* 63:1–436.

Jeannel, René. 1936. "Monographie des Catopidae." *Mémoires du Museum National d'Histoire Naturelle de Paris* 1:1–433.

Juberthie, C. 1983. "Le Milieu Souterrain: Étendue et Composition." *Mémoires de Biospéologie* 10:17–65.

Juberthie, Christian, Bernard Delay, and Michael Bouillon. 1980. Extension du Milieu Souterrain Superficiel en Zone Non-Calcaire: Description d'un Nouveau Milieu et de son Peuplement par les Coleopteres Troglobies. *Mémoires de Biospéologie* 7:19–52.

Kistner, David H. 1979. "Social and Evolutionary Significance of Social Insect Symbionts." *Social Insects* 1:339–413.

Latella, Leonardo. 1995. "La Fauna Cavernicola dei Monti Lepini." *Notiziario del Circolo Speleologico Romano* 6–7:77–119.

Latella, Leonardo. 2011. "Classificazione, Evoluzione e Adattamenti degli Organismi Ipogei." In *Atti 48° Corso di III Livello di Biospeleologia*, edited by Inguscio S., Maurano F. Pertosa 9–11 aprile 2010.

Latella, Leonardo. 2015a. "Coleotteri Leiodidi Colevini." *Quaderni del Museo delle Scienze* 3:165–175.

Latella, Leonardo. 2015b. "The Cholevinae Kirby, 1837 (Coleoptera, Leiodidae) of the Maritime Alps." *Zoosystema* 37:595–604.

Latella, Leonardo, and Valerio Sbordoni. 2002. "The Underground World." In *Wildlife in Italy,* edited by Alessandro Minelli, Claudio Chemini, Roberto Argano, and Sandro Ruffo, 339–358. Rome: Touring Club and Ministry for the Environment and Territory.

Latella, Leonardo, and Fabio Stoch. 2001. "Biospeleologia." In *Grotte e Fenomeno Carsico. La Vita nel Mondo Sotterraneo. Quaderni Habitat*, 53–86. Ed. Ministero dell'Ambiente, Museo Friulano di Storia Naturale.

Latella, Leonardo, Paola Bernabo, and Valeria Lencioni. 2008. "Distribution Pattern and Thermal Tolerance in Two Cave Dwelling Leptodirinae (Coleoptera, Cholevidae)." *Subterranean Biology* 6:81–86.

Latella, Leonardo, Anna Castioni, Laura Bignotto, Elisa Salvetti, Sandra Torriani, and Giovanna E. Felis. 2017. "Exploring Gut Microbiota Composition of the Cave Beetles *Neobathyscia pasai* Ruffo, 1950 and *Neobathyscia mancinii* Jeannel, 1924 (Leiodidae; Cholevinae)." *Bollettino del Museo Civico di Storia Naturale di Verona, Botanica Zoologia* 41:3–24.

Latella, Leonardo, Valerio Sbordoni, and Giuliana Allegrucci. 2017. "Three New Species of *Bathysciola* Jeannel, 1910 (Leiodidae, Cholevinae, Leptodirini) from Caves in Central Italy, Comparing Morphological Taxonomy with Molecular Phylogeny." *Insect Systematics and Evolution* 49:409–442.

Latreille, Pierre André. 1796. *Précis des Caractéres Génériques des Insectes Disposés dans un Ordre Naturel*. Bourdeaux, France: Brive.

Lencioni, Valeria, Paola Bernabò, and Leonardo Latella. 2010. "Cold Resistance in Two Species of Cave-Dwelling Beetles (Coleoptera: Cholevidae)." *Journal of Thermal Biology* 35:354–359.

Lindquist, Suzan, and Elizabeth A. Craig. 1988. "The Heat-Shock Proteins." *Annual Review of Genetics* 22:631–677.

Lucarelli, Marco, and Valerio Sbordoni. 1978. "Humidity Responses and the Role of Hamann's Organ of Cavernicolous Bathysciinae (Coleoptera Catopidae)." *International Journal of Speleology* 9:167–177.

Lucarelli, Marco, and Valerio Sbordoni. 1986. "Life History Traits in the Cave Beetle *Bathysciola sisernica* (Coleoptera, Bathysciinae)." *International Journal of Zoology* 53:105.

Lucarelli, Marco, Giuseppe Sgro, and Valerio Sbordoni. 1980. "Ciclo Biologico in Laboratorio di Tre Popolazioni di Cavernicole di *Bathysciola derosasi* Jeannel (Coleoptera, Bathysciinae)." *Mémorie de Biospéologie* 7:319–332.

Mammola, Stefano, Pedro Cardoso, David C. Culver, Louis Deharveng, Rodrigo L. Ferreira, Cene Fišer, et al. 2019. "Scientists' Warning on the Conservation of Subterranean Ecosystems." *BioScience* 69:641–650.

Mammola, Stefano, Pier Mauro Giachino, Elena Piano, Alexandra Jones, Marcel Barberis, Giovanni Badino, et al. 2016. "Ecology and Sampling Techniques of an Understudied Subterranean Habitat: The *Milieu Souterrain Superficiel* (MSS)." *The Science of Nature* 103:88.

Mammola, Stefano, Elena Piano, Pier Mauro Giachino, and Marco Isaia. 2015. "Seasonal Dynamics and Micro-Climatic Preference of Two Alpine Endemic Hypogean Beetles." *International Journal of Speleology* 44:239–249.

Nae, Ioana, and Raluca I. Băncilă. 2017. "Mesovoid Shallow Substratum as a Biodiversity Hotspot for Conservation Priorities: Analysis of Oribatid Mite (Acari: Oribatida) Fauna." *Acarologia* 57:855–868.

Nevo, Eviatar. 1979. "Adaptive Convergence and Divergence of Subterranean Mammals." *Annual Review of Ecology and Systematics* 10:269–308.

Newton, Alfred Francis 1998. "Phylogenetic Problems, Current Classification and Generic Catalog of World Leiodidae (including Cholevidae)." In *Phylogeny and Evolution of Subterranean and Endogean Cholevidae (=Leiodidae Cholevinae)*, edited by Pier Mauro Giachino and Stewart B. Peck, 41–177. Florence, Italy: Proceedings of the XX International Congress of Entomology.

Noh, Mi Young, Subbaratnam Muthukrishnan, Karl J. Kramer, and Yasuyuki Arakane. 2016. "Cuticle Formation and Pigmentation in Beetles." *Current Opinion in Insect Science* 17:1–9.

Paoletti, Maurizio G., Mattia Beggio, Angelo Leandro Dreon, Alberto Pamio, Tiziano Gomiero, Mauro Brilli, et al. 2011. "A New Foodweb Based on Microbes in Calcitic Caves: The *Cansiliella* (beetles) Case in Northern Italy." *International Journal of Speleology* 40:45–52.

Paoletti, Maurizio G., Luca Mazzon, Isabel Martinez-Saeudo, Mauro Simonato, Mattia Beggio, and Angelo Leandro Dreon. 2013. "A Unique Midgut-Associated Bacterial Community Hosted by the Cave Beetle *Cansiliella servadeii* (Coleoptera: Leptodirini) Reveals Parallel Phylogenetic Divergences from Universal Gut-Specific Ancestors." *BMC Microbiology* 13:1–16.

Parker, Joseph. 2016. "Myrmecophily in Beetles (Coleoptera): Evolutionary Patterns and Biological Mechanisms." *Myrmecological News* 22:65–108.

Patrizi, Saverio. 1956. "Introduzione ed Acclimatazione del Coleottero Catopide Bathysciola derosasi Dod. in una Grotta Laziale." *Le Grotte d'Italia* 1:303.

Paykull, Gustav. 1798. *Fauna Suecica. Insecta*. Tomus I. Upsaliae: Joh. F. Edman.

Peck, Stewart B. 1986. "Evolution of Adult Morphology and Life History Characters in Cavernicolous *Ptomaphagus* Beetles." *Evolution* 40:1021–1030.

Peck, Stewart B., and Joyce Cook. 2007. "Systematics, Distributions, and Bionomics of the *Neoeocatops* gen. nov. and *Nemadus* of North America (Coleoptera: Leiodidae: Cholevinae: Anemadini)." *The Canadian Entomologist* 139:87–117.

Perreau, Michel. 1989. "De la Phylogénie des Cholevidae et des Familles Apparentées (Coleoptera, Cholevidae)." *Archives des Sciences, Genéve* 39:579–590.

Perreau, Michel. 2000. "Catalogue des Coléoptères Leiodidae Cholevinae et Platypsyllinae." *Mémoires de la Société Entomologique de France* 4:1–460.

Perreau, Michel 2004. "Leiodidae." In *Catalogue of Palaearctic Coleoptera*. Vol. 2, *Hydrophiloidea, Histeroidea, Staphylinoidea*, edited by Ivan Löbl and Ales Smetana, 133–203. Stenstrup: Apollo Books.

Perreau, Michel. 2015. "Leiodidae." In *Catalogue of Palaearctic Coleoptera. Hydrophiloidea - Staphylinoidea Revised and Updated Edition*, edited by Ivan Löbl and Daniel Löbl, 180–290. Leiden/Boston: Brill.

Perreau, Michel. 2019. "Phylogeny of Bathysciotina Guéorguiev, 1974, Based on Morphology with a Special Emphasis to Italian Genera and with the Description of a New

Species of Halbherria (Coleoptera: Leiodidae: Cholevinae: Leptodirini)." *Zootaxa* 4590:367–381.

Rabasović, Mihailo D., Dejan V. Pantelić, Brana M. Jelenković, Srećko Ćurčić, Maja S. Rabasović, Maja Vrbica, et al. 2015. "Nonlinear Microscopy of Chitin and Chitinous Structures: A Case Study of Two Cave-Dwelling Insects." *Journal of Biomedical Optics* 20:016010.

Racovitza, Emil G. 1980. "Étude Écologique sur les Coléoptères Bathysciinae Cavernicoles." *Mémoire de Biospéologie* 6:1–199.

Rossi, Renato 1976. "Biology and External Morphology of the Larvae of *Boldoria (Pseudoboldoria) bergamasca* Jeannel and of *B. malanchinii* Pavan." *Annales de Spéléologie* 31:253–262.

Ruffo, Sandro. 1955. "Le Attuali Conoscenze sulla Fauna Cavernicola della Regione Pugliese." *Memorie di Biogeografia Adriatica* 3:1–143.

Ruffo, Sandro, and Augusto Vigna Taglianti. 2002. "Generalità sulla Fauna Italiana." In *La Fauna sulla Italia*, edited by Alessandro Minelli, Claudio Chemini, Roberto Argano, and Sandro Ruffo, 24–28. Rome: Ministero dell'Ambiente e della Tutela del Territorio.

Salgado, José M., Marina Blas Esteban, and Javier Fresneda 2008. "Coleoptera Cholevidae." In *Fauna Iberica*, *31*, edited by M. A. Ramos, 1–799. Madrid: Museo Nacional de Ciencias Naturales, CSIC.

Sbordoni, Valerio. 1980. "Strategie Adattative negli Animali Cavernicoli: Uno Studio di Genetica Ecologia di Popolazione." *Accademia Nazionale dei Lincei* 377:61–100.

Sbordoni, Valerio. 1982. "Advances in Speciation of Cave Animals." In *Mechanisms of Speciation*, edited by Claudio Barigozzi, 219–240. New York: Alan R. Liss.

Sbordoni, Valerio, Giuliana Allegrucci, and Donatella Cesaroni. 2012. "Population Structure." In *Encyclopedia of Caves*, 2nd ed., edited by David C. Culver and William B. White, 608–618. Amsterdam: Elsevier.

Sbordoni, Valerio, Giuliana Allegrucci, Adalgisa Caccone, Donatella Cesaroni, Marina Cobolli Sbordoni, and Elvira De Matthaeis. 1981. "A Preliminary Report of the Genetic Variability in Troglobitic Bathysciinae: *Leptodirus hohenwarti* and Two *Orostygia* Species." *Fragmenta Entomologica* 15:327–336.

Sbordoni, Valerio, Giuliana Allegrucci, and Donatella Cesaroni. 2004. "Population structure." In *Encyclopedia of Caves*, edited by William B. White and David C. Culver, 447–455. Amsterdam: Elsevier, Academic Press.

Sbordoni, Valerio, Mauro Rampini, and Marina Cobolli Sbordoni. 1982. "Coleotteri Catopidi Cavernicoli Italiani." *Biogeographia, Lavori della Società Italiana di Biogeografia* 7:253–336.

Silfverberg, Hans. 1990. "The Nomenclaturally Correct Names of Some Family-Groups in Coleoptera." *Entomologica Fennica* 1:119–121.

Sokolowsky, Kurt. 1942. "Die Catopiden der Nordmark (Col. Catopidae), eine Faunistisch-Oeokologische Studie (Catopiden Studien IV)." *Entomologische Blatter fur Biologie und Systematik der Kafer* 38:173–209.

Trewick, Steve A. 2017. "Plate Tectonics in Biogeography." In *International Encyclopedia of Geography: People, the Earth, Environment and Technology*, edited by Douglas Richardson, Noel Castree, Michael F. Goodchild, Audrey Kobayashi, Weidong Liu, and Richard A. Marston, 1–9. Hoboken, New Jersey: Wiley.

Trizzino, Mario, Paolo Audisio, Francesco Bisi, Alessandro Bottacci, Alessandro Campanaro, Giuseppe Maria Carpaneto, et al. 2013. "Gli Artropodi Italiani in Direttiva Habitat: Biologia, Ecologia, Riconoscimento e Monitoraggio." *Conservazione Habitat Invertebrati* 7:1–255.

Vailati, Dante. 1986. "Coleotteri Catopidi e Colonidi della Provincia di Brescia." *Natura Bresciana* 21:153–185.

Vailati, Dante. 1988. "Studi sui Bathysciinae delle Prealpi Centro-occidentali. Revisione Sistematica, Ecologia, Biogeografia della 'Serie Filetica di *Boldoria*' (Coleoptera Catopidae)." *Monografie di Natura Bresciana* 11:1–331.

Vailati, Dante. 2017a. "Revisione Tassonomica delle 'Serie Filetiche di *Dellabeffaella* e di *Boldoria*' con Descrizione di Quattro Nuovi Generi (Coleoptera Cholevidae Leptodirinae)." *Bollettino Della Società Entomologica Italiana* 149:3–32.

Vailati, Dante. 2017b. "Una Nuova Specie del Genere *Halbherria* Conci & Tamanini, 1951 del Massiccio del Monte Baldo e Considerazioni sulla 'Barriera' Biogeografica della Valle dell'Adige (Coleoptera Cholevidae Leptodirinae)." *Bollettino della Società Entomologica Italiana* 149:105–118.

van Husen, Dirk. 2000. "Geological Processes during the Quaternary." *Mitteilungen der Österreichischen Geologischen Gesellschaft* 92:135–156.

Walsh, David A., Marshall J. Edwards, and Murray Smith. 1997. "Heat Shock Proteins and Their Role in Early Mammalian Development." *Experimental & Molecular Medicine* 29:129–132.

Zoia, Stefano. 1986. "Il Genere *Parabathyscia* nell'Italia Settentrionale e in Toscana (Coleoptera, Catopidae, Bathysciinae)." *Fragmenta Entomologica* 18:329–418.

Zoia, Stefano. 1989. "Raccolte Entomologiche sul Monte Armetta (Italia: Liguria occidentale) per Mezzo di Trappole a Caduta: Catopidae e Silphidae (Coleoptera)." *Bollettino della Società Entomologica Italiana* 121:181–195.

Zoia, Stefano. 1996. "Considerations of the Present Knowledge of the Italian Cholevidae and Their Distribution, with Particular Reference to the Hypogean Species (Coleoptera)." In *Phylogeny and Evolution of Subterranean and Endogean Cholevidae (=Leiodidae Cholevinae)*, edited by Pier Mauro Giachino and Stewart B. Peck, 211–226. Museo Regionale di Scienze Naturali, Torino, Atti (Proceedings of) XX ICE, Firenze.

Zoia, Stefano, and Leonardo Latella. 2006. "Insecta Coleoptera Cholevidae and Platipsyllidae." In *Checklist and Distribution of the Italian Fauna*, edited by Sandro Ruffo and Fabio Stoch, 177–180. Memorie del Museo Civico di Storia Naturale di Verona, 2. Serie, Sezione Scienze della Vita, 17.

Zwick, Peter. 1979. "Contribution to the Knowledge of Australian Cholevidae (Catoptidae auct.: Coleoptera)." *Australian Journal of Zoology*, Supplementary Series 27:1–56.

5

Cave Trechine (Coleoptera: Carabidae) Radiation and Biogeography in Eastern North America

Karen A. Ober, Matthew L. Niemiller, and T. Keith Philips

Introduction

Caves are home to unique and diverse communities. Subterranean habitats are regarded as highly endangered ecosystems (Mammola et al. 2019), with 95% of obligate cave species in the United States considered to be "vulnerable" or "imperiled" by the Nature Conservancy (Culver et al. 2000). However, it is likely more are now imperiled, as this estimate is at least 20 years old. Species that complete their life cycles underground, cannot exist on the surface, and have morphological and physiological characters consistent with life underground are known as subterranean-adapted species (or troglobionts). These organisms are typically located within the deep cave zone, which is characterized by the absence of light, relatively constant temperature, and high relative humidity accompanied by a low rate of evaporation (Howarth 1980, 1982; Howarth and Wynne, Chap. 1). Therefore, troglobionts often exhibit traits correlated with these environmental factors, termed "troglomorphy" (Christiansen 1961), such as reduction or loss of eyes and pigmentation, elongated appendages and bodies, loss of flight wings and elongation of tactile setae in insects, and even slower metabolism compared with their surface relatives. These species typically show striking convergent evolution in morphological structures (Culver et al. 1990) that are often considered adaptations to a troglobiontic lifestyle (Darwin 1859; Culver 1982; Culver and Pipan 2019; Romero 2009). Troglobionts are frequently considered "super specialists" that cannot survive outside a narrow environmental threshold—characterized by the deep zone habitat (Culver and Pipan 2019).

Biologists have been interested in the origin and evolution of troglobionts for 250 years, beginning with the discovery of the first-known stygobiont, an aquatic salamander known as the olm (*Proteus anguinus*) described by Laurenti (1768). In addition to being valuable for understanding the processes of ecological and morphological adaptation, many troglobionts are excellent models for studying speciation and diversification (Kowalko et al. 2013; Soares and Niemiller 2013; Wynne et al., Chap. 2).

In this chapter, we review the diversity and distribution of troglobiontic trechine beetles of eastern North America in the context of regional geology and hydrology. Specifically, we (1) summarize the features of two major regions of cavernous karst in eastern North America, (2) introduce a speciose lineage of troglobiontic ground beetles that is an important component of subterranean ecosystems, (3) explore some biogeographic hypotheses that may have shaped the current distribution and diversity of trechine cave beetles, and (4) provide recommendations for areas of future research, as well as provide some ideas for the conservation of this lineage of subterranean beetles.

Eastern North America Karst Areas

In eastern North America, troglobionts are primarily associated with two major karst regions (Fig. 5.1), the Appalachian Ridge and Valley, also called Appalachians (APP), and the Interior Low Plateau (ILP; Barr and Holsinger 1985; Culver et al. 2000). The APP is approximately 37,000 km^2 extending from southeastern New York southwest to northeastern Alabama and consists primarily of a series of parallel sandstone ridges with intervening valleys of folded and faulted shales and carbonates. The ILP is approximately 61,000 km^2 encompassing a large region west of the Cumberland Plateau through southern Indiana and parts of Kentucky, Tennessee, and northern Alabama. The APP and ILP karstic regions are proximate to one another and come in contact near the junction of the borders of Tennessee, Alabama, and Georgia; however, the exact boundary between the APP and ILP is somewhat arbitrary. These karstic regions are highly fractured, support numerous caves (approximately 7,441 caves in the APP and 11,928 caves in the ILP; Culver et al. 2003), and thus support the greatest troglobiontic faunal diversity in the United States (Culver et al. 2003; Hobbs 2012; Niemiller et al. 2019).

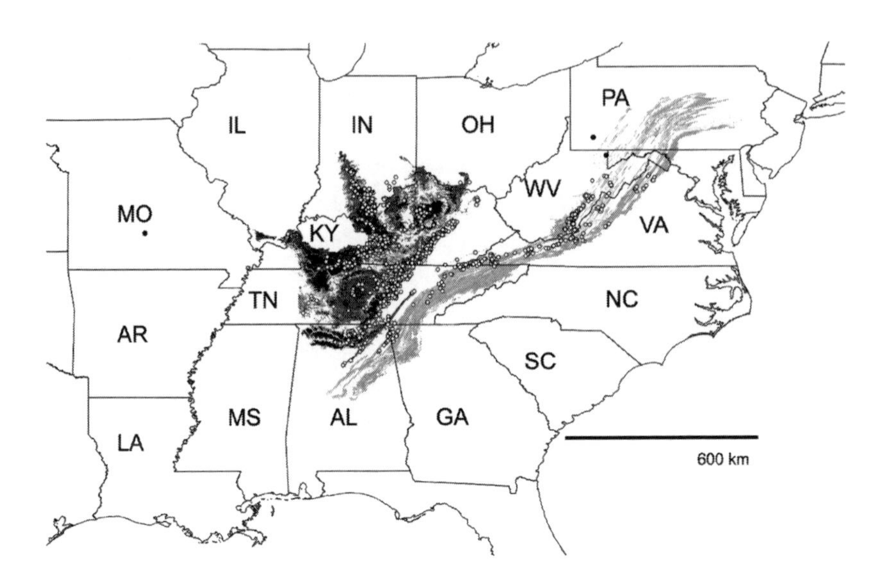

Figure 5.1. Relevant area of the Appalachian Ridge and Valley karst region (*gray region extending from northeastern Alabama into Pennsylvania*) and Interior Low Plateau karst region (*black area extending from northern Alabama north through Indiana and Ohio*) of eastern North America (e.g., Culver et al. 2003; Weary and Doctor 2014). White dots indicate locations of described cave trechine species of the genera *Pseudanophthalmus*, *Neaphaenops*, *Darlingtonea*, *Ameroduvalius*, and *Nelsonites* (e.g., Barr 2004), as well as unpublished occurrence records. Black dots indicate undescribed *Pseudanophthalmus* species from Maryland, Pennsylvania, and Missouri.

The geologic character of the APP is quite different from the ILP. APP sedimentary rock strata are folded and faulted in intricate complex patterns. As a result, narrow linear bands of cavernous limestone (primarily Cambrian and Ordovician in age with some Silurian, Devonian, and Mississippian strata) stretch out in parallel valleys separated by sandstone, shale, or noncarbonate ridges that were uplifted during the Permian period (Hack 1969; Barr 1985a). Caves in the APP are structured with major passages oriented along the narrow limestone bands (Barr 1967b). ILP geology is characterized by vast expanses of thick, flat-bedded, relatively pure, highly soluble limestone, which is mostly Mississippian age but also contains Ordovician, Silurian, and Devonian strata (Barr 1967b). Largely honeycombed with caves characterized by sinuous branch-like passages (Barr 1967b), the ILP is comprised of four subregions: (1) the Bluegrass Region in

Kentucky, (2) the western escarpment of the Cumberland Plateau from northeastern Kentucky to northern Alabama, (3) the Central Basin in Tennessee, and (4) the Pennyroyal Plateau from southern Indiana near Bloomington extending westward into Kentucky and north-central Tennessee. The Pennyroyal Plateau, Central Basin, and the western escarpment of the Cumberland Plateau are highly caverniferous with extensive cave systems that may have facilitated dispersal of terrestrial troglobionts (Barr 1985a). In contrast, the Bluegrass Region in northern Kentucky near the Ohio and Indiana borders contains small and isolated caves (Barr 1967b), which may limit subterranean dispersal of troglobionts.

Historic geologic and climatic processes in eastern North America influence the diversity and distribution of troglobiontic fauna. Troglobionts are confined to their subterranean habitat and inevitably their distributions reflect the area's paleogeography and paleoecology. For example, large rivers, such as the Ohio River along the Indiana-Ohio-Kentucky border, serve as a biogeographic barrier, which has facilitated the divergence, isolation, and speciation of troglobiontic populations (Niemiller et al. 2013). Other barriers, such as unsuitable habitat (i.e., substratum not suitable for troglobiontic fauna) may be major factors driving isolation between species of troglobionts, resulting in disjunct distributions. Within cavernous limestone that is porous and permeable, small-bodied troglobionts may disperse to surrounding subterranean habitats through networks of subterranean fissures or cavities (i.e., mesocaverns; Howarth 1982, 1983).

Cave Trechine Beetles

Morphology and Biology

Cave trechines in eastern North America are tiny beetles (approximately 3.0 to 7.5 mm long) that lack eyes and flight wings (Plate 5.1). Relative to other trechines, they are typically depigmented with long, slender bodies, elongated appendages, and sensory setae. These beetles generally resemble "aphaenopsian" cave trechines from Europe and Asia (Jeannel 1943; Barr and Holsinger 1985; Culver et al. 1990).

Most cave trechines are difficult to differentiate, as many species have converged on a shared set of morphological adaptations. Historically, taxonomists relied on the morphological characteristics of the male genitalia to distinguish between species (e.g., Valentine 1948; Barr 1965, 1981, 2004). Barr (2004) and Barr and Holsinger (1985) arranged the species and species groups of *Pseudanophthalmus*, as well as *Neaphaenops tellkampfii* (Erichson, 1844), *Darlingtonea kentuckensis* Valentine, 1952, *Ameroduvalius jeanneli* Valentine, 1952, *Nelsonites jonesei* Valentine, 1952, and *Nelsonites walteri* Valentine, 1952 based primarily on morphological features, but also on their distributions in karst regions. Morphological characteristics included differentiation of the male genitalia and meso- and abdominal ventrites, as well as the dorsal surface puncture and setal shapes and positions. Although these characters have not been tested in any rigorous phylogenetic study, some of these groups likely represent natural or monophyletic clades.

All cave trechines (like many carabid beetles) are predatory, likely feeding on arthropod eggs and juveniles, annelids, and other macroarthropods. However, little is known about the prey and feeding behavior of cave trechines. Barr (1968) observed members of *Neaphaenops tellkampfii* preying on cave cricket eggs, which they extract from damp silt on ledges and the cave floor with their elongated mandibles. Species within the genus *Pseudanophthalmus* and *A. jeanneli* feed on minute tubificid and enchytraeid worms that burrow in the silt along cave streams (Barr 1968). These beetles are often abundant among worm castings and have been observed carrying partially eaten worms in their mandibles (Barr 1968).

Diversity and Distribution

The vast majority of described cave beetles belong to the family Carabidae, tribe Trechini. About half of the more than 2,000 species of trechines (in numerous lineages) worldwide have colonized subterranean habitats and occur primarily in temperate and subtropical climates (Jeannel 1926, 1927, 1928, 1930, 1941; Barr 1965; Casale et al. 1998; Moravec et al. 2003; Uéno 2005, 2006; Townsend 2010; Fang et al. 2016). In North America, more than two-thirds of all trechine beetle species are troglobionts.

Cave trechine beetles represent the dominant terrestrial troglobionts in eastern North America. Six genera and 155 species are considered cave-obligate inhabitants; the largest genus, *Pseudanophthalmus*, contains 148 described species and is arranged into 26 species groups of “apparently related” species (Barr 2004). However, there are likely more than 80 undescribed species (Peck 1998; Barr 2004). *Pseudanophthalmus* species are found in caves of the APP and ILP karst regions in Alabama, Georgia, Illinois, Indiana, Kentucky, Maryland (based on a single undescribed species), Ohio, Pennsylvania (based on a single undescribed species), Tennessee,

Virginia, and West Virginia (Barr 2004), as well as in the Ozark karst region of Missouri (based on a single undescribed species; Elliott 2007). Five additional trechine genera composed entirely of troglobiontic species exist in eastern North America, but unlike *Pseudanophthalmus*, diversity is low across these genera. Three genera contain a single species each, *Neaphaenops tellekampfi* (Erichson, 1844), *D. kentuckensis*, and *A. jeanneli*. *Nelsonites* and *Xenotrechus* each contain two species (*Nelsonites jonesei* (Valentine, 1952), *Nelsonites walteri* (Valentine, 1952), *Xenotrechus condei* (Barr & Krekeler, 1967), and *Xenotrechus denticollis* (Barr & Krekeler, 1967)). *Neaphaenops tellkampfii*, *D. kentuckensis*, *A. jeanneli*, *N. jonesei*, and *N. walteri* have been assumed to be closely related to *Pseudanophthalmus* (Valentine 1952; Barr 1972, 1980, 1981, 1985b; Maddison et al. 2019; Fig. 5.2). However, the phylogenetic relationships among these five genera are unclear, in part, due to limited sampling of species within *Pseudanophthalmus*. Preliminary molecular phylogenetic studies of eastern North American cave trechines suggest these genera are derived from *Pseudanophthalmus* (Niemiller, unpublished data).

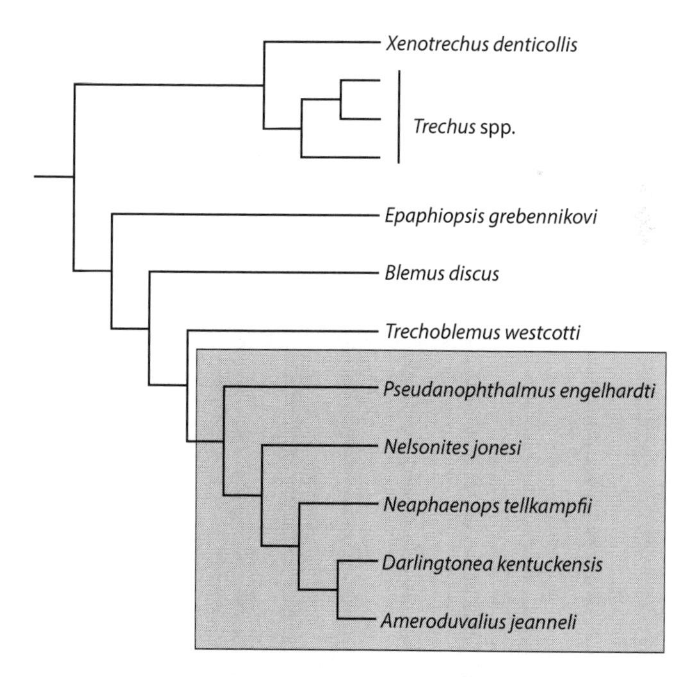

Figure 5.2. Hypothesized phylogenetic tree for the five genera of eastern North America cave trechines (*gray box*) in the *Trechus* assemblage. Redrawn from Maddison et al. (2019).

The ranges of the aforementioned four species-poor genera are relatively widespread and include the Pennyroyal Plateau (*N. tellkampfii*), the western escarpment of the Cumberland Plateau (*D. kentuckensis*, *A. jeanneli*, *N. walteri*, and *N. jonesei*) in the ILP of eastern Kentucky and north-central Tennessee. Notably, all these genera have distributions overlapping with different *Pseudanophthalmus* species and often co-occur in the same cave systems. In contrast, the genus *Xenotrechus*, known from a few caves in eastern Missouri west of the Mississippi River (Barr and Krekeler 1967), is thought to be taxonomically distant from *Pseudanophthalmus* and other related genera. Specifically, *Xenotrechus* is hypothesized to be related to the eastern European genera *Chaetoduvalius* and *Geotrechus* but lacks any close relatives in North America (Barr and Krekeler 1967).

The *Pseudanophthalmus* group, consisting of five troglobiontic genera, belongs to the *Trechoblemus* series in the *Trechus* assemblage (Jeannel 1926, 1927, 1928, 1930, 1949). However, *Trechoblemus* is a predominately Eurasian genus of surface-dwelling, winged trechines; only a single species, which occurs only in the Willamette Valley of Oregon (Barr 1972), is known from North America. Aside from the possibility of *Pseudanophthalmus* (with *P. sylvaticus* occurring within the deep soil of a high elevation spruce-fir forest in West Virginia; Barr 1967a), there are no obvious surface-dwelling sibling group or ancestral taxon for *Trechoblemus* in North America.

Biogeographic and Evolutionary Patterns

Biogeography

The pattern of cave trechine species ranges differ in the APP versus the ILP. Broadly speaking, the species groups of *Pseudanophthalmus* are restricted to either the APP or the ILP karst region, while *N. tellkampfii*, *D. kentuckensis*, *A. jeanneli*, *N. jonesei*, and *N. walteri* occur only in the ILP (Table 5.1). In the APP, many *Pseudanophthalmus* species are restricted to a single cave or cave system (Barr 1965, 1981). Of the 64 described *Pseudanophthalmus* species from the APP, 22 are considered single-cave endemics. APP caves supporting these species typically occur within narrow bands of karst separated by shales and sandstone. This stands in contrast to the 12 of 84 ILP species endemic to single caves—as the ILP region typically contains much more contiguous karst and a presumed higher level of subterranean connectivity than the APP.

Table 5.1. Number of described cave trechine species including individually described species, with *Pseudanopthalmus* species groups presented *sensu* Barr (2004). Karst regions are defined by Culver et al. (2003) and watershed basins were identified using the US Geological Survey's watershed boundary map (Seaber et al. 1987). Eighty-one cave trechine species occur in the Interior Low Plateau (ILP) karst, while 72 species are found in the Appalachian Ridge and Valley (APP) karst.

Genus, species, and/or species group	Species (#)	Region	Watershed basin(s)
Ameroduvalius			
A. jeanneli Valentine, 1952	1	ILP	Upper Cumberland
Nelsonites			
N. jonesi Valentine, 1952	1	ILP	Upper Cumberland
N. walteri Valentine, 1952	1	ILP	Upper Cumberland
Darlingtonea			
D. kentuckensis Valentine, 1952	1	ILP	Upper Cumberland
Neaphaenops			
N. tellkampfii (Erichson, 1844)	1	ILP	Green
Pseudanopthalmus			
P. alabamae group	2	APP	Coosa-Tallapoosa
P. audax group	3	ILP	Middle Ohio–Raccoon; Patoka-White
P. barri group	2	ILP	Lower Ohio–Salt
P. cumberlandus group	10	ILP	Upper & Lower Cumberland
P. engelhardti group	20	APP & ILP Central Basin	Upper & Middle Tennessee; Coosa-Tallapoosa; Black Warrior-Tombigbee
P. eremita group	2	ILP	Lower Ohio–Salt
P. gracilis group	2	APP	James & Kanawha
P. grandis group	9	APP	Kanawha; Monongahela
P. hirsutus group	7	APP	Upper & Middle Tennessee
P. horni group	13	ILP	Middle Ohio–Little Miami; Kentucky; Patoka-White
P. hubbardi group	7	APP	Upper Tennessee
P. hubrichti group	6	APP	Upper Tennessee; Kanawha; French Broad–Holston
P. hypolithos group	5	APP	Kentucky; Upper Tennessee; Big Sandy; Upper Cumberland
P. inexpectatus group	6	ILP	Lower Ohio-Salt; Green
P. intermedius group	4	ILP	Middle Tennessee–Elk; Upper Cumberland
P. jonesi group	8	APP	Upper Tennessee

(*continued*)

Table 5.1. (continued)

Genus, species, and/or species group	Species (#)	Region	Watershed basin(s)
P. leonae group	1	ILP	Lower Ohio–Salt
P. menetriesi group	8	ILP	Green
P. petrunkevitchi group	3	APP	Kanawha; French Broad–Holston; Potomac
P. pubescens group	5	ILP	Green; Lower Cumberland
P. pusio group	6	APP	James; Roanoke
P. rittmani group	3	ILP	Kentucky
P. robustus group	4	ILP	Upper Cumberland
P. simplex group	2	ILP	Upper Cumberland
P. tennesseensis group	4	APP	Upper Tennessee
P. tenuis group	6	ILP	Lower Ohio–Salt; Patoka-White

An important caveat to mention is the limited accessibility of these subterranean habitats for collecting cave beetles. Importantly, cave trechines may be using widespread networks of underground voids and fissures that are inaccessible to humans. Researchers can only access this habitat through a limited number of caves. As a consequence, the designation of these animals as "single-cave endemics" may be due to limited collecting and sampling bias, rather than actual biogeography.

Barr (1985a) suggested the major ridges in the APP separated limestone valleys and prevented dispersal and subsequent further radiation of cave trechines. This was evidenced by morphologically similar allopatric species occurring on opposite sides of ridges within adjacent limestone valleys. For example, different species of the *P. hypolithos* species group appear restricted to parallel limestone bands separated by the Cumberland Mountain ridge. Specifically, *P. praetermissus* (Barr, 1981) occurs to the south with *P. scholasticus* (Barr, 1981), *P. calcareus* (Barr, 1981), *P. frigidus* (Barr, 1981), and *P. hypolithos* (Barr, 1981) distributed in the north (Fig. 5.3A). Similarly, the four species in the *P. tennesseensis* group of the APP of the eastern Tennessee region are not only restricted to separate bands of limestone (Weary and Doctor 2014) but are separated by the Clinch River (Fig. 5.3B).

Some new discoveries challenge the hypothesis that species distributions are typically small and restricted to the APP region. A recently discovered population of *P. hortulanus* (Barr, 1965) in Tazewell County, Virginia (Ober, unpublished data) suggests that conspecifics were encountered in

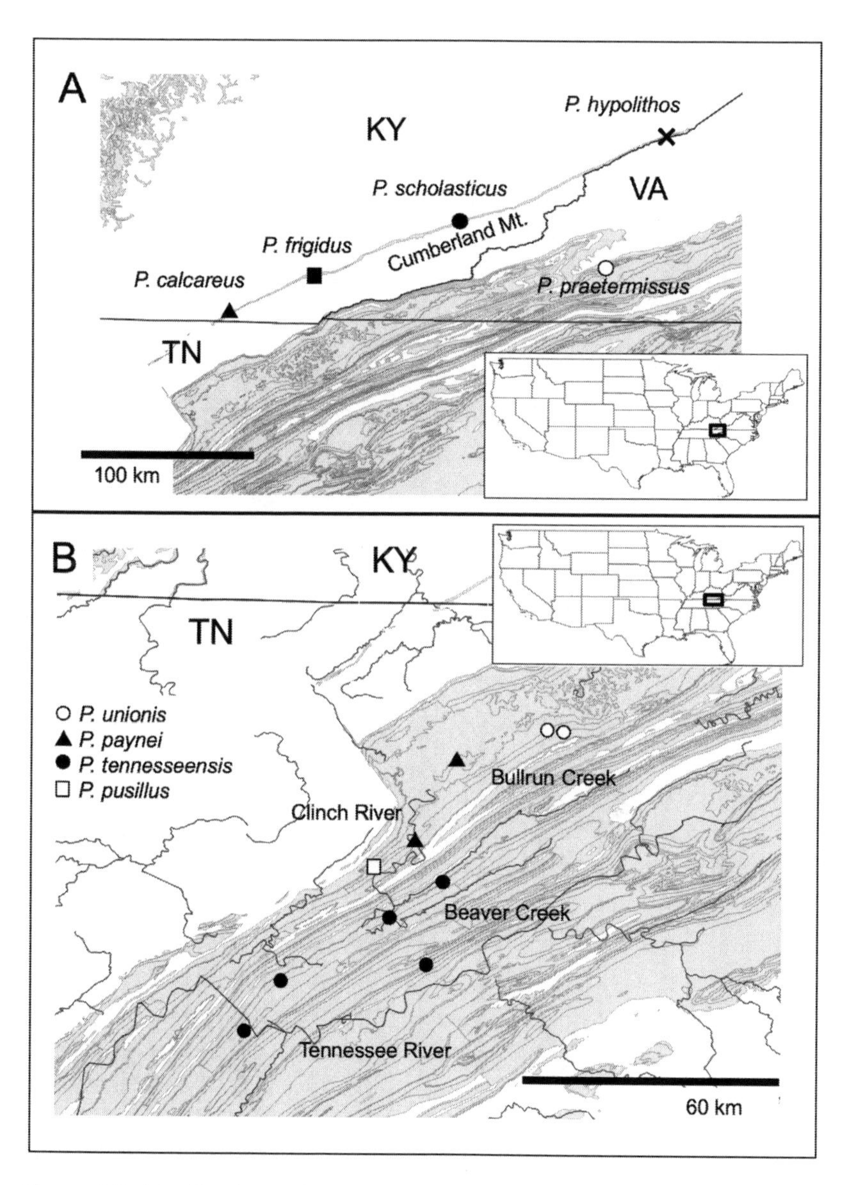

Figure 5.3. Distributions of cave trechines in the Appalachian karst region. [A] Members of the *Pseudanophthalmus hypolithos* group are found in narrow bands of limestone (*gray*). Separate bands of limestone are denoted with dark gray lines. Kentucky species are separated from Virginia species by an intervening ridge of nonkarst (Cumberland Mountains). [B] Species of the *Pseudanophthalmus tennesseensis* group occur in separate bands of limestone (*dark gray lines*) with species divided by the Clinch River and Bullrun Creek (*thin gray lines*).

the same APP karst formation. Previously known from only one cave in the Kanawha River watershed, a new population of *P. hortulanus* was found approximately 26 km to the west in the Upper Tennessee basin. While this population is separated by the Little River, it is within the same Moccasin or Bays karst formation as the previously known population (Weary and Doctor 2014). Conversely, another recently discovered population of *P. deceptivus* (Barr, 1981) in Scott County, Virginia (Ober, unpublished data) expands this species range approximately 21 km to the southeast to a second cave in Lee County, Virginia. These two populations occur in karst formations (Weary and Doctor 2014) separated by the Powell River.

The isolated and disjunct nature of karst in the APP (Barr 1981; Culver 1982) led Barr (1967b) to hypothesize that more cave trechine species occur per unit area of karst than in the ILP—although the ILP karst formation is more contiguous. In fact, Barr claimed the APP contained more than two to four times as many cave trechine species as the ILP per unit area. Based on karst area estimates of the APP and ILP (Culver et al. 2003) and the list of recognized trechine species (from Barr 2004 and Bousquet 2012), we estimated there are approximately 2.0×10^{-3} species per km^2 in the APP compared to 1.3×10^{-3} species per km^2 in the ILP. This estimate does not account for the many undescribed species identified by Barr (1967b, 2004), but it does suggest a higher species density in the APP where limestone exposures are more discontinuous, which may limit dispersal and gene flow across these isolated karst units.

The ILP has lower cave trechine species diversity and larger distributional ranges compared to the APP. ILP species inhabit extensive cave systems in large areas of relatively flat, highly caverniferous limestone (Barr 1967b, 1968, 1981, 1985a). More heterogeneous karst areas exist in the ILP, such as the Bluegrass Region, where there are small, isolated caves. Several species have either small ranges, including *P. packardi* (Barr, 1959) and *P. troglodytes* (Krekeler, 1973) (Barr 2004), or are considered single-cave endemics (e.g., *P. parvus* (Krekeler, 1973), *P. exoticus* (Krekeler, 1973), and *P. pholeter* (Krekeler, 1973) (Barr 2004)).

Overall, cave trechines in the ILP have more extensive ranges, and fewer species have been identified as endemic to a single cave. ILP species found in more than one cave typically have significantly larger ($p < 0.01$) distributions ($\bar{x} = 629.12$ km^2, SD $= 942.77$) than species found in more than one cave in the APP ($\bar{x} = 159.43$ km^2, SD $= 364.47$). For example, *Pseudanoph-*

thalmus bendermani (Barr, 1959), one of the few single-cave endemics in the Central Basin of the ILP, has recently been discovered in a second cave 93 km east of the type locality (Niemiller and Ober, unpublished data). However, both localities occur within the Nashville karst formation (Weary and Doctor 2014). This suggests that ILP troglobiontic species, exclusive of karst islands, may have relatively large distributions. Barr (1967b, 1968, 1981, 1985a) suggested the geologic structure and stratigraphy of the ILP, which has large areas of highly fractured caverniferous limestone, permitted more subterranean dispersal among cave populations and species. For example, *D. kentuckensis* is distributed over a large area (approximately 3,728 km^2) along the western escarpment of the Cumberland Plateau in southeastern Kentucky and adjacent northern Tennessee (Fig. 5.4). In addition, more cave trechine species are found in sympatry in the ILP than in the APP. This likely occurred as different species colonized the subterranean realm at different times (Barr 1967b) due to the extensive, highly fractured nature of the ILP.

Nonetheless, there are potential hydrologic and geologic barriers to dispersal that separate species and species groups in the ILP. For example, the *P. tenuis* group is located primarily in the Mitchell Plain karst area in southern Indiana, but one species, *P. barberi* (Jeannel, 1928), from northern Kentucky is separated from the other five species by the Ohio River (Barr 2004). Moreover, two species of the *P. barri* group, *P. barri* (Krekeler, 1973) and *P. troglodytes* (Krekeler, 1973), occur on opposite sides of the Ohio River (Barr 2004). These distributions of sibling species suggest that the pre-Kansan Ohio River was not a barrier for ancestral populations (Barr 1985a). Although identified as initially shallow, the modern Ohio River has deeply incised the cave-bearing strata along its course through the Crawford-Mammoth Cave Uplands and Mitchell Plain, reaching nearly 75 m below the bottom of the ancient, pre-Pleistocene Teays River valley (Teller and Goldthwait 1991). This depth extends below the strata where caves occur. In some reaches, the Ohio River has completely bisected the cave-bearing Ste. Genevieve and St. Louis limestones, likely isolating cave faunas on either side of the river, including stygobiontic species (Niemiller et al. 2013). The Cumberland River is a barrier for troglobionts in both the Pennyroyal Plateau Highland Rim karst area and the Central Basin karst area (Barr 2004); for example, it divides two species of the genus *Nelsonites*, with *N. jonesi* occurring to the north and *N. walteri* to the south (Barr 1985a).

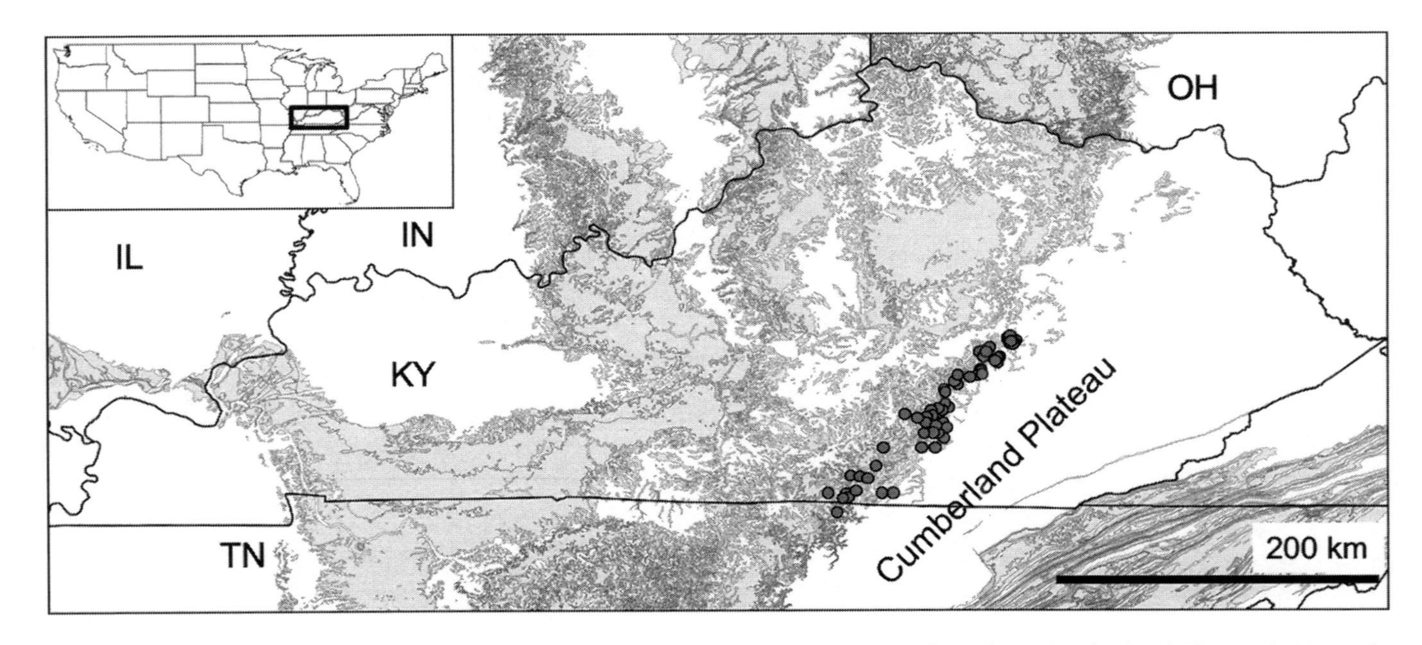

Figure 5.4. Distribution of *Darlingtonea kentuckensis* (*dots*) along the western edge of the Cumberland Plateau in Kentucky and northern Tennessee. Interior Low Plateau karst areas shown in gray.

In contrast, smaller streams and rivers are not necessarily dispersal barriers for all cave trechines (Barr and Peck 1965). Some species occur in subterranean habitats along both sides of smaller water courses. For example, *P. ciliaris* (Valentine, 1937) occurs along both sides of the Red River, as does *P. loganensis* (Barr, 1959) (Fig. 5.5), while *Neaphaenops tellkampfii* has been documented in caves along both sides of Sinking Creek and the Green River.

Barr and Peck (1965) hypothesized that flooding may wash beetles out of their subterranean habitats, transporting them downstream to other subterranean areas on either bank. Barr (1985a) examined the "meander frequencies" of rivers by dividing the distributions of isolated versus widespread species or populations. He concluded that the higher the meander frequency, the more often troglobionts were washed out of first-known habitats and transported to other subterranean habitats downstream.

As such, it appears distributional patterns of groups of related species often resemble regional drainage patterns. While most cave trechine species are restricted to a single watershed basin, several ILP species occur in more than one present-day watershed basin. These include *P. templetoni* (Valentine, 1948) and *P. macradyi* (Valentine, 1948) of the Upper Cumberland and Middle Tennessee-Elk basins (Barr 2004) and the widespread species *P. stricticollis* (Jeannel, 1931) and *P. youngi* (Krekeler, 1958) within both the Lower Ohio-Salt and Patoka-White basins (Barr 2004). Current distributions of these animals may reflect a Pleistocene or older drainage pattern, rather than the contemporary watershed basin.

Generally, cave trechine species groups are restricted to a single karst area in the ILP. For example, *N. jonesei*, *N. walteri*, *D. kentuckensis*, and the *P. intermedius* group are found only in caves along the western escarpment of the Cumberland Plateau. The *P. pubescens* and *P. menetriesi* species groups are restricted to Pennyroyal Plateau subterranean habitats, while the *P. simplex* group seems restricted to the Central Basin of the ILP.

In some cases, *Pseudanophthalmus* species groups occupy more than one ILP karst area. The *Pseudanophthalmus inexpectatus* group is found in the Bluegrass Region (*P. parvus*, *P. puteanus* (Krekeler, 1973), and *P. umbratilis* (Krekeler, 1973)) and the Pennyroyal Plateau (*P. cnephosus* (Krekeler, 1973), *P. inexpectatus* (Barr, 1959), and *P. orientalis* (Krekeler, 1973)), while the *P. robustus* group occurs in the Central Basin (*P. farrelli* (Barr, 1959)) and the western escarpment of the Cumberland Plateau (*P. beakleyi* (Valentine, 1937), *P. robustus* (Valentine, 1931), and *P. valentinei* (Jeannel, 1949)). In addition, while

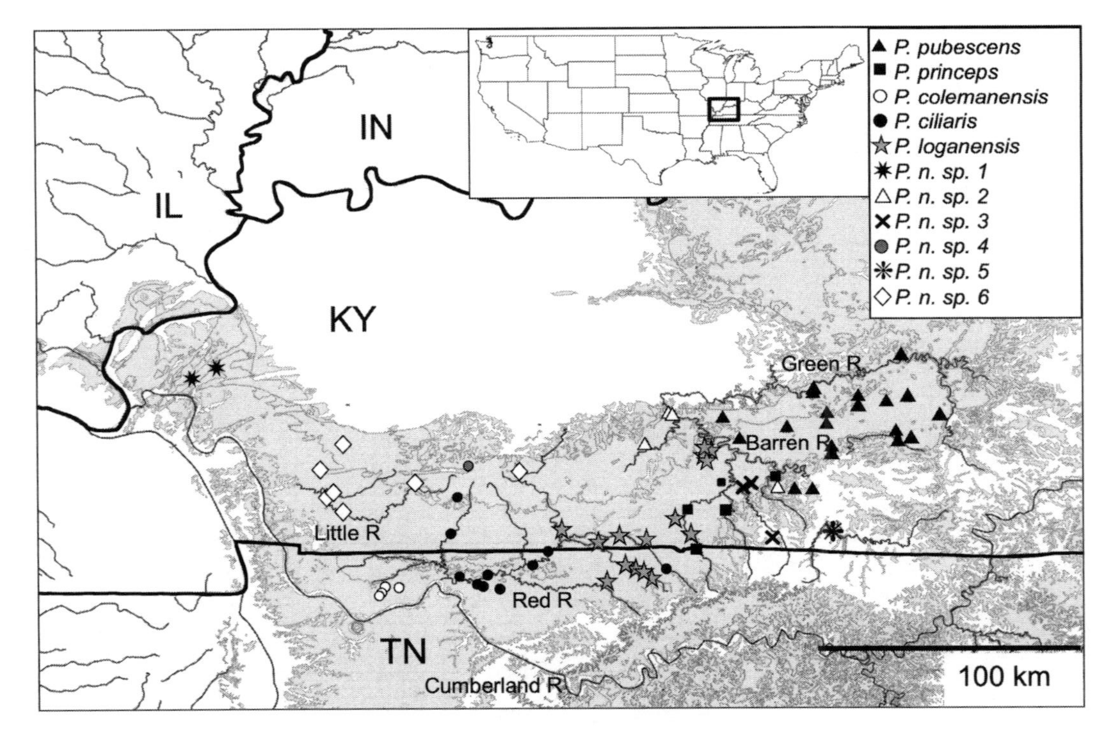

Figure 5.5. Distributional range of the *Pseudanophthalmus pubescens* group within the Pennyroyal Plateau in Kentucky and northern Tennessee. Five recognized species and six undescribed species are denoted by symbology in the legend. Species designations *sensu* Barr (2004). Karst areas shown in gray. Rivers and streams are shown with thin black lines.

the *P. engelhardti* group is distributed mostly in the APP, one species (*P. hesperus*) occurs in the Central Basin.

Thus, these distributional patterns suggest these species groups may be monophyletic. Although the phylogeny of cave trechines in this region is incomplete, Barr's (2004) species groups provided us with some insights. He combined species into species groups based on shared characteristics of the shape of a groove at the apex of the elytra and features of the male genitalia. Additionally, Barr (1981, 1985a) claimed that species identified as closely related based upon morphological characters were generally either co-occurring or close geographically, which suggested common ancestry.

In the ILP, most closely related cave trechine species are distributed on the landscape as a patchwork of adjacent but nonoverlapping ranges. These include many species in the *P. tenuis*, *P. menetriesi*, *P. robustus*, and *P. pubescens* species groups (Barr 1967b, 2004). Specifically, the *P. pubescens* group (composed of five described species and six undescribed species) ranges from the Mammoth Cave region south to Allen County, Kentucky, southwest to the margin of the Central Basin, and west into the Pennyroyal Plateau and adjacent Highland Rim of Tennessee; its range encompasses approximately 13,000 km^2. Most of the species within the *P. pubescens* group are allopatric or parapatric with respect to other species in the group, although *P. ciliaris* and *P. loganensis* are co-located within one cave (i.e., limestone unit). However, they can be found in caves containing other more distantly related cave trechines, such as *Neaphaenops tellkampfii* and *P. cerberus* (Horn, 1871).

As with other cave trechines, hydrologic or geologic barriers separate some *P. pubescens* group species. For example, the Barren River separates *P. pubescens* from *P. loganensis* and a likely undescribed species (*Pseudanophthalmus* n. sp. 2; T. Barr, unpublished notes). Some subterranean areas at the margins of the Pennyroyal Plateau karst are considered "karst islands" isolated from other areas of the main karst plain (Barr 1985a). Two potentially new species of the *P. pubescens* group (*Pseudanophthalmus* n. sp. 2 and *Pseudanophthalmus* n. sp. 4; T. Barr, unpublished notes) occupy small areas in these more isolated karst islands (T. Barr, unpublished notes). Two additional potentially undescribed species (*Pseudanophthalmus* n. sp. 1 and *Pseudanophthalmus* n. sp. 6; T. Barr, unpublished notes) may extend the range of the *P. pubescens* group west to the edge of the Pennyroyal Plateau karst area toward the Illinois border, while another purportedly undescribed species (*Pseudanophthalmus* n. sp. 3; T. Barr, unpublished notes) occurs at close proximity (i.e., less than 20 km) to both *P. princeps* (Barr, 1979) and *P. pubescens* (Horn, 1868) in the Barren River subbasin.

Evolutionary History

Biogeographic explanations of cave trechine distribution patterns must include how these species diverged from their epigean ancestors, and the evolutionary processes that occurred once they colonized subterranean habitats. In many respects, cave trechine species are like island species, playing an important role as models for speciation, colonization, and evolution of specialized phenotypes. However, the apparent absence of North American surface-dwelling relatives that share a recent common ancestor in eastern North America has provoked considerable speculation on the origin and evolutionary history of this tribe.

Contrasting hypotheses have been proposed to explain the speciation and biogeography of troglobiontic trechines. These hypotheses fall into two general categories: (1) a single event or a few cave colonization events with subterranean dispersal with subsequent isolation and diversification (single or few evolutionary origin[s] of cave trechines; Fig. 5.6A–D), and (2) multiple independent subterranean colonizations with subsequent isolation and diversification (multiple, independent evolutionary origins of cave trechines, Fig. 5.6E–H).

We recognize these hypotheses are not mutually exclusive and both most likely contributed to the diversity of eastern North American cave trechines. Previous authors have assumed that once trechine populations became restricted to subterranean habitats (due to climate and/or phenotypic adaptations), troglobionts typically would become isolated from other subterranean habitats and could not migrate between subterranean habitats, unless they were interconnected (Krekeler 1959; Barr 1967b; Barr and Holsinger 1985). Thus, allopatric species evolved in different subterranean areas and likely occur throughout a given limestone stratum but are reproductively isolated from other species by geographic barriers that prevent dispersal (Barr 1967b).

In support of the multiple, independent evolutionary origins hypothesis, Barr (1981, 1985a) proposed only trechine populations near karst areas were able to colonize subterranean habitats and thus survive the warm, dry conditions during successive interglacial periods of the Pleistocene. Distributions of cave trechine species in the APP eastward and ILP west-

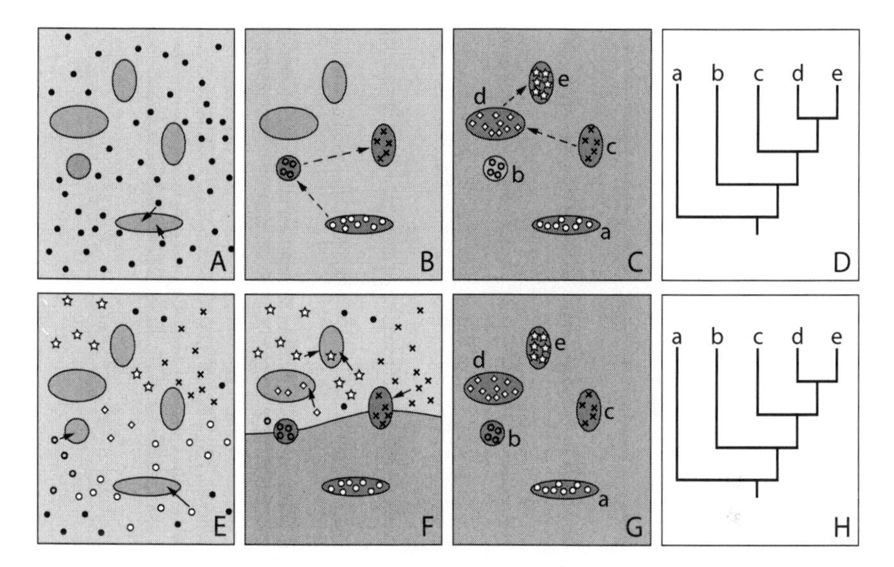

Figure 5.6. Schematic of the evolution and biogeography of cave trechines via the single (or few) origin hypothesis [A–D] and the multiple independent origins hypothesis [E–H]. [A] As the climate grew warmer and drier during interglacial periods, a trechine population colonized available subterranean habitat, which is represented by gray ovals. [B] Epigean trechine ancestors in Appalachia went extinct, while hypogean trechine populations became isolated underground and evolved into troglobiontic forms. [C] These species ultimately dispersed to other subterranean habitats where they became isolated and further diverged from their troglobiontic sibling species (*b* and *c*). [D] The phylogenetic relationships of cave species reflect a history of new subterranean colonization and speciation, yielding the same pattern as the vicariance hypothesis. [E] The multiple independent origins hypothesis proposes the ancestor of cave trechines was a surface-dwelling, widespread polymorphic species living in cool, moist habitats in Appalachia during a glacial period when the climate was cooler and wetter. [F] As the climate changed and became warmer and drier, local trechine variants colonized cave habitats and ultimately became isolated. [G] Surface populations went extinct, leaving only isolated subterranean relicts. [H] Phylogenetic relationships of cave species reflect the recency of isolation from other species. Ancestral surface conditions in the single (or few) origin hypothesis for cave trechines is identical to the multiple independent origins hypothesis.

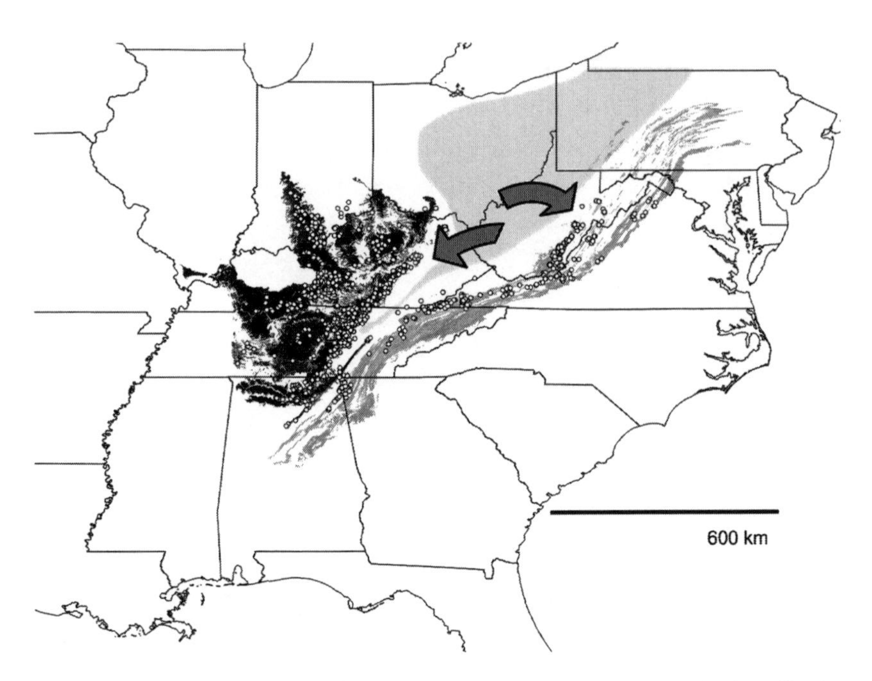

Figure 5.7. Barr (1981, 1985a) suggested that epigean trechines in the Allegheny and Cumberland Plateaus (*light gray shading*) were the ancestral source of cave trechines (*dispersal indicated with arrows*) in the Interior Low Plateau (*black*) to the west and Appalachians (*gray*) to the east prior to changes during the last glacial-interglacial period. Circles represent individual detections of cave trechines.

ward suggest that epigean trechine populations of the Allegheny and Cumberland Plateaus (between the APP and ILP) were a primary ancestral source from which troglobiontic forms evolved (Barr 1981, 1985a; Fig. 5.7). Barr (1981) also suggested the trechine surface-dwelling ancestor in the Appalachian region had considerable intraspecific and geographic variation prior to subterranean colonization. Regardless, the distinctly different lineages occupying subterranean habitats of the APP and ILP suggest substantial local differentiation of trechine populations prior to subterranean colonization (Barr 1981). Furthermore, geographic clustering of related species implies vicariance among cave descendants of these locally differentiated surface-dwelling ancestors (Barr 1965, 1981, 1985a).

While Barr (1965) and Jeannel (1926, 1927, 1928, 1930) suggested cave trechines evolved after the most recent ice age in the Pleistocene, the precise age of cave trechine lineages in eastern North America is unknown. Given the large number (more than 150) of cave trechine species and the evolution of specialized troglobiontic characters, it seems unlikely that cave colonization and diversification occurred after the most recent Pleistocene ice age. Additionally, the presence of two or more cave trechine species occurring in a single cave or cave system (e.g., the Mammoth Cave assemblage contains six cave trechine species) suggests multiple cave colonization events occurred over multiple interglacial periods—perhaps as early as the late Pliocene when some of these caves were forming.

Divergence-time studies based on molecular sequence data for European cave beetles suggest diversification dates much earlier than the Pleistocene. Faille et al. (2011) found one lineage of European trechines colonized caves about 10 million years ago (mya). For other beetle families, the major lineages of western Mediterranean cave beetles (family Leiodidae) diverged ~30 mya (Ribera et al. 2010), while Texas cave carabid beetles (tribe Platynini) in the United States diversified about 5 mya (Gómez et al. 2016).

Barr (1962) proposed that closely related species of cave trechines found in different subterranean systems, but in the same geographic regions, descended from a common, surface-dwelling ancestor. For example, species of the *P. robustus* group occupy adjacent but mutually exclusive ranges in the Central Basin and western escarpment of the Cumberland Plateau (Barr 1962). Similarly, species in the *P. intermedius* group also occupy caves in a limestone escarpment along the western edge of the Cumberland Plateau. Barr (1985a) cited these distributions as evidence of a gradual restriction of a much larger ancestral range encompassing the Cumberland and Allegheny Plateaus, which resulted in relicts within subterranean habitats (see Fig. 5.6E–H).

One difficulty in finding support for the multiple independent origins hypothesis in cave trechine evolution is the lack of an extant surface-dwelling trechine ancestor or close relative in eastern North America. If there was a polymorphic ancestral species (or even several closely related trechine species) that gave rise to the diverse cave trechine fauna in eastern North America, there most likely would be remnants of the ancestral species, perhaps in the cool, moist habitats of Appalachian mountain summits. We would expect to find multiple examples of trechine species in eastern North America with "intermediate" phenotypes (e.g., partially depigmented bodies, reduced eyes, and reduced or vestigial flight wings) interspersed among the highly troglobiontic cave trechine species. This pattern is described as adaptive shifts (Howarth 1980, 1987; Peck and Finston 1993; Desutter-Grandcolas and Grandcolas 1996), which have been identified primarily in tropical regions (Hoch and Howarth 1999; Wessel et al. 2007). Adaptive shifts have been implicated in the *Trechus* radiation in the Canary Islands, where troglobiontic species have closely related epigean congeners (Borges et al. 2007; Contreras-Díaz et al. 2007). Moreover, in the western Mediterranean, several independent cave trechine lineages evolved from extant epigean ancestors (Faille et al. 2011).

European trechine fauna includes several examples of morphological "intermediates" occupying deep soil and interstices in addition to a diversity of troglobiontic taxa (Borges et al. 2007; Contreras-Díaz et al. 2007; Faille et al. 2011). It is unclear why intermediate forms do not exist in North America; however, one explanation may be the relative lack of suitable cool, moist, high elevation habitats in Appalachia (Barr 1969).

One widely accepted version of the multiple independent origins hypothesis is the climatic relict hypothesis (Jeannel 1943; Peck and Finston 1993; Holsinger 2000). North American cave trechines are considered relicts of a now-extinct, late Pliocene or Pleistocene surface-dwelling ancestor, which was likely once widely distributed (Jeannel 1926, 1927, 1928, 1930, 1949; Krekeler 1959; Barr 1965, 1967b). It is possible that restriction to the subterranean realm occurred as early as the Miocene or Oligocene, as evidenced by European cave species—as these trechines show cave lineage origins as early as ~30 mya (Ribera et al. 2010; Faille et al. 2011). Regardless, it is plausible to envisage a scenario where now-extinct surface-dwelling ancestors that once lived in cool, wet leaf litter, moss, or moss-covered rocky debris of Appalachian conifer forests (i.e., preadapted to subterranean life) ultimately colonized subterranean habitats. The result of widespread climatic or other environmental changes associated with alternating glacial-interglacial periods would have restricted ancestral trechine species to cooler, wet microhabitats in shaded ravines, sinkholes, and underground habitats. Continued warming and drying during more recent interglacial periods would have led to the extinction of all remaining epigean populations. Hence, the subterranean-restricted populations became geographically and reproductively isolated and evolved into subterranean forms (Jeannel 1926, 1927, 1928, 1930, 1949; Krekeler 1959; Barr 1965, 1967b).

An alternative to the multiple independent origins hypothesis is the single (or few) evolutionary origin(s) of cave trechines. Over time, epigean trechines colonized caves, their subterranean populations became isolated and geographically localized, which led to vicariance of numerous species and an ultimate radiation of cave trechines. Barr (1967b) suggested distributional patterns of eastern North American cave trechines were indicative of the single (or few) cave trechine origin hypothesis. Each major lineage within a karst region represents a successful single-cave colonization by a surface-dwelling ancestor. Subterranean dispersal, diversification, and speciation, followed by concurrent extinction of the surface populations, were likely due to a climatic shift (see Fig. 5.6A–D). Howarth (1983) posited that most troglobionts disperse solely through continuous fissures and mesocaverns within the host rock. In the ILP, the expansive, honeycomb-like structure of caves and the associated network of small subterranean cracks and crevices make subterranean dispersal by cave trechines likely. Dispersal interstitially is also possible, as other trechine species are deep soil-dwellers (Jeannel 1937; Barr 1962; Arnett and Thomas 2000; Ortuño and Gilgado 2011; Schuldt and Assmann 2011).

Strong evidence for the dispersal potential of cave trechines is currently unknown. The relatively large distributions of some species (e.g., *D. kentuckensis*, *P. robustus*, and *P. stricticollis*) and species sympatry have been documented in several distantly related cave trechines in the ILP. This suggests these animals may have dispersed via mesocaverns. Dispersal mechanism(s) of *P. farrelli* and *P. tiresias* (Barr, 1959) from Fox Cave in Tennessee, *P. caecus* (Krekeler, 1973) and *P. umbratilis* (Krekeler, 1973) in Clifton Cave in Kentucky, and *N. jonesei*, *A. jeanneli*, and *D. kentuckensis* in Wind Cave in Kentucky (Barr 2004) are less clear. We are uncertain whether these troglobiontic species radiated via dispersal and subsequent vicariance through mesocaverns or if each species represents a different colonization event from a surface-dwelling ancestor.

Conversely, the distribution of closely related species in the *P. pubescens* and *P. menetriesi* species groups of the Pennyroyal Plateau further supports the single (or few) cave trechine origin hypothesis. *P. pubescens* and *P. menetriesi* (Moltschulsky, 1862) are widely distributed in the region, and related species exhibit an archipelago-like pattern of ranges across the Pennyroyal Plateau (Barr 2004). This suggests speciation by dispersal, subsequent isolation, and divergence. Incidentally, leiodid cave beetles in the western Mediterranean (Ribera et al. 2010) and cave trechines in the Pyrenees (Faille et al. 2010) display a similar pattern of a single subterranean colonization event with subsequent underground dispersal and diversification.

At present, there is no evidence to support a single (or few) origin versus a multiple independent origins for eastern North American cave trechines. In fact, some combination of both modes of cave colonization and evolution may have occurred in separate karst regions (ILP vs. APP) or even within the same karst area at different geologic times during the late Pliocene and Pleistocene. A complication with attempting to disentangle the evolutionary history of cave trechines in eastern North America and distinguish between these two competing hypotheses is that both hypotheses can yield similar phylogenetic patterns among cave species (see Fig. 5.6D and H)—that is, the extinction of a close surface-dwelling relative(s), as is intimated by the climatic relict hypothesis (Jeannel 1943; Peck and Finston 1993; Holsinger 2000).

Discussion

Carbonate karst regions of eastern North America support a remarkable diversity of cave trechine beetles, which is the result of an explosive species diversification and adaptation of this tribe to the cave environment. These trechines converged on some general biogeographic patterns shaped by historical, stratigraphic, and fluvial factors. Caves are generally smaller and more isolated in the fragmented karst of the APP, where carbonate rock is patchy and discontinuous. The APP karst region supports a high diversity of endemic cave beetles per unit area with many species limited in their known distribution to one or a few caves. Conversely, the large and highly interconnected cave systems that developed in the broad and more continuous carbonate exposures in the ILP karst region host cave trechine species with larger ranges and more instances of sympatry.

Furthermore, cave trechines likely represent relict lineages descended from epigean ancestors that occurred in cool, moist surface habitats during the glacial maxima and colonized subterranean habitats during warmer, drier interglacial periods. These colonization events may have occurred as early as the Oligocene. The lack of any surface-dwelling ancestral trechines provides support for the climatic relict hypothesis as the explanation for the current distribution of subterranean trechines. Lacking eyes, flight wings, and the physiology to exist outside a narrow range of temperature and humidity requirements, subterranean-adapted species are limited in their dispersal ability through surface habitats, and closely related taxa are allopatrically isolated. On the other hand, subterranean habitats are often connected through a network of small fissures and voids—that is, meso-caverns (Howarth 1983)—which facilitate dispersal of troglobionts underground. Thus, the evolutionary history and colonization of subterranean habitats of eastern North American trechine beetles is complex with many questions remaining unanswered.

Conservation and Future Research

There is a strong potential to advance our understanding of cave trechine diversity patterns, whereby we can use this tribe as a model group for studying the evolution of subterranean life. More research is needed to better characterize the diversification and origin of cave trechines and to understand their phylogeographic patterns in eastern North America. Specifically, accurate information about their systematics, genetics, distribution, life history, and habitat requirements is needed to develop appropriate recommendations to preserve and protect these troglobionts.

Molecular phylogenetic analyses (Niemiller, Ober, and Phillips, unpublished data), with relaxed molecular clock dating of cave trechines, reconstruction of ancestral ranges, and speciation models, are essential to better define evolutionary history of this group. With time-calibrated phylogenies, we can ultimately connect the evolutionary history of species diversification to models of paleoclimatic, paleogeologic, and paleohydrologic changes. Genetic analysis will enable us to infer demographic changes in populations. In addition, research on homologous or analogous changes in loss of function in wings or eyes and other troglobiontic characters can aid in clarifying the phylogeography and evolutionary history of cave trechines. Molecular systematics and biogeography of these beetles offer a model for other comparative evolutionary and ecological studies of troglobionts to further our understanding of factors driving speciation and biogeographic patterns.

Cave ecosystems and their biodiversity face mounting anthropogenic threats. As humans impact the surface environment, these changes will negatively affect subterranean ecosystems and cave trechines. Deforestation can change hydrologic regimes and nutrient input from the surface (Trajano 2000; Souza-Silva et al. 2015; Zhao et al., Chap. 7)—effecting the food web in which cave trechines serve as one of the primary predators of eastern North American cave ecosystems. Development along rivers and streams results in hydrologic alterations, including flooding or desiccation of subterranean systems (Elliott 2000; Souza-Silva et al. 2015; Šugai et al. 2015). These changes can lead to the loss of cave trechine habitat. Mining activities contribute to the destruction of karst and destroy cave trechine habitat by altering groundwater tables and/or polluting groundwater via sedimentation and contamination (Mammola et al. 2019; Zhao et al., Chap. 7). Furthermore, global climate change is expected to drastically impact subterranean organisms (Badino 2004; Mammola et al. 2018, 2019). It has been proposed that most troglobionts will be unable to adapt to rising temperatures and desiccation of

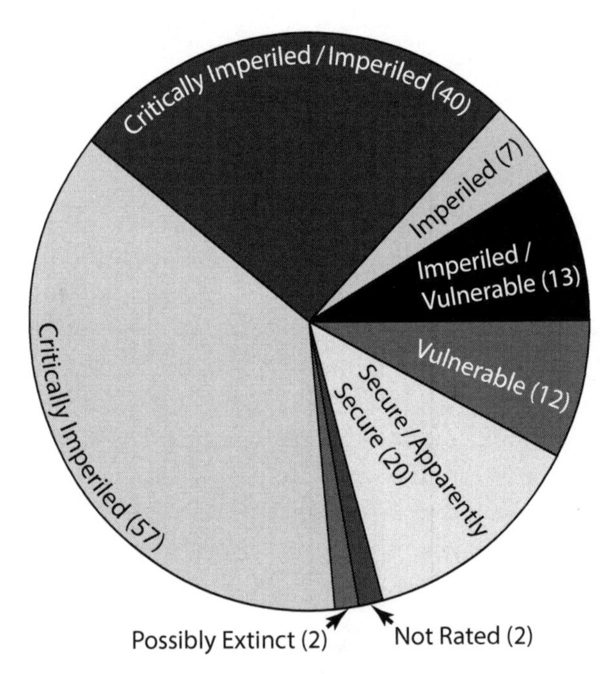

Figure 5.8. NatureServe conservation status (NatureServe 2022) of the 153 eastern North American cave trechines. Proportion of described species classified as "Critically Imperiled" (37.3%), "Critically Imperiled/Imperiled" (26.1%), "Imperiled" (4.6%), "Imperiled/Vulnerable" (8.5%), "Vulnerable" (7.8%), "Secure/Apparently Secure" (13.1%), "Possibly Extinct" (or "Extinct") (1%), and "Not Rated" (1%).

the cave environment due to climate change (Mammola et al. 2018). As cave trechines in the major karstic regions of eastern North America are troglobionts, these populations are expected to be particularly vulnerable to increased cave temperatures and decreased relative humidity.

Caves and subterranean habitats include some of the most unique, inaccessible, and understudied environments on earth. They support extraordinary forms of life and represent crucial habitats to be preserved and prioritized through conservation policies. Furthermore, subterranean-adapted organisms remain among the least documented faunas on the planet. Because of their restricted distributions and life history traits, many populations and species of troglobionts are considered at risk of extinction or critically imperiled; thus, these groups represent high-priority targets for conservation and management (Culver et al. 2000; Niemiller and Zigler 2013; Niemiller et al. 2017). In general, because of extremely restricted distribution and anthropogenic threats to both habitat and species, most (84.3%) eastern North American cave trechines are at an elevated risk of extinction, while at least two species are either extinct or possibly extinct (NatureServe 2022; Fig. 5.8).

Surprisingly, none of these species have been assessed for the International Union for Conservation of Nature's *IUCN Red List of Threatened Species*. An updated IUCN Red List and NatureServe conservation assessment for each taxon will be important for elevating the global importance of this tribe as well as guiding cave conservation policy in the karst regions of eastern North America.

ACKNOWLEDGMENTS

Robert Davidson and Kirk Zigler provided thoughtful comments leading to the improvement of this chapter.

REFERENCES

Arnett, Ross H., and Michael C. Thomas. 2000. *American Beetles*. Vol. I. *Archostemata, Myxophaga, Adephaga, Polyphaga: Staphyliniformia*. Boca Raton: CRC Press.

Badino, Giovanni. 2004. "Cave Temperatures and Global Climate Change." *International Journal of Speleology* 33:103–114.

Barr, Thomas C., Jr. 1959. "New Cave Beetles (Carabidae) from Tennessee and Kentucky." *Journal of the Tennessee Academy of Science* 18:195–196.

Barr, Thomas C., Jr. 1961. "Caves of Tennessee." *Tennessee Division of Geology Bulletin* 64:1–567.

Barr, Thomas C., Jr. 1962. "The *robustus* Group of the Genus *Pseudanophthalmus* (Coleoptera: Carabidae)." *Coleopterists Bulletin* 16:109–118.

Barr, Thomas C., Jr. 1965. "The *Pseudanophthalmus* of the Appalachian Valley (Coleoptera: Carabidae)." *American Midland Naturalist* 73:41–72.

Barr, Thomas C., Jr. 1967a. "A New *Pseudanophthalmus* from an Epigean Environment in West Virginia (Coleoptera: Carabidae)." *Psyche* 74:166–172.

Barr, Thomas C., Jr. 1967b. "Observations on the Ecology of Caves." *American Naturalist* 101:475–491.

Barr, Thomas C., Jr. 1968. "Cave Ecology and the Evolution of Troglobites." In *Evolutionary Biology,* edited by Theodosius Dobzhansky, Max K. Hecht, and William C. Steere, 35–102. Boston: Springer.

Barr, Thomas C., Jr. 1969. "Evolution of the (Coleoptera) Carabidae in the Southern Appalachians." In *The Distributional History of the Biota of the Southern Appalachians. Part I. Invertebrates* (Research Division Monograph, 1), edited by Perry C. Holt, 67–92. Blacksburg: Virginia Polytechnic Institute.

Barr, Thomas C., Jr. 1972. "*Trechoblemus* in North America, with a Key to North American Genera of Trechinae (Coleoptera: Carabidae)." *Psyche* 78:140–149.

Barr, Thomas C., Jr. 1979. "The Taxonomy, Distribution, and Affinities of *Neaphaenops*, with Notes on Associated Species of *Pseudanophthalmus* (Coleoptera, Carabidae)." *American Museum Novitates* 2682:1–20.

Barr, Thomas C., Jr. 1980. "New Species Groups of *Pseudanophthalmus* from the Central Basin of Tennessee (Coleoptera: Carabidae: Trechinae)." *Brimleyana* 3:85–96.

Barr, Thomas C., Jr. 1981. "*Pseudanophthalmus* from Appalachian Caves (Coleoptera: Carabidae): The *engelhardti* Complex." *Brimleyana* 5:37–94.

Barr, Thomas C., Jr. 1985a. "Pattern and Process in Speciation of Trechine Beetles in Eastern North America (Coleoptera: Carabidae: Trechinae)." In *Taxonomy, Phylogeny, and Biogeography of Beetles and Ants,* edited by George E. Ball, 350–407. Dordrecht: Springer.

Barr, Thomas C., Jr. 1985b. "New Trechine Beetles (Coleoptera: Carabidae) from the Appalachian Region." *Brimleyana* 11:119–132.

Barr, Thomas C., Jr. 2004. *A Classification and Checklist of the Genus* Pseudanophthalmus *Jeannel (Coleoptera: Carabidae: Trechinae).* Martinsville: Virginia Museum of Natural History, Special Publication 11.

Barr, Thomas C., Jr., and John Holsinger. 1985. "Speciation in Cave Faunas." *Annual Review of Ecology and Systematics* 16:313–337.

Barr, Thomas C., Jr., and Carl H. Krekeler. 1967. "*Xenotrechus*, a New Genus of Cave Trechines from Missouri (Coleoptera: Carabidae)." *Annals of the Entomological Society of America* 60:1322–1325.

Barr, Thomas C., Jr., and Stewart B. Peck. 1965. "Occurrence of a Troglobiotic *Pseudanophthalmus* outside a Cave (Coleoptera: Carabidae)." *American Midland Naturalist* 73:73–74.

Borges, Paulo A.V., Pedro Oromi, Artur R. M. Serrano, Isabel R. Amorim, and Fernando Pereira. 2007. "Biodiversity Patterns of Cavernicolous Ground-Beetles and Their Conservation Status in the Azores, with the Description of a New Species: *Trechus isabelae* n. sp. (Coleoptera: Carabidae: Trechinae)." *Zootaxa* 1478:21–31.

Bousquet, Yves. 2012. "Catalogue of Geadephaga (Coleoptera, Adephaga) of America, North of Mexico." *ZooKeys* 245:1–1722.

Boyd, Olivia F. 2015. "Phylogeography of *Darlingtonea kentuckensis* and Molecular Systematics of Kentucky Cave Trechines." Master's thesis and Specialist Project. Paper 1477. Western Kentucky University.

Caccone, Adalgisa. 1985. "Gene Flow in Cave Arthropods: A Qualitative and Quantitative Approach." *Evolution* 39:1223–1234.

Casale, Achille, Augusto Vigna Taglianti, and Christian Juberthie. 1998. "Coleoptera Carabidae." In *Encyclopaedia Biospeologica* 2, edited by Christian Juberthie and Vasil Decu, 1047–1081. Moulis-Bucarest: Société Internationale de Biospéologie.

Christiansen, Kenneth. 1961. "Convergence and Parallelism in Cave Entomobryinae." *Evolution* 15:288–301.

Contreras-Díaz, Hermans G., Oscar Moya, Pedro Oromi, and Carlos Juan. 2007. "Evolution and Diversification of the Forest and Hypogean Ground-Beetle Genus *Trechus* in the Canary Islands." *Molecular Phylogenetics and Evolution* 42:687–699.

Culver, David C. 1982. *Cave Life: Evolution and Ecology*. Cambridge, MA: Harvard University Press.

Culver, David C., and Tanja Pipan. 2019. *The Biology of Caves and Other Subterranean Habitats*. Oxford: Oxford University Press.

Culver, David C., Mary Christman, William R. Elliott, Horton H. Hobbs, and James R. Reddell. 2003. "The North American Obligate Cave Fauna: Regional Patterns." *Biodiversity and Conservation* 12:441–468.

Culver, David C., Thomas C. Kane, Daniel W. Fong, Ross Jones, Mark A. Taylor, and S. C. Sauereisen. 1990. "Morphology of Cave Organisms: Is It Adaptive?" *Mémoires de Biospéologie* 17:13–26.

Culver, David C., Lawrence L. Master, Mary C. Christman, and Horton H. Hobbs. 2000. "Obligate Cave Fauna of the 48 Contiguous United States." *Conservation Biology* 14:386–401.

Darwin, Charles. 1859. *On the Origin of Species by Means of Natural Selection, or the Preservation of Favoured Races in the Struggle of Life*. London: John Murray.

Desutter-Grandcolas, Laure, and Philippe Grandcolas. 1996. "The Evolution toward Troglobiotic Life: A Phylogenetic Reappraisal of Climatic Relict and Local Habitat Shift Hypotheses." *Mémoires de Biospéologie* 23:57–63.

Elliott, William R. 2000. "Conservation of the North American Cave and Karst Biota." In *Ecosystems of the World: Subterranean Ecosystems*, edited by Horst Wilkens, David C. Culver, and William F. Humphreys, 665–689. Amsterdam: Elsevier.

Elliott, William R. 2007. "Zoogeography and Biodiversity of Missouri Caves and Karst." *Journal of Cave and Karst Studies* 69:135–162.

Erichson, Wilhelm F. 1844. Footnote in Article by Tellkampf, Theodore. 1844. "Ueber den blinden Fisch der Mammuthhöhle in Kentucky, mit Bemerkungen über einige andere in dieser Höhle lebende Thiere." *Archiv für Anatomie, Physiologie und Wissenschaftliche Medicin*, 381–394.

Faille, Arnaud, Achille Casale, and Ignacio Ribera. 2011. "Phylogenetic Relationships of West Mediterranean Troglobiotic Trechini Ground Beetles (Coleoptera: Carabidae)." *Zoologica Scripta* 40:282–295.

Faille, Arnaud, Ignacio Ribera, Louis Deharveng, Charles Bourdeau, Lionel Garnery, Eric Queinnec, et al. 2010. "A Molecular Phylogeny Shows the Single Origin of the Pyrenean Subterranean Trechini Ground Beetles (Coleoptera: Carabidae)." *Molecular Phylogenetics and Evolution* 54:97–106.

Fang, Jie, Wenbo Li, and Mingyi Tian. 2016. "Occurrence of Cavernicolous Ground Beetles in Anhui Province, Eastern China (Coleoptera, Carabidae, Trechinae)." *ZooKeys* 625:99.

Gómez, R. Antonio, James Reddell, Kipling Will, and Wendy Moore. 2016. "Up High and Down Low: Molecular Systematics and Insight into the Diversification of the Ground Beetle Genus *Rhadine* LeConte." *Molecular Phylogenetics and Evolution* 98:161–175.

Hack, John T. 1969. "The Area, Its Geology: Cenozoic Development of the Southern Appalachians." In *The Distributional History of the Biota of the Southern Appalachians. Part I. Invertebrates* (Research Division Monograph, 1), edited by Perry C. Holt, 1–17. Blacksburg: Virginia Polytechnic Institute.

Hobbs, Horton H., III. 2012. "Diversity Patterns in the United States." In *Encyclopedia of Caves*, 2nd ed., edited by William B. White and David C. Culver, 251–264. New York: Elsevier, Academic Press.

Hoch, Hannelore, and Frank G. Howarth. 1999. "Multiple Cave Invasions by Species of the Planthopper Genus *Oliarus* in Hawaii (Homoptera: Fulgoromorpha: Cixiidae)." *Zoological Journal of the Linnean Society* 127: 453–475.

Holsinger, John R. 2000. "Ecological Derivation, Colonization, and Speciation." In *Ecosystems of the World: Subterranean Ecosystems*, edited by Horst Wilkens, David C. Culver, and William F. Humphrey, 399–415. Amsterdam: Elsevier.

Horn, George H. 1868. "Catalog of Coleoptera from South-West Virginia." *Transactions of the American Entomological Society* 2:123–128.

Horn, George H. 1871. "Descriptions of New Coleoptera of the United States." *Transactions of the American Entomological Society* 3:325–344.

Howarth, Francis G. 1980. "The Zoogeography of Specialized Cave Animals: A Bioclimatic Model." *Evolution* 34:394–406.

Howarth, Francis G. 1982. "Bioclimatic and Geologic Factors Governing the Evolution and Distribution of Hawaiian Cave Insects." *Entomologia Generalis* 8:17–26.

Howarth, Francis G. 1983. "Ecology of Cave Arthropods." *Annual Review of Entomology* 28:365–389.

Howarth, Francis G. 1987. "The Evolution of Non-relictual Tropical Troglobites." *International Journal of Speleology* 16:1–16.

Jeannel, René. 1926. "Monographie des Trechinae. Morphologie Comparée et Distribution Géographique d'un Group de Coléoptères. Première Livraison." *L'Abeille* 32:221–550.

Jeannel, René. 1927. "Monographie des Trechinae. Morphologie Comparée et Distribution Géographique d'un Group de Coléoptères. Deuxième Livraison." *L'Abeille* 33:1–592.

Jeannel, René. 1928. "Monographie des Trechinae. Morphologie Comparée et Distribution Géographique d'un Group de Coléoptères. Troisième Livraison." *L'Abeille* 34:1–808.

Jeannel, René. 1930. "Monographie des Trechinae. Morphologie Comparée et Distribution Géographique d'un Group de Coléoptères. Quatrième Livraison. Supplément." *L'Abeille* 34:59–122.

Jeannel, René. 1931. "Révision des Trechinae de l'Amérique du Nord." *Archives de Zoologie Expérimentale et Générale* 71:403–499.
Jeannel, René. 1937. "Les Bembidiides Endogés (Col. Carabidae). Monographie d'une Linée Gondwanienne." *Revue Française d'Entomologie* 3:241–396.
Jeannel, René. 1941. "L'isolement, Facteur de l'Evolution." *Revue Française d'Entomologie* 8:101–110.
Jeannel, René. 1943. "Nouveaux Hénicocephalides sudaméricains." *Bulletin de la Société Entomologique de France* 48:125–128.
Jeannel, René. 1949. "Les Coléoptères Cavernicoles de la Région des Appalaches. Étude Systématique." *Notes Biospéologiques* 4:37–104.
Kowalko, Johanna E., Nicolas Rohner, Tess A. Linden, Santiago B. Rompani, Wesley C. Warren, Richard Borowsky, et al. 2013. "Convergence in Feeding Posture Occurs through Different Genetic Loci in Independently Evolved Cave Populations of *Astyanax mexicanus*." *Proceedings of the National Academy of Science* 110:16933–16938.
Krekeler, Carl H. 1958. "Speciation in Cave Beetles of the Genus *Pseudanophthalmus* (Coleoptera, Carabidae)." *American Midlands Naturalist* 59:167–189.
Krekeler, Carl H. 1959. "Dispersal of Cavernicolous Beetles." *Systematic Zoology* 8:119–130.
Krekeler, Carl H. 1973. "Cave Beetles of the Genus *Pseudanophthalmus* (Coleoptera, Carabidae) from the Kentucky Bluegrass and Vicinity." *Fieldiana (Zoology)* 62:35–83.
Laurenti, Josephus N. 1768. *Specimen Medicum, Exhibens Synopsin Reptilium Emendatam Cum Experimentis Circa Venenaet Antidota Reptilium Austriacorum*. Wien: Joan Thomae.
Maddison, David R., Kojun Kanda, Olivia F. Boyd, Arnaud Faille, Nick Porch, Terry L. Erwin, et al. 2019. "Phylogeny of the Beetle Supertribe Trechitae (Coleoptera: Carabidae): Unexpected Clades, Isolated Lineages, and Morphological Convergence." *Molecular Phylogenetics and Evolution*. 132:151–176.
Mammola, Stefano, Pedro Cardoso, David C. Culver, Louis Deharveng, Rodrigo L. Ferreira, Cene Fišer, et al. 2019. "Scientists' Warning on the Conservation of Subterranean Ecosystems." *BioScience* 69:641–650.
Mammola, Stefano, Sara L. Goodacre, and Marco Isaia. 2018. "Climate Change May Drive Cave Spiders to Extinction." *Ecography* 41:233–243.
Moravec, P., Shun-Ichi. Uéno, and Igor A. Belousov. 2003. "Tribe Trechini." In *Catalogue of Palaearctic Coleoptera, Archostemata, Myxophaga, Adephaga*, vol. 1, edited by Ivan Löbl and Ales Smetana, 288–346. Stenstrup: Apollo Books.
Motschulsky, Victor. 1862. "Entomologie Speciale. Remarques sur la Collection d'Insectes de V. de Motschulsky. Coléoptères." *Etudes Entomologiques* 11:15–55.
NatureServe. 2022. "NatureServe Explorer: An Online Encyclopedia of Life." Accessed January 27, 2022. http://explorer.natureserve.org.
Niemiller, Matthew L., and Kirk S. Zigler. 2013. "Patterns of Cave Biodiversity and Endemism in the Appalachians and Interior Plateau of Tennessee, USA." *PLOS ONE* 8:e64177.
Niemiller, Matthew L., James R. McCandless, R. Graham Reynolds, James Caddle, Thomas J. Near, Christopher R. Tillquist, et al. 2013. "Effects of Climatic and Geological Processes during the Pleistocene on the Evolutionary History of the Northern Cavefish, *Amblyopsis spelaea* (Teleostei: Amblyopsidae)." *Evolution* 67:1011–1025.

Niemiller, Matthew L., Steven J. Taylor, Michael E. Slay, and Horton H. Hobbs III. 2019. "Diversity Patterns in the United States and Canada." In *Encyclopedia of Caves*, 3rd ed., edited by William B. White, David C. Culver, and Tanja Pipan, 163–176. Amsterdam: Elsevier, Academic Press.

Niemiller, Matthew L., Kirk S. Zigler, Karen A. Ober, Evin T. Carter, Annette S. Engel, Gerald Moni, et al. 2017. "Rediscovery and Conservation Status of Six Short-Range Endemic *Pseudanophthalmus* Cave Beetles (Carabidae: Trechini)." *Insect Conservation and Diversity* 10:495–501.

Ortuño, Vicente M., and Jose D. Gilgado. 2011. "Historical Perspective, New Contributions and an Enlightening Dispersal Mechanism for the Endogean Genus *Typhlocharis* Dieck 1869 (Coleoptera: Carabidae: Trechinae)." *Journal of Natural History* 45:1233–1256.

Peck, Stewart B. 1998. "A Summary of Diversity and Distribution of the Obligate Cave Inhabiting Faunas of the United States and Canada." *Journal of Cave and Karst Studies* 60:18–26.

Peck, Stewart B., and Terrie L. Finston. 1993. "Galapagos Islands Troglobites: The Question of Tropical Troglobites, Parapatric Distributions with Eyed Sister-Species, and Their Origin by Parapatric Speciation." *Mémoires de Biospéologie* 20:19–37.

Ribera, Ignacio, Javier Fresneda, Ruxandra Bucur, Ana Izquierdo, Alfried P. Vogler, Jose M. Salgado, et al. 2010. "Ancient Origin of a Western Mediterranean Radiation of Subterranean Beetles." *BMC Evolutionary Biology* 10:29.

Romero, Aldemaro. 2009. *Cave Biology: Life in Darkness*. Cambridge: Cambridge University Press.

Schuldt, Andreas, and Till Assmann. 2011. "Below Ground Carabid Beetle Diversity in the Western Palaearctic—Effects of History and Climate on Range-Restricted Taxa (Coleoptera, Carabidae)." *ZooKeys* 100:461–474.

Seaber, Paul R., F. Paul Kapinos, and George L. Knapp. 1987. "Hydrologic Unit Maps. U.S. Geological Survey Water-Supply Paper 2294." Accessed July 25, 2019. http://pubs.usgs.gov/wsp/wsp2294/.

Soares, Daphne, and Matthew L. Niemiller. 2013. "Sensory Adaptations of Fishes to Subterranean Environments." *BioScience* 63:274–283.

Souza-Silva, Marconi, Rui P. Martins, and Rodrigo L. Ferreira. 2015. "Cave Conservation Priority Index to Adopt a Rapid Protection Strategy: A Case Study in Brazilian Atlantic Rain Forest." *Environmental Management* 55:279–295.

Sugai, Larissa S. M., Jose M. Ochoa-Quintero, Richieri Costa-Pereira, and Fabio O. Roque. 2015. "Beyond Above Ground." *Biodiversity and Conservation* 24:2109–2112.

Teller, James T., and Richard P. Goldthwait. 1991. "The Old Kentucky River; a Major Tributary to the Teays River." *Geological Society of America Special Papers* 258:29–42.

Townsend, James I. 2010. "Trechini (Insecta: Coleoptera: Carabidae: Trechinae)." *Fauna of New Zealand* 62:1–101.

Trajano, Eleonora. 2000. "Cave Faunas in the Atlantic Tropical Rain Forest: Composition, Ecology, and Conservation." *Biotropica* 32:882–893.

Uéno, Shun-Ichi. 2005. "A New Genus and Species of Cave Trechine (Coleoptera, Trechinae) from a Deep Pothole in Northwestern Guangxi, South China." *Journal of the Speleological Society of Japan* 29:1–11.

Uéno, Shun-Ichi. 2006. "Cave Trechines from Southwestern Guizhou, South China, with Notes on Some Taxa of the *Guizhaphaenops* Complex." *Journal of the Speleological Society of Japan* 31:1–27.

Valentine, Joseph M. 1937. "Anophthalid Beetles (Fam. Carabidae) from Tennessee Caves." *Journal of the Elisha Mitchell Scientific Society* 53:93–100.

Valentine, Joseph M. 1948. "New Anophthalmid Beetles from the Appalachian Region." *Geological Survey of Alabama Museum Paper* 27:1–20.

Valentine, Joseph M. 1952. "New Genera of Anophthalmid Beetles from Cumberland Caves (Carabidae, Trechinae)." *Geological Survey* of *Alabama Museum Paper* 34:1–41.

Weary, David J., and Daniel H. Doctor. 2014. "Karst in the United States: A Digital Map Compilation and Database." US Geological Survey Open-File Report 2014–1156. Accessed July 15, 2019. http://pubs.usgs.gov/of/2014/1156/.

Wessel, Pål, Petra Erbe, and Hannelore Hoch. 2007. "Pattern and Process: Evolution of Troglomorphy in the Cave-Planthoppers of Australia and Hawai'i—Preliminary Observations (Insecta: Hemiptera: Fulgoromorpha: Cixiidae)." *Acta Cardiologica* 36:199–206.

6

Subterranean Colonization and Diversification of Cave-Dwelling Salamanders

Matthew L. Niemiller, Evin T. Carter, Danté B. Fenolio, Andrew G. Gluesenkamp, and John G. Phillips

Introduction

Only two groups of vertebrates have successfully colonized caves and other subterranean habitats, having adapted to constant darkness with generally limited energy resources: cavefishes (class Actinopterygii) and cave salamanders (class Amphibia, order Caudata). Cavefishes are rather diverse, with more than 300 species reported to use caves and more than 170 species (across 10 orders and 22 families) considered subterranean-adapted (Proudlove 2006, 2010; Soares and Niemiller 2013, 2020; Niemiller and Soares 2015; Zhao et al., Chap. 7). For salamanders, 14 species in two families are considered true subterranean inhabitants (Gorički et al. 2012, 2019; J. G. Phillips et al. 2017). This is a fraction of the over 760 salamander species described worldwide (AmphibiaWeb 2021). Troglobiontic (terrestrial, subterranean-adapted) and stygobiontic (aquatic, subterranean-obligate) salamanders exhibit a suite of morphological, physiological, and behavioral traits associated with a permanent subterranean existence—termed troglomorphy (Christiansen 1961). Several salamander species are known to use subterranean habitats on a temporary to semipermanent basis for shelter/refuge, foraging opportunities, and reproduction (e.g., Baird et al. 2006; Camp and Jensen 2007; B. T. Miller et al. 2008; Niemiller and Miller 2009). These species vary in the amount of time spent underground and the degree to which they are adapted for survival in such habitats.

Most cave-dwelling salamander diversity is represented by the lungless salamanders of the family Plethodontidae, particularly the subfamily Spelerpinae. A total of 14 species are recognized globally. This includes 12 cave-obligate species known from karst regions of eastern and central North

America—nine species in the genus *Eurycea* and three species in the genus *Gyrinophilus*. Additionally, one cave-obligate species in the genus *Chiropterotriton* occurs in central Mexico, while another troglobiont in the genus *Proteus* (family Proteidae) occurs in eastern Europe. Many of these species have restricted distributions, as they are endemic to a specific karst region. Consequently, these species are also of conservation concern. Our review focuses primarily on cave-obligate species, but we also note the other troglophilic/stygophilic and trogloxenic/stygoxenic salamander species where applicable.

Troglomorphy in Salamanders

The lack of light is an obvious sensory constraint in underground habitats. Subterranean colonization and inhabitation have led to the evolution of a suite of traits associated with living in perpetual darkness. In particular, energy limitation in subterranean ecosystems is hypothesized to be a major selection pressure driving the evolution of troglomorphy, including physiological adaptations such as lower metabolic and growth rates, decreased reproductive output, slower development, and longer life spans, compared to surface relatives (Brandon 1971; Culver and Pipan 2009). Cave salamanders often display several degenerate troglobiontic traits, such as reduced eyes and pigmentation. Many species also have evolved constructive adaptations associated with nonvisual sensory modalities (Plate 6.1). Here, we briefly discuss some of the major degenerate and constructive traits that have evolved in subterranean salamanders.

Degeneration of Eyes and Pigmentation

Similar to cavefishes (Soares and Niemiller 2013; Niemiller and Soares 2015), the olm (*Proteus anguinus* Laurenti, 1768) has initial eye development that is normal, but then development slows, involution of the cornea occurs, and the lens undergoes severe lytic processes; ultimately, this results in a greatly reduced eye that is sunken into the orbit and beneath the skin (Durand 1971, 1973, 1976; Rétaux and Casane 2013). Photoreceptor cells in the retina exhibit signs of degeneration, with the outer segments completely missing in many populations (Durand 1971, 1973; Kos and Bulog 1996b; Kos et al. 2001). Also, photoreceptors are degenerate in the pineal body, which is a photoreceptive organ in the roof of the posterior part of the forebrain (Kos and Bulog 1996a, 1996b, 2000; Kos et al. 2001).

The eyes of the Georgia blind salamander (*Eurycea wallacei* (Carr, 1939)) are more degenerated than those of other stygobiontic salamanders (Brandon 1968). Small eyes are embedded in a mass of adipose tissue with no extrinsic eye muscles present, which is common in many cavefish and other cave salamander species. In addition, the retina lacks rods and cones, an outer plexiform layer, and a subdivided nuclear layer. A rudimentary lens is present in some individuals, and the retina is undifferentiated. Likewise, the eyes of *E. rathbuni* (Stejneger, 1896) are quite degenerate, lack a lens and eye muscles, and have undifferentiated rods and cones (Eigenmann 1900, 1909). In other species, such as *Gyrinophilus palleucus* McCrady, 1954, eyes are reduced but optomotor responses persist (Besharse and Brandon 1974). Comparative observations of the eyes of a grotto salamander species (*E. nerea* (Bishop, 1941)) and several surface plethodontid species demonstrated that postembryonic degenerative changes involve eyelids, cornea, retina, outer plexiform layer, and pigmented epithelium (Besharse and Brandon 1974). Interestingly, these changes are initiated only after metamorphosis, although late-stage larvae can also have degenerating retinas. Ultimately, the eyelids will fuse completely (or nearly completely), and the eyes degenerate during, or up to a year after, metamorphosis (Brandon 1971; Durand et al. 1993). However, there is considerable variation in the degree and timing of eye degeneration, perhaps related to light exposure (Noble and Pope 1928; Besharse and Brandon 1974, 1976).

Although several of the 14 subterranean-adapted salamanders are depigmented, most depigmented species still retain scattered melanophores that are greatly reduced in number and size (Gorički et al. 2012, 2019; Fenolio et al. 2013). Melanin synthesis can be induced by exposure to light, as in *Proteus anguinus* Laurenti, 1768 (Prelovšek et al. 2008; Gorički et al. 2012) and larvae of the *Eurycea spelaea* (Stejneger, 1892) complex; in both cases, these species appear to retain pigment when isolated in surface springs and streams (Fenolio and Trauth 2005). Conversely, prolonged light exposure does not induce changes in skin pigmentation in *Gyrinophilus palleucus* (McCrady 1954; Gorički et al. 2012). In other cases, extensive pigmentation has been retained, as in *G. gulolineatus* Brandon, 1965. It is possible that the same mechanisms that govern melanin reduction in cavefishes (Protas et al. 2006; Gross and Wilkens 2013) are responsible for depigmentation in salamanders.

Craniofacial and Other Skeletal Modifications

Troglomorphy also involves skeletal changes in subterranean-adapted salamanders. Subterranean-restricted lineages often have more teeth, fewer trunk vertebrae, and attenuated limbs compared to surface relatives (Brandon 1971). Likewise, cranial morphology in these lineages seem to converge on a longer head with a flat snout. This duckbill-like shaped head is shared across troglomorphic lineages where the rostral end of the head is flattened on the dorsal-ventral dimension and broadened (Brandon 1971), as in *Eurycea rathbuni*, *E. wallacei*, and *Proteus anguinus*. All subterranean-adapted *Eurycea* possess a single premaxilla, while *Gyrinophilus* have two premaxillae separated for their entire length (Wake 1966). Several anterior cranial bones, as well as cervical vertebrae, are often elongated, and the teeth are more numerous than in related surface species (Brandon 1971). The flattening and broadening of the head are not restricted to stygobiontic taxa, although this is most commonly observed in aquatic salamanders (larvae and paedomorphic species). However, these cranial modifications have been taken to the extreme in several stygobiontic species. For example, the premaxillae, frontal, vomers, and palatopterygoids of *E. rathbuni* are the most modified relative to surface-dwelling *Eurycea* (Mitchell and Smith 1965; Wake 1966; Potter and Sweet 1981; Clemen et al. 2009). This head shape might be driven through ontogenetic change, possibly in association with the degeneration of eyes. For example, eye degeneration in cavefishes is known to impact craniofacial development (Yamamoto et al. 2003; Dufton et al. 2012). The role of ontogeny in craniofacial development has been well studied in fishes (e.g., Franz-Odendaal et al. 2007; Kapralova et al. 2015), but similar studies are lacking for cave salamanders. This deficiency likely stems from the challenges of collecting adequate sample sizes while maintaining viable populations.

There is also a trend toward attenuation of limbs and digits (i.e., longer, thinner limbs and digits) in some stygobiontic salamanders, such as *E. rathbuni* and *E. wallacei*; for these species, limbs can be so slender that they appear frail (Valentine 1964; Brandon 1971). Limb attenuation has also been noted in metamorphosed *G. gulolineatus* (Niemiller and Miller 2010). Interestingly, limb attenuation in metamorphosed *G. gulolineatus* may be a function of developmental issues during metamorphosis and possibly due to starvation given that metamorphosis is abnormal in this species, and metamorphosed individuals do not feed or survive long after gill resorption (Niemiller, unpublished data).

Nonvisual Senses

Nonvisual sensory modalities, including mechanosensation, chemosensation, and electroreception, may be enhanced in troglobiontic species to compensate for the lack of visual stimuli in complete darkness. These senses play a critical role in orientation and detection of prey and predators, as well as the detection of and communication with conspecifics. The lateral-line system associated with mechanosensation is well developed in stygobiontic salamanders, as it is in aquatic salamanders in general. In the genus *Gyrinophilus*, mechanosensory pores are more numerous on the head of stygobiontic species than in epigean species (e.g., *G. porphyriticus* (Green, 1827); Niemiller, unpublished data). Yet the degree of troglomorphy in mechanosensation observed in stygobiontic salamanders pales in comparison with that exhibited by many cavefishes (Soares and Niemiller 2013). It remains uncertain if the ability to detect and catch prey is significantly enhanced in stygobiontic species compared to related surface-dwelling taxa.

In lieu of using the lateral-line system, direct tactile stimulation likely plays a primary role in feeding for metamorphosed adult grotto salamanders (Stone 1964). During an experiment where adult *E. spelaea* were fed beef liver after an acclimation period, Stone (1964) observed the animals did not orient themselves prior to tactile stimulation. Although beef liver is not part of the natural diet of *E. spelaea*, these findings suggested these salamanders were not relying on chemical cues. Aquatic *E. spelaea* larvae employ suction feeding (Fenolio et al. 2006; Grigsby 2009), whereas adult *E. spelaea* can feed both aquatically and terrestrially (P. W. Smith 1948; Brandon 1971; Fenolio and Trauth 2005) and may use jaw or tongue projection to feed aquatically (G. Roth 1987; Grigsby 2009).

Chemosensation is important during courtship in salamanders (Houck and Reagan 1990; Houck et al. 1998; Rollmann et al. 1999), but our knowledge of cave-dwelling species is limited to the subterranean-adapted *Proteus anguinus* and troglophilic *Calotriton asper* (Dugès, 1852) (Guillaume 2000). However, comparative studies are needed to examine chemosensation between related troglomorphic and surface taxa to determine if sensory trade-offs exist—as is observed in fishes (e.g., Soares et al. 2004; Jeffery 2008; Yoshizawa et al. 2010).

Electroreception is a sensory modality that should be favorable in aquatic, subterranean habitats to detect weak electrical fields generated by prey, conspecifics, and potential predators, as well as to assist in navigation and orientation. *P. anguinus* possesses ampullary organs (Istenic and Bulog 1984) and responds to electrical fields both behaviorally and electrophysiologically (A. Roth and Schlegel 1988; Schlegel 1997; Schlegel and Roth 1997). This behavior is widespread in salamanders and has been observed in troglophilic *Eurycea lucifuga* Rafinesque, 1822 (J. B. Phillips 1977) and newts (J. B. Phillips 1986, 1987). Subterranean-adapted *Eurycea* and *Gyrinophilus* may also possess the ability to magnetically orient. Unfortunately, few other stygobiontic salamanders have been studied to date, and thus little is known regarding the overall advantages and how species exploit this sensory modality.

Life History

In addition to morphology, several physiological and life history traits are associated with troglomorphy in cave-obligate salamanders. While we recognize that life history has not been well studied for most species, inferences may be made based on limited or anecdotal evidence and life history theory. Most troglomorphic salamanders are completely aquatic and exhibit larval form paedomorphosis (i.e., sexual maturity reached in the larval stage whereby external gills and tail fins are retained from larval stage through adulthood)—exceptions include the *Eurycea spelaea* complex and *Chiropterotriton magnipes* Rabb, 1965. The latter species belongs to a clade of terrestrial, direct-developing species (subfamily Bolitoglossinae).

Multiple species are considered obligate or facultative paedomorphs across several salamander families. For example, *Proteus anguinus* is an obligate paedomorph and stygobiont. Yet, all other members of the family Proteidae that live in aquatic surface environments are also obligate paedomorphs. Additionally, several other families of epigean salamanders (i.e., Amphiumidae, Cryptobranchidae, and Sirenidae) contain strictly paedomorphic lineages, although the extent of paedomorphosis differs among these families. Amphiumids lose their gills and retain a pair of gill slits, whereas cryptobranchids lose their gills, eyelids do not develop, and the tail remains largely laterally compressed. Interestingly, all stygobiontic salamanders retain external gills and a well-developed tail fin throughout their lives.

When compared to related surface-dwelling species, many subterranean-adapted organisms exhibit traits related to reduced resting metabolism, such as decreased growth rates, delays in reaching sexual maturity, and increased lifespan (Poulson 1963; Hervant et al. 2000; Culver and Pipan 2009; Niemiller and Poulson 2010). For example, average lifespan is 68.5 years in *P. anguinus*, with a maximum predicted lifespan in excess of 100 years (Voituron et al. 2011). Sexual maturity is also significantly delayed and not reached until 15 years. However, it is unclear whether all cave-obligate salamanders exhibit delays in sexual maturity and increased longevity when compared to epigean relatives. In studied populations of *Gyrinophilus palleucus*, lifespan has been estimated to be at least six years (Huntsman et al. 2011) or between 9 to 14 years (Niemiller et al. 2016)—neither lifespan nor sexual maturity in *G. palleucus* appear to differ significantly when compared to life history data of surface-dwelling *G. porphyriticus*. This supports the hypothesis that *G. palleucus* is a relatively young subterranean species (Niemiller et al. 2008, 2009). Fenolio et al. (2014) estimated *E. spelaea* live at least nine years, which is similar to epigean *Eurycea*. In some individuals, sexual maturity is delayed and may not be reached until six years. However, these and other cave-obligate salamanders may live considerably longer than these estimates, as few species and populations have been studied to date.

Although reproduction has not been well studied for most cave-dwelling salamanders, it appears to be influenced by low resting metabolic rates and limited food resources. Reproductive strategies, such as egg or clutch size, ultimately depend on both maternal and offspring environments at both ecological and evolutionary scales. Limited evidence and working hypotheses suggest that stygobiontic salamanders produce fewer and/or larger eggs compared to their surface relatives, because low energy resources must be balanced with slow metabolism. This low energy budget must, in turn, be balanced across processes such as growth, reproductive maturity, and gametogenesis (Burton et al. 2011; Pettersen et al. 2016). Notably, females are unlikely to reproduce annually due to energy limitations. In fact, intervals between reproductive events might be particularly long for some species, such as *P. anguinus*, which may reproduce every 12.5 years on average with a mean clutch size of 35 eggs (Voituron et al. 2011). In contrast, one of its epigean relatives, *Necturus maculosus* (Rafinesque, 1818), produces up to 193 eggs in a single clutch (Petranka 1998) and is capable of oviposition at least every 24 months. Their eggs can be produced annually but are often retained for up to one year and deposited at a later time or reabsorbed, as was observed in captive individuals (Calatayud et al. 2018; M. Stoops, pers. Comm., October 21, 2018). Meanwhile, *Necturus beyeri* Viosca, 1937 does not exhibit the seasonality of activity and food intake observed in *N. maculosus*. Additionally, it has an average clutch size of 30 eggs, which rivals *P. anguinus* (Shoop 1965; Sever and Bart 1996) but appears to oviposit annually (Calatayud et al. 2018; M. Stoops, pers. Comm., October 21, 2018).

Perhaps balancing the production of fewer eggs, larger egg size provides several advantages for stygobiontic salamanders. Larger eggs yield larger larvae (Nussbaum 1987), which may be more resistant to starvation due to, in part, lower mass-specific metabolism (Glazier 2014). Although salamanders are often recognized as top predators in cave ecosystems, eggs and larvae can be vulnerable to predation by cave-dwelling invertebrates, such as crayfishes and flatworms, as well as conspecifics. Larger size of both eggs and larvae may be driven by selection to escape predation. However, lower activity levels or lower predator density (including density of conspecifics) suggest lower encounter rates with predators. This may reduce the advantage of larger size, which has been demonstrated in food webs involving spiders with either active or sit-and-wait foraging modes (Verdeny-Vilalta et al. 2015). Additionally, because larger eggs give rise to larger-bodied larvae, these larvae are more likely to possess a wider mouth gape to accommodate a greater range of prey sizes—another potential adaptation for coping with limited food resources in subterranean habitats.

Some taxa use an especially active foraging strategy when prey is abundant. For example, *P. anguinus* has an array of mechanical and chemical adaptations for spatial orientation and prey acquisition (Uiblein et al. 1992; Uiblein and Parzefall 1993; Schlegel et al. 2009) that enable it to capitalize on abundant prey. Both *Gyrinophilus* and *Proteus* (and likely other stygobiontic species) often exhibit extended periods of inactivity punctuated by episodes of foraging when prey is abundant and at close proximity. Such behavior may serve to optimize energy budgets (Uiblein et al. 1992; Gorički et al. 2012; M. L. Niemiller and E. T. Carter, pers. Obs.). This is in contrast to the highly active foraging modes observed in *G. porphyriticus* and *Necturus* species; individuals may emerge from cover and forage throughout low-light periods (Resetarits 1991; Petranka 1998). Overall, there remains a paucity of information on activity, foraging mode, and degree of predation for most other troglomorphic salamanders. Elucidating these subjects will be central to understanding the ecological and evolutionary implications of energy limitations and reproductive strategies in subterranean-adapted salamanders.

Evolution of Paedomorphosis

Most subterranean-adapted salamanders are completely aquatic and paedomorphic. Bayesian reconstructions of stygobiontic taxa suggest that ancestral spelerpine salamanders were metamorphic and at least four independent shifts to paedomorphosis have occurred (Bonett, Steffen, Lambert et al. 2014; Bonett, Steffen, and Robison 2014). These include (1) an ancient shift in the ancestor of Edwards Plateau *Eurycea* in Texas, (2) a shift in the ancestor of *E. wallacei* in the Upper Floridan Aquifer, (3) possibly multiple shifts among populations of *G. palleucus* and *G. gulolineatus* in the Interior Low Plateau and Appalachians karst regions, and (4) a shift in *Gyrinophilus subterraneus* Besharse & Holsinger, 1977 in the Appalachians karst region (although current evidence suggests the population is not entirely paedomorphic). Interestingly, shifts to paedomorphosis do not appear to be irreversible, even after several millions of years. Bonett, Steffen, Lambert, et al. (2014) found support for the re-evolution of metamorphosis in *Eurycea troglodytes* Baker, 1957 from the Edwards Plateau. The shift to paedomorphosis from a metamorphic ancestor is highly correlated with a concurrent shift from surface to subterranean habitats (Bonett, Steffen, Lambert, et al. 2014).

Paedomorphosis has been hypothesized to evolve when surface conditions became relatively inhospitable, and selection favored a longer or permanent aquatic stage because environmental conditions in aquatic habitats were more suitable (Wilbur and Collins 1973). In aquifers, limited terrestrial habitat around subterranean water bodies may have limited the opportunity to metamorphose for species colonizing those habitats. In caves, paedomorphosis may be advantageous due to either limited food resources or harsher abiotic conditions in terrestrial habitats than aquatic habitats (Brandon 1971; Wilbur and Collins 1973; Ryan and Bruce 2000). Brandon (1971) suggested paedomorphosis via extension of the larval stage in the troglobiontic *Gyrinophilus* is a response to lower food availability in caves compared to surface habitats. Bruce (1979) extended this hypothesis to account for the quality, and not just quantity, of available prey. In surface populations of *G. porphyriticus*, metamorphosis occurs near the optimal body size, enabling this species to switch from smaller invertebrates to larger invertebrates and salamanders. In caves where abundances of larger invertebrates and other salamanders are low, selection favors extension of the larval period, which ultimately leads to paedomorphosis in the stygobiontic species (Bruce 1979). Some *G. porphyriticus* cave populations have reduced growth rates and may remain in the larval stage up to 10 years before metamorphosing (Niemiller, unpublished data). This is nearly three to five times longer than surface-dwelling populations.

It is unclear which occurred first, evolution of paedomorphosis or subterranean colonization, or if both occurred in concert. In Edwards Plateau *Eurycea*, paedomorphosis appears to have evolved before the transition into subterranean habitats (Bonett, Steffen, Lambert, et al. 2014). Many nontroglomorphic species of *Eurycea* continue to use subterranean habitats (Chippindale et al. 2000; Bendik and Gluesenkamp 2013), which is likely in response to inhospitable surface conditions. Likewise, it is highly probable that paedomorphosis preceded subterranean colonization in *P. anguinus*, as all proteids are paedomorphic (Sket and Arntzen 1994; Gorički and Trontelj 2006). The absence of metamorphosis in *Proteus* may be due to loss of function in thyroid hormone-dependent genes required for tissue transformation (Safi et al. 2006). In Edwards Plateau *Eurycea*, paedomorphosis followed by subterranean colonization appears to be driven by the climate becoming progressively more arid during the Miocene (Sweet 1977; Bonett, Steffen, Lambert, et al. 2014).

Paedomorphosis is also associated with reduced distributional range size in spelerpine salamanders including subterranean species (Bonett, Steffen, Lambert, et al. 2014). While paedomorphosis may limit dispersal and gene flow among populations, leading to greater risk of extinction, it may also facilitate divergence and ultimately speciation (e.g., Niemiller et al. 2008, 2009). In fact, Bonett, Steffen, Lambert, et al. (2014) found that paedomorphic and cave-dwelling spelerpine lineages had both higher speciation rates and higher extinction rates than metamorphic and surface-dwelling lineages.

Colonization, Diversification, and Evolutionary History

In addition to *Proteus* and *Chiropterotriton*, there have been at least six cave colonization events of spelerpine salamanders resulting in subterranean-adapted forms. (1) At least two known events resulted in stygobionts on the Edwards Plateau—the entire subgenus *Typhlomolge* (i.e., *Eurycea rathbuni* (Stejneger, 1896), *E. robusta* (Longley, 1978), *E. waterlooensis* Hillis, Chamberlain, et al., 2001, and *E. latitans* [inclusive of *E. tridentifera*, *E. pterophila* Burger, Smith & Potter, 1950, and *E. neotenes* Bishop & Wright, 1937] in the *Blepsimolge* clade) (Chippindale et al. 2000; Hillis et al. 2001; Wiens et al. 2003; Bonett, Steffen, Lambert, et al. 2014). (2) *Eurycea wallacei* evolved into a troglomorphic form in the Upper Floridan Aquifer. (3) The *E. spelaea* complex radiated across the Ozark Plateau (Bonett and Chippindale 2004; J. G. Phillips et al. 2017). (4) The stygobiontic *G. subterraneus* evolved in the Appalachian Mountains. (5) Possibly multiple colonization events occurred in *G. palleucus* and *G. gulolineatus* (both stygobionts) of the Interior Low Plateau and Appalachian Mountains. Here we review subterranean colonization, diversification, and the evolutionary history of the major cave-obligate lineages.

Olm (*Proteus anguinus*)

To date, the genus *Proteus* consists of one recognized stygobiontic species distributed throughout the Dinaric Karst of Italy, Slovenia, Croatia, Bosnia and Herzegovina, and northwest Montenegro (Kletečki et al. 1996; Sket 1997; Gorički et al. 2017). Its closest relatives are surface-dwelling North American waterdogs (*Necturus* species), which together form the monophyletic Proteidae family (Trontelj and Gorički 2003; Wiens et al. 2005). Incidentally, all members of Proteidae are obligately paedomorphic.

The monophyly of Proteidae has been debated, most notably owing to the large degree of geographic separation between extant populations. Noble (1931) assigned European cave-dwelling *Proteus* and their surface-dwelling North American relatives, *Necturus*, to Proteidae according to morphology and traits related to paedomorphosis (Trontelj and Gorički 2003). Hecht (1957) later suggested the morphological likeness of *Proteus* and *Necturus* resulted simply from neoteny and thus convergence. The disparate ranges of present-day *Necturus* and *Proteus* apparently precluded connections between North American and European forms during the Tertiary—when brackish or marine water would have imposed a significant barrier to dispersal for these aquatic forms. However, Larsen and Guthrie (1974) asserted the common ancestor of *Proteus* and *Necturus* could have preceded the Eocene epoch, when land connections and freshwater would have facilitated migration. Moreover, from a morphological perspective, both *Necturus* and *Proteus* share the unusual pattern among caudates of having a haploid number of 19 chromosomes (Kezer et al. 1965), as well as similarities in the hyobranchial apparatus, teeth, ossification between the stylus and squamosal, and a similar columellar process (i.e., a bony projection of the squamosal bone involved in hearing). Early on, Hecht and Edwards (1976) argued against this placement, but morphological and mtDNA (*12S* ribosomal) analyses support monophyly of Proteidae (Trontelj and Gorički 2003; Wiens et al. 2005; Shen et al. 2013).

Various texts and depictions dating back to around the tenth century fit the description of European *Proteus*. Intricate carvings that date from the tenth or eleventh century occur on a wellhead in Venice, Italy (see also Shaw 1999). Other seventeenth-century observations appear to be related to flood events at surface springs (e.g., Valvasor 1689). It was not until 1768 that the first specimen was described as *Proteus* (Laurenti 1768) from a surface spring at Cerknica Lake (near Postojna caves) in southwestern Slovenia. Subsequent repeated surveys of Laurenti's Cerknica site yielded no additional specimens. As this site is separated hydrologically from all known *Proteus* localities, it is now believed that Laurenti's specimen originated from the springs near the village of Sticna in central Slovenia (Sket 1997; Shaw 1999). The first specimens collected inside a cave were discovered in 1797 by Jersinovic von Loewengreif at the Pivka River at Črna Jama (formerly Magdalenen-Grotte). Mertens and Müller (1940) officially recognized Črna Jama (Magdalenen-Grotte) in Slovenia as the type locality of *P. anguinus*—owing in part to the initial work of Fejervary (1926).

Additional species of *Proteus* (*Hypochthon*), described by Fitzinger (1850), have generally been discounted (Mertens and Müller 1940; Sket 1997). To date, only two subspecies have been widely recognized based on the degree of troglomorphy (Sket and Arntzen 1994; Arntzen and Sket 1996, 1997) and mtDNA (Gorički and Trontelj 2006). However, more recent evidence suggests the existence of at least six monophyletic lineages (see below). These two forms can differ markedly in morphology, but the more stygobiontic form typically exhibits greater inter- and intrapopulation variation (Arntzen and Sket 1996, 1997; Gorički et al. 2012; Ivanović et al. 2013). White proteus (*P. anguinus anguinus*) represents the more ubiquitous, lightly pigmented form (Plate 6.1A), although melanin synthesis can be induced by exposure to light.

Throughout much of the Dinaric Karst from Italy to Montenegro (Sket 1997; Gorički et al. 2017), this species occurs in ~300 springs and groundwater systems to depths of 113 m (Šarić and Konrad 2017). It was introduced to a subterranean research laboratory in Moulis, Saint-Girons, France in 1952, which is where much of our current knowledge on life history and behavior originates (Schlegel et al. 2006; Voituron et al. 2011). The black proteus (*P. a. parkelj*), as its name suggests, represents the less common, more darkly pigmented form. This animal was first discovered by the Slovenian Karst Research Institute in 1986 in the Dobličica spring near Črnomelj in Bela Krajina and later confirmed in Na trati near Jelševnik, southeast Slovenia (Aljančič et al. 1986; Sket 2017). The current distribution, confirmed by living specimens from five caves, comprises an area less than 2 km^2 (Gorički et al. 2017).

Black proteus is thought to resemble the common surface ancestor of *P. anguinus*. Its eyes are well developed and covered by a thin layer of skin rather than being reduced and embedded as in white proteus. Limbs, tail, and cranial bones are shorter and appear more robust, while the trunk is more elongate. Additionally, the permanently pigmented skin is thicker than white proteus (Gorički et al. 2012). Although black proteus lacks many of the troglomorphic traits of white proteus, Sket (2017) considers both forms to be stygobionts inhabiting the same or similar subterranean systems.

Gorički et al. (2017) suggest the distributions of black and white proteus may overlap along the Dobličica River, Slovenia. Both forms were detected syntopically via environmental DNA (eDNA) analysis of water samples at Šprajcarjev Zdenec spring. Their thin, elongate body and strong swimming ability may enable them to disperse greater distances. This is further supported to some degree by microsatellite data from the Postojna and Planina cave systems, separated by more than 10 km, which exhibit apparently high gene flow and near panmixia ($F_{ST} = 0.0024$; Zakšek et al. 2018). Based upon recent eDNA analysis, Gorički et al. (2017) suggest reproductive barriers may exist and thus each form may represent distinct species.

Six independent *P. anguinus* lineages with nonoverlapping geographic ranges have been identified (Trontelj et al. 2009; Gorički et al. 2017). These lineages include the (1) Istra clade in Croatia, (2) Lika clade in Croatia, (3) Southeastern Coastal Region clade in southeastern Croatia and southern Herzegovina, (4) Bosanska Krajina clade in Bosanska Krajina and Bosnia and Herzegovina, (5) Southwestern Slovenia clade, and (6) Southeastern Slovenia clade (Trontelj et al. 2009).

Gorički and Trontelj (2006) suggested each *Proteus* lineage evolved troglomorphic traits multiple times independently. Interestingly, caves did not form in the Dinaric area until after colonization by early proteids. Thus, an early surface proteid likely colonized the region from the east or west during the Oligocene or Miocene when land connections existed (Sket 1997). Recent divergence estimates for *Euproctus* (using *12S* and *16S* rDNA clock) suggest most lineages split during the Pliocene or Miocene between 16 and 8.8 million years ago (mya). An additional northwestern split from the "non-troglomorphic" lineage probably occurred approximately 5 to 4 mya (Trontelj et al. 2007). Various estimates placed the split between the non-troglomorphic lineage and the last troglomorphic lineage between 4.5 and 0.5 mya (Sket and Arntzen 1994; Gorički and Trontelj 2006). Mitochondrial rDNA provided a more tightly constrained estimate of 0.6 to 0.5 mya (Trontelj et al. 2007). Thus, the cave colonization event and the subsequent evolution of a full suite of subterranean-adapted characters occurred within the past 500,000 years (Gorički and Trontelj 2006; Trontelj et al. 2007). Sket (1997) suggested that warmer climatic conditions following the last glaciation (<10,000 years ago) likely drove this lineage northward, where it ultimately retreated into the cooler subterranean environment. The northward movement would have been halted by the northern extent of the Dinaric Karst.

Bigfoot Splayfoot Salamander (*Chiropterotriton magnipes*)

The bigfoot splayfoot salamander, *Chiropterotriton magnipes* Rabb, 1965, is unique among cave-dwelling salamanders because it is a terrestrial direct-developer, while other species have an aquatic larval stage. This species is known from the Sierra Madre Oriental in the Mexican states of San Luis Potosí and Querétaro. Unlike other subterranean-adapted salamanders discussed in this chapter, *C. magnipes* has well-developed eyes and typically occurs from the twilight to transition zone rather than in the deep cave zone (refer to Howarth and Wynne, Chap. 1, for more information on cave zonal environments). To date, this salamander has not been detected in surface forests. However, other members of the genus *Chiropterotriton* (*C. arboreus*, *C. cieloensis*, *C. cracens*, *C. infernalis*, *C. mosaueri*, and *C. multidentatus*) are known from both hypogean and epigean habitats in Mexico and are considered troglophiles.

All *Chiropterotriton* species have traits that would serve them well in the terrestrial cave environment. For example, *C. magnipes* has extremely large feet, enabling it to traverse cave ceilings (Plate 6.1B). In addition, this species has the unexpected combination of reduced pigment and large eyes; the latter trait renders it a successful hunter in the twilight zone. Moreover, *C. magnipes* has attenuated limbs and a wide, flattened head. We consider this species a troglobiont, despite the limited degree of troglomorphy when compared with other subterranean-adapted salamanders. However, some may classify this species as an obligate troglophile (*sensu* Peck 1970), as no surface populations are known to date.

A time-calibrated molecular analysis of two mtDNA regions and three nuclear genes indicated that *Chiropterotriton* diverged from its nearest common ancestor approximately 30 to 20 mya (Rovito et al. 2015). Their study built upon previous work by Wiens et al. (2007) that suggested a divergence of ~15 mya for the common ancestor of *C. magnipes* from its congener, *C. arboreus*.

Spring Salamanders (*Gyrinophilus* spp.)

The genus *Gyrinophilus* consists of four large (up to 24 cm total length), semi- to permanently aquatic species in the highlands of eastern North America. These species are associated with caves to varying degrees. Of these, three species are considered stygobionts (*G. palleucus*, *G. gulolineatus*, and *G. subterraneus*); these species occur in the Interior Low Plateau and Appalachians karst regions of Alabama, Tennessee, Georgia, and West Virginia. They are closely related to a fourth species (with four subspecies)—a surface and facultative cave-dwelling congener, the spring salamander (*G. porphyriticus*).

The Tennessee cave salamander (*G. palleucus*; Plate 6.1C) was the first stygobiontic *Gyrinophilus* described (McCrady 1954). It was discovered in Franklin County, Tennessee in 1944, but McCrady delayed formal description in hopes of confirming paedomorphosis. In 1954, H. C. Yeatman confirmed this trait when he collected a paedomorphic male (from the type locality) with a spermatophore protruding from its cloaca (Lazell and Brandon 1962). Historically, two subspecies were recognized—the pale salamander (*G. p. palleucus*) and the Big Mouth Cave salamander (*G. p. necturoides*). Although these subspecies differ in pigmentation and number of trunk vertebrate, genetic work (Niemiller et al. 2008, 2009; Bonett, Steffen, Lambert, et al. 2014; Kuchta et al. 2016) and the presence of populations with intermediate phenotypes (Brandon 1966; B. T. Miller and Niemiller 2008) does not support recognition of these taxa as distinct subspecies (Niemiller and Niemiller 2020). Populations that lack noticeable spotting and are pale in pigmentation occur in caves along the eastern escarpment of the Cumberland Plateau in southern Franklin and Marion Counties, Tennessee, and Jackson County, Alabama (B. T. Miller and Niemiller 2008). More heavily pigmented populations with numerous small to large spots on the dorsum are known from the western escarpment of the Cumberland Plateau, Eastern Highland Rim, and Central Basin of Tennessee (B. T. Miller and Niemiller 2008). Several additional populations are known from portions of northern Alabama and northwestern Georgia. The range of *G. palleucus* is parapatric with respect to the range of *G. porphyriticus*, although the two species may intergrade in a couple of cave systems in northwestern Alabama.

The Berry Cave salamander (*G. gulolineatus*) was described originally as a subspecies of *G. palleucus* (Brandon 1965) and is known from 10 caves in the Clinch River and Tennessee River watersheds of the Appalachians karst region of east Tennessee. The specific status of *G. gulolineatus* had been long debated (e.g., Brandon et al. 1986; Petranka 1998), but is a distinct species based on osteological evidence, morphology, physiography, and phylogenetic analyses (B. T. Miller and Niemiller 2008; Niemiller et al. 2008; Niemiller and Miller 2010). The distributions of *G. gulolineatus* and *G. porphyriticus* overlap, and these species occur syntopically at five caves. However, *G. gulolineatus* does occur in greater abundance at these sites.

Finally, the West Virginia spring salamander (*G. subterraneus*) is endemic to a single cave system, General Davis Cave in Greenbrier County, West Virginia. Unlike the other two stygobiontic congeners, *G. subterraneus* regularly metamorphoses, albeit at a very large size (>90 mm snout to vent length; Besharse and Holsinger 1977; Niemiller, Osbourn, et al. 2010). Its large larvae are likely paedomorphic (Besharse and Holsinger 1977), as metamorphosed adults are not regularly observed (Niemiller, Osbourn, et al. 2010). Both larvae and adults can be differentiated morphologically from *G. porphyriticus*, which also occurs abundantly at General Davis Cave.

A few studies have examined the phylogenetic relationships of species and subspecies long recognized in the genus *Gyrinophilus*. Distinction of *G. porphyriticus* and the three stygobiontic species currently recognized is based primarily on the morphological analyses of Brandon (1966). Allozyme data (Addison Wynn, pers. comm., 2006) supports the recognition of *G. palleucus* and *G. gulolineatus* as distinct lineages from *G. porphyriticus*, while *G. gulolineatus* Brandon, 1965 possesses three unique alleles not shared with *G. palleucus*. Allozyme electrophoresis conducted by Howard et al. (1984) found six unique alleles in *G. subterraneus* not shared with *G. porphyriticus*. Although sample sizes were small, the authors believed *G. subterraneus* was likely a valid species and distinct from *G. porphyriticus*. Baldwin (2002) found that the two subspecies of *G. palleucus* form a monophyletic group sibling to *G. porphyriticus*, which was based on sequences of the *nd4* and *cytb* mitochondrial genes and the *rag1* nuclear gene.

The most comprehensive phylogenetic study of cave-dwelling *Gyrinophilus* was conducted by Niemiller et al. (2008), who inferred phylogenetic relationships from two mitochondrial loci (*12S* ribosomal and *cytb*) and the nuclear *rag1* locus from 27 populations of *G. palleucus* and *G. gulolineatus* and 15 populations of *G. porphyriticus* in Tennessee, Alabama, and Georgia. In a later study, Niemiller et al. (2009) included *G. subterraneus* in the phylogenetic analysis of *cytb* that included data from Niemiller et al. (2008). Both studies revealed the stygobiontic species were phylogenetically nested within *G. porphyriticus*, a result also observed in subsequent phylogenetic studies of *Gyrinophilus* (Bonett, Steffen, Lambert, et al. 2014; Kuchta et al. 2016). In general, these analyses supported the current taxonomy, including recognition of *G. palleucus* and *G. gulolineatus*, and potentially multiple lineages in *G. porphyriticus*. However, *G. subterraneus* may not represent a valid species but rather a variant of *G. porphyriticus* (see below). Although populations of the two subspecies of *G. palleucus* from and near the type localities are morphological distinct, intergrade populations are known from parts of northern Alabama. Moreover, mitochondrial divergence is low between these subspecies and does not fall into distinct genealogical clusters (Niemiller et al. 2008, 2009). Consequently, the continued recognition of these two distinct subspecies may not be warranted (refer to Niemiller et al. 2009). Despite geographic overlap with *G. porphyriticus*, *G. gulolineatus* maintains genetic distinctness from its congener, albeit with some gene flow. This distinction is similar to *G. palleucus* with *G. porphyriticus*.

In *Gyrinophilus*, subterranean colonization appears to have occurred in the late Pliocene and Pleistocene (Niemiller et al. 2008; Bonett, Steffen, Lambert, et al. 2014). This is reflected in the poorer phylogenetic resolution of nuclear loci (Niemiller et al. 2008). Brandon (1971) hypothesized that *G. palleucus* and *G. gulolineatus* arose during the Pleistocene from a surface-dwelling, metamorphosing ancestor similar to present-day *G. porphyriticus*. He further suggested that cave colonization was driven by fluctuating climatic surface conditions that forced preadapted populations at the peripheral range of the ancestral species underground; this may have ultimately resulted in speciation (i.e., the climate-relict model; Vandel 1964; Holsinger 2000; Wynne et al., Chap. 2). Although mitochondrial and nuclear gene genealogies and time-calibrated phylogenetic analyses primarily support Brandon's (1971) hypothesis, evidence of gene flow suggests that the stygobiontic species of *Gyrinophilus* did not diverge from their surface ancestor in strict allopatry.

Coalescent-based analyses by Niemiller et al. (2008) strongly suggest divergence and ultimately speciation of stygobiontic species occurred despite either low levels of continuous or periodic gene flow with *G. porphyriticus* (i.e., an adaptive shift; Howarth 1973; Rivera et al. 2002; Howarth et al. 2019). Niemiller et al. (2009) favored a scenario with more episodic gene flow between *Gyrinophilus* cave- and surface-dwelling populations, which was prompted by changing climate from the late Pliocene through the Pleistocene. Under this scenario (termed the "periodic-isolation hypothesis"), cave-dwelling populations were isolated from surface populations during warm interglacial periods but broadly overlapped during glacial periods when favorable cool, moist conditions were more prevalent at lower elevations. This facilitated secondary contact and gene flow. Given that interglacial periods were generally shorter than glacial periods and epigean and hypogean *Gyrinophilus* are not completely geographically isolated presently, periods of true isolation were likely brief relative to periods favorable to geographic overlap. Nonetheless, *G. palleucus*, which is parapatrically distributed relative to *G. porphyriticus* and *G. gulolineatus*, was isolated from *G. porphyriticus* for a long enough period to cause reproductive isolation. However, evidence of contemporary gene flow with *G. porphyriticus* in some populations of *G. palleucus* and *G. gulolineatus* suggests that reproductive isolation is not entirely complete throughout their ranges (Niemiller, Miller, et al. 2010; Niemiller et al. 2018).

Gyrinophilus subterraneus, which occurs much further north in West Virginia, is syntopic with *G. porphyriticus* at General Davis Cave, Greenbrier County, West Virginia. It occurs within the range of *G. porphyriticus* in the Appalachians. Thus, periods of isolation were likely much shorter than those experienced by *G. palleucus* and *G. gulolineatus* to the south. Like other stygobiontic *Gyrinophilus*, *G. subterraneus* is phylogenetically nested within *G. porphyriticus* (Niemiller et al. 2009; Bonett, Steffen, Lambert, et al. 2014). Gene networks indicate that mitochondrial haplotypes (*cytb*) and nuclear alleles (*rag1* and *rhodopsin*) are closely related to, and shared with, those of *G. porphyriticus* at General Davis Cave and other caves in West Virginia and Virginia (Fig. 6.1). Moreover, although *G. subterraneus* and *G. porphyriticus* occur syntopically, and *G. porphyriticus* adults are commonly observed at General Davis Cave, no larvae have been morphologically identified as *G. porphyriticus*—despite more than 30 years of semiregular surveys (Niemiller, Osbourn, et al. 2010).

Niemiller, Osbourn, et al. (2010) suggested the absence of observed *G. porphyriticus* larvae (as well as smaller size classes of *G. subterraneus* larvae) may be due to size-based detection bias, size-based habitat segregation, or low survivorship of eggs and smaller larvae. We propose another possible explanation in light of recent genetic data. *Gyrinophilus* at General Davis Cave represents a single admixed population, and consequently, *G. subterraneus* should be synonymized with or treated as a subspecies of *G. porphyriticus*. Past authors have also disputed the validity of *G. subterraneus* as a distinct species. For example, Blaney and Blaney (1978) suggested *G. porphyriticus* is a polymorphic species, and the population at General Davis Cave is a transitional cave form with varying levels of paedomorphosis and troglomorphy. Rather than representing a distinct taxon, larval salamanders at General Davis Cave currently recognized as *G. subterraneus* may be a unique local variant of *G. porphyriticus* with smaller eyes, wider heads, more premaxillary and prevomerine teeth, and distinct patterning compared to larvae in populations from other localities in the Virginias. If so, the larvae at General Davis Cave ultimately metamorphose at a large body size (>90 mm snout to vent length) into the large *G. porphyriticus* adults commonly observed in this cave. The uniqueness of this population may be attributed to the past isolation from other *G. porphyriticus* populations, but secondary contact occurred before reproductive isolation developed. Therefore, *Gyrinophilus* at General Davis Cave may represent an example of failed speciation.

But what do we make of the adult form of *G. subterraneus* that is morphologically distinct from *G. porphyriticus*? Metamorphosed *G. subterraneus* have a gaunt appearance, reduced eyes, an indistinct canthus rostralis, and

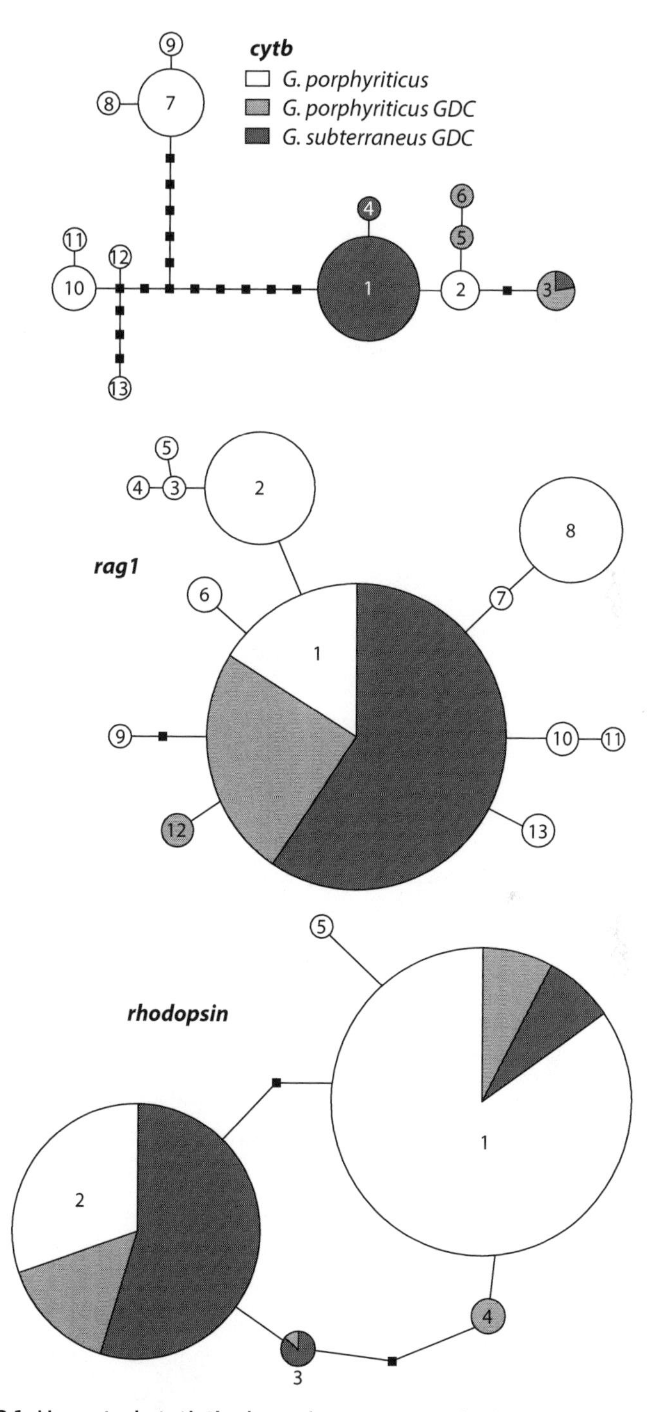

Figure 6.1. Unrooted statistical parsimony networks for the mitochondrial *cytb* and nuclear *rag1* and *rhodopsin* loci for 28 *Gyrinophilus subterraneus* and 12 *G. porphyriticus* sampled at General Davis Cave, Greenbrier County, West Virginia, and other *G. porphyriticus* from West Virginia and Virginia. Area of circles in each network is proportional to the number of individuals with that haplotype. Black squares represent unsampled or extinct haplotypes.

an undivided premaxilla, and they retain much of the reticulate patterning observed in larvae (Besharse and Holsinger 1977; Niemiller, Osbourn, et al. 2010). In several respects, these metamorphosed salamanders resemble metamorphosed *G. palleucus* and *G. gulolineatus* (Simmons 1976; Brandon et al. 1986; B. T. Miller and Niemiller 2008; Niemiller, Miller, et al. 2010). Since 1975, 39 metamorphosed *G. subterraneus* have been documented; of these, only 15 metamorphosed individuals have been recorded since 1985. Under our hypothesis of a single admixed population, this unusual adult form may represent individuals that possess deleterious alleles in loci associated with traits that change during metamorphosis. Although the fate of these individuals with this phenotype has not been tracked, metamorphosed *G. gulolineatus* (some may have hybridized with *G. porphyriticus*) die within a few weeks in captivity, possibly due to starvation (as a result of not feeding) or developmental issues (Niemiller, Miller, et al. 2010). In contrast, if *G. subterraneus* is distinct, the adult form may represent individuals induced to metamorphose in association with environmental stressors, such as flooding, starvation, pollution, and/or disease (B. T. Miller 1995; B. T. Miller and Niemiller 2008; Niemiller, Miller, et al. 2010). Which of these two competing hypotheses is correct will remain unclear until larvae are reared through metamorphosis in the lab or tracked in situ in General Davis Cave. A robust population genomic study would also elucidate whether one or two diagnosable genetic clusters exist. Although *G. subterraneus* is closely related to *G. porphyriticus* at General Davis Cave, the larvae possess both a unique *cytb* haplotype and distinctive allozyme alleles not found in adult *G. porphyriticus* (Howard et al. 1984).

Finally, what facilitates divergence between related cave and surface populations of *Gyrinophilus* that may have experienced episodic gene flow? Several factors may contribute to genetic distinctness, including assortative mating and selection against hybrids, as well as parapatric divergence. As subterranean-adapted species become better adapted to conditions underground, they may disperse deeper into cave systems and ultimately shift to paedomorphosis (Niemiller et al. 2008, 2009).

Paedomorphosis may contribute to premating isolation (Niemiller et al. 2008). Courtship between paedomorphs must occur in water, whereas courtship of metamorphs can occur in water or on land. In *Gyrinophilus*, only the latter has been observed (Beachy 1997), perhaps leading to reproductive isolation associated with breeding habitat. Moreover, paedomorphosis may interfere with mate recognition if paedomorphs do not develop the same courtship pheromones as adults (Bonett and Chippindale 2006). However, successful courtship between paedomorphs and metamorphs have been observed in other salamander species (e.g., Semlitsch and Wilbur 1989; Whiteman et al. 1999).

Moreover, availability of breeding habitat free from competitors may be a primary ecological advantage of subterranean colonization in *Gyrinophilus* (Niemiller et al. 2008) and perhaps other hypogean species. However, permanent existence in caves requires special sensory, metabolic, and life history adaptations for efficient foraging and resource use. These adaptations likely involve life history trade-offs necessary for ecological speciation (Coyne and Orr 2004).

Edwards Plateau *Eurycea*

Nearly all Edwards Plateau *Eurycea* (*Paedomolge*) species of central southwestern Texas include at least one subterranean population, and many are stygobionts. We estimate there may have been as many as 21 independent cave colonization events in this group (Plate 6.2), as not all subterranean populations within a given species are each other's closest relatives (Bendik et al. 2013).

The taxonomy of this group is complicated, principally because its evolutionary history includes confounding classifications based on traditional morphology. Molecular approaches have largely resolved these issues, but shallow divergence times and ongoing gene flow among epigean and hypogean populations (and even among species) challenge delimitating species' distributional ranges (Devitt et al. 2019).

Approximately 22 mya, Edwards Plateau *Eurycea* (*Paedomolge*) shared a paedomorphic ancestor. This likely set the stage for high levels of speciation and colonization of subterranean environments (Bonett, Steffen, Lambert, et al. 2014). *Paedomolge* includes three clades (Hillis et al. 2001): *Typhlomolge*, *Blepsimolge*, and *Septentriomolge*. All *Typhlomolge* are highly specialized stygobionts inhabiting aquifers at the eastern margin of the Edwards Plateau in Comal, Hays, and Travis Counties. The most recognizable of these species is *E. rathbuni* (Plate 6.1D). Two other members of the subgenus *Typhlomolge* (*E. robusta* and *E. waterlooensis*) have been described, while a fourth species from New Braunfels awaits description. Mitchell and Reddell (1965) proposed placing *Typhlomolge* in the genus *Eurycea*. Wake

(1966) proposed retention of the genus *Typhlomolge* (consisting of *E. rathbuni*, *E. robusta*, and *E. tridentifera* [= *E. latitans*]) based on morphological similarities shared among stygobiontic members. Chippindale et al. (2000) demonstrated that *E. tridentifera* (= *E. latitans*) is deeply nested within the clade now referred to as *Blepsimolge*. However, molecular studies have demonstrated that *Typhlomolge* forms a clade with *Blepsimolge* (*Notiomolge* = *Typhlomolge* + *Blepsimolge*), and the Colorado River has separated *Notiomolge* from *Septentriomolge* since the Miocene (Hillis et al. 2001; Devitt et al. 2019). The split within *Notiomolge* reflects an ancient cave colonization event by *Typhlomolge*, resulting in an adaptive shift. This pattern was also observed throughout more recent lineages within *Paedomolge* (Bendik et al. 2013).

The taxonomic history of other *Paedomolge* is more complex. Most epigean populations were lumped into a single species (e.g., *E. neotenes*). Molecular studies revealed high levels of species diversity and a deep divergence separating southern and northern species about 10 mya (Chippindale et al. 2000; Bonett et al. 2013). Combined molecular and morphological studies also illustrate contrasting patterns of morphological conservatism among epigean populations, as well as considerable variation across hypogean populations (Bendik et al. 2013). Devitt et al. (2019) used genome-wide sequence data to reveal cryptic species, as well as to define distributional boundaries of several species within this group.

The *Blepsimolge* clade contains seven described species (with three additional species awaiting description) that occur in springs and caves from Hays County to Val Verde County (Devitt et al. 2019). With the exception of *E. nana*, all Edwards Plateau *Eurycea* species include both epigean and hypogean populations. Some populations of *E. latitans*, *E. neotenes*, and *E. pterophila* may be considered highly specialized stygobionts. Recent studies suggest that *E. tridentifera* is a stygobiontic ecomorph of *E. latitans* (Bendik et al. 2013; Devitt et al. 2019). Individuals at the type locality are not uniformly stygobiontic, although they represent a single interbreeding population. While most individuals are blind with reduced pigment, individuals from the twilight zone of Honey Creek Cave are pigmented with functional eyes (Bendik et al. 2013). Genetic distances between epigean and hypogean populations within *Blepsimolge* are generally small, and morphologically similar populations are not necessarily one another's closest relatives. Thus, because of rapid, parallel phenotypic evolution, the current taxonomy may not provide an accurate reflection of species boundaries (Bendik et al. 2013).

Eurycea chisholmensis Chippindale, Price, Wiens & Hillis, 2000, *E. naufragia* Chippindale, Price, Wiens & Hillis, 2000, and *E. tonkawae* Chippindale, Price, Wiens & Hillis, 2000 form the clade *Septentriomolge*, which occurs north of the Colorado River in Travis, Williamson, and Bell Counties (Chippindale et al. 2000). These species are typically associated with springs, although individuals may spend most of their time underground. In addition, obligate subterranean populations of all three species have been documented. The most extreme stygobiontic forms are populations of *E. tonkawae* that occur in the Buttercup Creek Karst caves of northern Travis and southern Williamson Counties.

As mentioned above, paedomorphosis likely facilitated the colonization of subterranean habitats in this group. Spring-dwelling species often retreat underground during periods of drought, and it is thought that epigean species lay their eggs underground. Additionally, some populations may spend months or years without access to the surface. At least one species (*E. tonkawae*) is known to undergo reversible body-length shrinkage due to low food availability during drought (Bendik and Gluesenkamp 2013). This suite of traits greatly increases the ability of epigean salamanders to colonize subterranean environments, as evidenced by the high frequency of cave colonization. However, stygobiontic physical traits do not appear to be a requirement for cave colonization. Although epigean forms may occur in a wide range of subterranean environments, stygobiontic forms only occur in extensive deep zone habitats. In this environment, climatic conditions are relatively stable over time and characterized by reduced seasonal fluctuation (e.g., Howarth and Wynne, Chap. 1).

Georgia Blind Salamander (*Eurycea wallacei*)

One of the least studied hypogean salamanders is the stygobiontic Georgia blind salamander (*Eurycea wallacei* (Carr, 1939); Plate 6.1E). Restricted to the Upper Floridan Aquifer in the Dougherty Plain of southwestern Georgia and adjacent northwestern Florida, this species was initially placed in the monotypic genus *Haideotriton* (Carr, 1939) but later synonymized in the genus *Eurycea* (Frost et al. 2006). Recent analysis of *E. wallacei*'s evolutionary history demonstrated significant mitonuclear discordance, which has been observed in other plethodontids (e.g., Edgington et al. 2016). *Eurycea wallacei* is composed of two highly divergent mtDNA lineages. Both exhibit high within-population diversity and haplotype sharing between distant localities, suggesting subterranean connectivity (J. G. Phillips, unpublished data). Although mitochondrial evolution in *E. wallacei* may follow two separate trajectories, the nuclear genome suggests an alternative hypothesis—that there is more recent coalescence and high levels of admixture throughout the species range. The presence of mitochondrially divergent individuals suggest that gene flow among *E. wallacei* lineages occurred following a period of isolation and may represent secondary contact. Because karst can be dynamic over geologic time scales and the fracturing and fragmentation of this substrate is likely (Palmer 2007; de Waele 2017), the development of new subterranean conduits connecting previously isolated habitats plausibly explains the high levels of genetic diversity observed in *E. wallacei*.

Multilocus estimations of phylogenetic structure of *E. wallacei* indicate this species likely diverged from two-lined salamanders (*E. bislineata/E. cirrigera* complex) between 25 mya (based upon mtDNA) and 12 mya (based upon nuclear DNA; J. G. Phillips, unpublished data). Using single nuclear and 2–3 mitochondrial genes, the nuclear estimate is much closer to previous estimates from other spelerpine phylogenies (Bonett, Steffen, Lambert, et al. 2014; Steffen et al. 2014; Wray and Steppan 2017). Although the phylogeny cannot definitively identify a location for the origin of the *E. wallacei* clade, we suggest this species likely originated in central Georgia near the population in the Lower Flint River drainage, Radium Springs area. Interestingly, Radium Springs and the surrounding aquifer were likely inundated with salt water until the late Oligocene to early Miocene (J. A. Miller 1986). This shallow sea was believed to have receded in the middle Miocene, which aligns with the approximate divergence date of *E. wallacei* from a surface ancestor and its subsequent colonization of subterranean habitats. This species may have dispersed from the present-day Upper Floridan Aquifer to the Chipola River drainage around the middle Miocene, but permanent colonization and further dispersal to the Choctawhatchee River drainage was unlikely until the Pliocene. Both the Chipola and Choctawhatchee drainages failed to form a clade (J. G. Phillips, unpublished data), which suggests multiple dispersal events and potentially ongoing movement of individuals between drainages through the associated karst aquifer.

Unfortunately, this relationship is difficult to evaluate because molecular samples are scarce from the northern part of *E. wallacei*'s range. If diversification occurred after the colonization of groundwater habitats and the divergence from an epigean ancestor, *E. wallacei* lineages could have become separated and genetically isolated in the northern part of their current range. As the aquifer to the south became accessible, both lineages could have dispersed along similar routes to their contemporary habitats. This would account for the high diversity within localities and haplotype sharing among distant localities. Additionally, these lineages would be reproductively isolated, thus preserving the high levels of genetic diversity observed.

Grotto Salamander (*Eurycea spelaea*) Complex

The grotto salamander species complex is endemic to the Springfield and Salem Plateaus in the Ozark Mountains of southern Missouri, extreme southeastern Kansas, and adjacent areas in Arkansas and Oklahoma. Until recently, all Ozark populations were considered a single species. Initially described as *Typhlotriton spelaeus* (Stejneger, 1892), this species remained in a monotypic genus and was based on morphological distinctness. However, Wake (1993) suggested the taxonomic status warranted reevaluation. This complex had been erroneously separated into two other species, *T. nereus* (Bishop, 1944) and *T. braggi* (C. C. Smith, 1968), but were later synonymized with *T. spelaeus* (Brandon 1966; Brandon and Black 1970). Genetic analysis later revealed *Typhlotriton* was nested within other species of *Eurycea* from the Ozark Plateau (Bonett and Chippindale 2004).

Three highly divergent lineages with strong support and concordance between mitochondrial and nuclear DNA were recently uncovered (J. G. Phillips et al. 2017). Coalescence of these lineages likely occurred during the middle Miocene (~15–10 mya) with the deepest divergence separating populations from the western and eastern portions of its range (Fig. 6.2). Divergence among these clades follows the divide between the Middle Arkansas River and ancestral White River paleodrainages. The south-flowing White River paleodrainage included the Gasconade and Osage Rivers, but both rivers were captured by the Lower Missouri River drainage during the Pleistocene (where they still flow today). The eastern clade (which occurs entirely within the White River paleodrainage) contained an early divergence (13–9 mya) corresponding to the northern and southern clades, and predominantly overlapping with the Salem and eastern Springfield

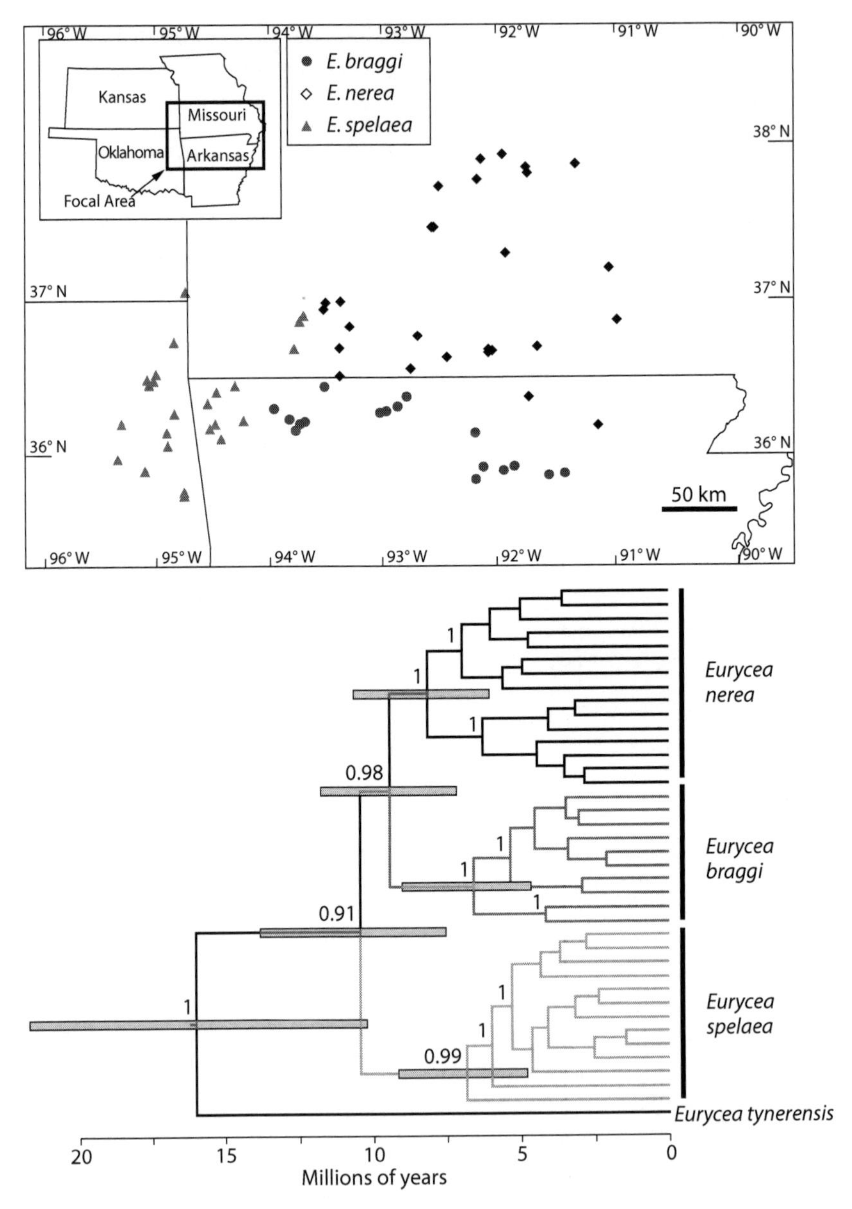

Figure 6.2. Phylogenetic relationships and geographic distributions for the grotto salamander (*Eurycea spelaea*) species complex in the Ozark Highlands of Arkansas, Missouri, and Oklahoma. Modified from J. G. Phillips et al. (2017).

Plateaus (J. G. Phillips et al. 2017). A similar line of evidence to support the separation of these clades is the presence of several species of fishes that demonstrate morphological similarities across the White and Lower Missouri River drainages (e.g., R. V. Miller 1968; Cross et al. 1986).

Rapid divergence of the three *Eurycea* lineages indicates a nonadaptive radiation resulting in multiple cryptic species. Consequently, J. G. Phillips et al. (2017) recommended resurrecting the two former synonyms of *E. spelaea* (now *E. braggi* and *E. nerea*). This change now places *E. braggi* in the eastern Springfield Plateau and the eastern portion of the West Springfield Plateau of northern Arkansas, while *E. nerea* (Bishop, 1944) occurs in the Salem Plateau and eastern West Springfield Plateau of south-central Missouri and northern Arkansas (J. G. Phillips et al. 2017). *Eurycea spelaea* (Plate 6.1F) is distributed across in the West Springfield Plateau of southwestern Missouri, extreme southeastern Kansas, northwestern Arkansas, and northeastern Oklahoma. Around 25 mya, the ancestor of the endemic *Eurycea* (*E. spelaea* complex and its sibling taxon *E. tynerensis*) likely dispersed to the Ozark Mountains (Martin et al. 2016), with rapid radiations occurring shortly thereafter. Deep phylogenomic divergences suggest grotto salamanders have been subterranean-adapted since the middle Miocene, which is older than some cave-obligate lineages (e.g., *Gyrinophilus gulolineatus* and *G. palleucus*; Niemiller et al. 2008). This may be attributable to the unique ontogenetic shift—visually adept larvae capable of living on the surface and subterranean-adapted adults. These larvae may also exhibit behavior of longer inhabitance of subterranean habitats by way of omnivory and coprophagy of bat guano (Fenolio et al. 2006). Middle Miocene divergence among grotto salamander lineages provides a minimum estimate for colonization of hypogean environments and subsequent evolution of subterranean adaptation.

Conservation and Future Research

Cave salamanders provide remarkable examples of adaptation to extreme environments. Subterranean colonization occurred multiple times in three distinct lineages (bolitoglossines, proteids, and spelerpines), offering numerous opportunities to study patterns of vertebrate evolution, ecological and morphological convergence, and drivers of speciation. In addition, cave salamanders represent an excellent group to compare patterns and processes observed in the only other vertebrate group to colonize the subterranean realm, cavefishes. Although the study of cave salamanders currently lags behind that of cavefishes in both breadth and depth, both groups promise to be mutually informative, and various researchers are working on developing cave salamanders as model organisms, as has been done with some cavefish (e.g., Jeffery 2008; Zhao et al., Chap. 7).

Exciting advances in captive husbandry have also been made. For example, long-term lab studies of *Proteus* have begun to inform us of longevity and reproductive cycles (e.g., Aljančič 1993). Research on captive populations can provide important information related to reproductive biology and reproductive output, as well as provide specimens for enhanced research in genetics, environmental toxicology, and behavioral biology.

Unfortunately, most species of hypogean salamanders are considered vulnerable or threatened with extinction. All species are sensitive to human activities. For example, both overexploitation of groundwater and contamination of the Floridan Aquifer across the range of *E. wallacei* threatens this species' persistence (Fenolio et al. 2013). Texas *Eurycea* species are at risk due to current declines in groundwater availability and increased air temperature caused by global climate change (Devitt et al. 2019). Populations of *G. gulolineatus* are threatened by habitat degradation and groundwater pollution associated with urbanization, agriculture, and previous quarrying operations in and around the Knoxville metropolitan area in eastern Tennessee (Niemiller et al. 2018).

Proteus anguinus is identified as "Vulnerable" (IUCN 2021) by the International Union for Conservation of Nature for the *IUCN Red List of Threatened Species* and listed in Appendix II of the Berne Convention and Annex II and Annex IV of the EU Habitats Directive (Arntzen et al. 2009). Additionally, this species is considered part of the cultural heritage for many regions throughout its native range in Bosnia and Herzegovina, Croatia, Italy, and Slovenia. This likely aided in the legislative protection of this species across much of its range—particularly in Slovenia, Croatia, and Italy. Both *Proteus* subspecies lack protection in Bosnia and Herzegovina, and Montenegro. Given their endemism and restricted distribution, both *P. anguinus* and black proteus (*P. a. parkelj*) are perhaps in greatest need of protection and further assessment (Arntzen et al. 2009; Gorički et al. 2017). While multiple evidenced-based conservation initiatives are in place at various scales, including the PROTEUS Olm Genome Project (POGP 2021) and the work of Zakšek and Trontelj (2017), more research is needed.

Most salamanders have not received the attention awarded to *Proteus*, although most Edwards Plateau *Eurycea* (10 of 14 species) are state or federally listed as threatened or endangered (TPWD 2019). All are listed as "Vulnerable" or "Data Deficient" by IUCN (Table 6.1)—although both ranks reflect a general lack of information about the biological status of the species in question. Because we now know more about these species than when the assessments were conducted, we recommend these species be reassessed. The *E. spelaea* complex was only recently split. The new nomenclature has not been implemented, nor have either of the newly delimited lineages been evaluated for state-level protection in Missouri or Arkansas. However, this complex has been listed as "endangered" in Kansas (KDWP 2021) and as a species of conservation need or concern in Missouri (MNHP 2019) and Oklahoma (ODWC 2016). Also, *Eurycea wallacei* is listed as IUCN "Vulnerable" (IUCN 2021), "threatened" in Florida, and "endangered" in Georgia (Fenolio et al. 2013). *Chiropterotriton magnipes* is listed as IUCN "Endangered" (IUCN 2021), and Mexico lists the species under their "special protection" category (Parra-Olea et al. 2008). *Gyrinophilus subterraneus* and *G. gulolineatus* are both listed as IUCN "Endangered" (Hammerson 2004; Hammerson and Beachy 2004b; IUCN 2021). *Gyrinophilus subterraneus* probably faces the greatest risk, as it occurs in a single system in West Virginia and is not protected. The cave system itself is at least partially protected by a conservation easement. Additionally, *Gyrinophilus gulolineatus* is protected under Tennessee state law, and the status of the species is currently being evaluated for listing under the US Endangered Species Act (Withers 2016; Niemiller et al. 2018). *Gyrinophilus palleucus* is considered "Vulnerable" by the IUCN (Hammerson and Beachy 2004a; IUCN 2021) and is protected in Alabama (under "priority 2" status; ADCNR 2015), Georgia ("Threatened"; GDNR 2015), and Tennessee ("Threatened"; Withers 2016).

Despite recent advances in next-generation sequencing (NGS), the analysis of genome-scale data in salamanders has lagged behind other vertebrate groups. This is primarily due to large genomes with low nuclear variation (Jockusch 1997; Sun et al. 2012; Sun and Mueller 2014; Keinath et al. 2015) that can be 14 to 120 Gb, and up to 40 times larger than the human genome (Nowoshilow et al. 2018). Recently, the number of studies on salamanders using NGS data has increased on several fronts, including high-throughput sequencing of parallel-tagged amplicons (O'Neill

Table 6.1. Stygobiontic and troglobiontic salamander diversity in North America and Europe and IUCN Red List conservation status and associated criteria. Troglobiontic species are denoted by an asterisk (*). All other species are stygobionts. IUCN Red List categories include Data Deficient (DD), Least Concern (LC), Near Threatened (NT), Vulnerable (VU), Endangered (EN), Critically Endangered (CR), Extinct in the Wild (EW), and Extinct (EX). For criteria see IUCN (2019, 2021).

Species	Common name	Region	IUCN Red List
Family Plethodontidae			
**Chiropterotriton magnipes* Rabb, 1965	Bigfoot splayfoot salamander	Sierra Madre Oriental, Mexico	CR B2ab (iii, v)
**Eurycea braggi* (Smith, 1968)	Southern grotto salamander	Ozark Highlands, USA	
**Eurycea nerea* (Bishop, 1944)	Northern grotto salamander	Ozark Highlands, USA	
Eurycea rathbuni (Stejneger, 1896)	Texas blind salamander	Edwards Plateau, USA	VU D2
Eurycea robusta (Longley, 1978)	Blanco blind salamander	Edwards Plateau, USA	DD
**Eurycea spelaea* (Stejneger, 1892)	Western grotto salamander	Ozark Highlands, USA	LC
Eurycea tonkawae Chippindale, Price, Wiens & Hillis, 2000	Jollyville Plateau salamander	Edwards Plateau, USA	EN B1ab (iii, v)
Eurycea tridentifera Mitchell & Reddell, 1965 [= *E. latitans* Smith & Potter, 1946]	Comal blind salamander	Edwards Plateau, USA	VU D2
Eurycea wallacei (Carr, 1939)	Georgia blind salamander	Floridan Aquifer, USA	VU B1ab (iii) + 2ab (iii)
Eurycea waterlooensis Hillis, Chamberlain, et al. 2001	Austin blind salamander	Edwards Plateau, USA	VU D2
Gyrinophilus gulolineatus Brandon, 1965	Berry Cave salamander	Appalachian Mountains, USA	EN B1ab (iii) + 2ab (iii)
Gyrinophilus palleucus McCrady, 1954	Tennessee cave salamander	Interior Low Plateau, USA	VU B2ab (iii, v)
Gyrinophilus subterraneus Besharse & Holsinger, 1977	West Virginia spring salamander	Appalachians Mountains, USA	EN D
Family Proteidae			
Proteus anguinus Laurenti, 1768	Olm	Dinaric Karst, Bosnia and Herzegovina, Croatia, Italy, and Slovenia	VU B2ab (ii, iii, v)

et al. 2013; Wielstra et al. 2014; Zielinski et al. 2014), anchored hybrid enrichment (Newman and Austin 2016; J. G. Phillips et al. 2017), and RADseq (Nunziata et al. 2017; Nunziata and Weisrock 2018; Devitt et al. 2019). In 2018, 15 years after the human genome was sequenced, the first salamander genome (*Ambystoma mexicanum*) was published (Nowoshilow et al. 2018). This study may serve to further advance methods used to study salamanders' genetics (refer to Weisrock et al. 2018 for a full review). Additionally, eDNA will enable researchers to identify the presence of organisms with low detection probabilities. This provides an alternative to typical survey methods by permitting the sampling of habitats previously inaccessible (i.e., wellhead water samples may now be analyzed for salamander presence). Gorički et al. (2017) recently used this approach successfully to detect and monitor *Proteus anguinus*.

These genetic methods could be integrated with physiology and life history data, especially in relation to surface relatives, to better understand past, present, and future colonization of subterranean environments. Importantly, more studies are required to better describe the basic life history characteristics of cave-dwelling salamanders. For most species, we have limited information on the most important life history traits, including growth rates, age of sexual maturity, lifespan, fecundity, and parental care.

A continual theme in studying the evolution of subterranean-adapted organisms is that cave inhabitants are subjected to reduced food resources. While some cave systems have high allochthonous input, energy limitations in general may form a predominant selection pressure on subterranean-adapted species. However, variance in energy limitation, rather than sheer reduction, may be a more important driver. Accordingly, greater attention should be placed on actual levels of historical and contemporary energy resources when considering subterranean colonization and evolutionary patterns of troglomorphism.

Studies to determine geographic distribution and population estimates will vastly improve our ability to conserve these unique amphibians. We also need improved environmental toxicology research that clearly defines concentrations of common agrochemicals (common to the regions where these salamanders exist) known to interfere with the health of these populations. Additionally, research is needed to determine the minimum aquifer levels that retain enough water to sustain viable salamander populations, and the retention of said levels should be incorporated into the environmental regulations of local municipalities (e.g., Stamm et al. 2015).

Community education and outreach efforts should also be enhanced to connect the health of salamander populations with good groundwater quality—a shared goal for both salamanders and humans, inasmuch as healthy salamanders living within the groundwater upon which humans are reliant can be used as indicators for water quality. Activities that clearly highlight the importance of this linkage should become a priority of researchers, land managers, and speleological organizations. As the human population increases, exploitation of the natural resources that these organisms inhabit will be exploited, intensified, and contaminated beyond current levels, challenging the persistence of these taxa.

Most species are threatened due to groundwater quality and quantity, while some species face direct impacts due to human activities including climate change (Stamm et al. 2015; Devitt et al. 2019; NatureServe 2019). An overarching issue challenging the persistence of subterranean salamanders (their subterranean habitats and most cave-dwelling fauna, in general) is best summarized as "out of sight, out of mind." Most people are unaware that such organisms exist (often near expansive human population centers). Caves, aquifers, and the organisms that inhabit them are often considered mysterious and are difficult to access and explore—especially for nonspecialists, leading to significant knowledge gaps influencing our ability to conserve and manage the communities inhabiting these areas. We hope this chapter serves to underscore the importance of these amazing animals. We hope readers will gain a sense of appreciation that will lead to increased study, awareness, and policies that help to safeguard cave-dwelling salamanders and their habitats.

ACKNOWLEDGMENTS

The authors thank Chris Beachy and Brian Miller for reviewing this chapter. Jerry Fant provided the image of *Chiropterotriton magnipes* shown in Plate 6.1, and Nathan F. Bendik contributed images of *Eurycea* species provided in Plate 6.2.

REFERENCES

[ADCNR] Alabama Department of Conservation and Natural Resources. 2015. *Alabama's Wildlife Action Plan*. Montgomery, Alabama, USA: Alabama Department of Conservation and Natural Resources, Division of Wildlife and Fisheries.

Aljančič, Marko, Boris Bulog, Andrej Kranjc, Drasko Josipovič, Boris Sket, and Peter Skoberne. 1993. *Proteus—the Mysterious Ruler of Karst Darkness*. Ljubljana, Slovenia: Vitrum.

Aljančič, Marko, Peter Habič, and Andrej Mihevc. 1986. "Črni močeril iz Bele krajine [The Black Olm from White Carniola]." *Naše Jame* 28:39–44.

AmphibiaWeb. 2021. Accessed February 2, 2021. https://amphibiaweb.org.

Arntzen, Jan W., and Boris Sket. 1996. "Speak of the Devil: The Taxonomic Status of *Proteus anguinus parkelj* Revisited (Caudata: Proteidae)." *Herpetozoa* 8:165–166.

Arntzen, Jan W., and Boris Sket. 1997. "Morphometric Analysis of Black and White European Cave Salamanders, *Proteus anguinus*." *Journal of Zoology* 241:699–707.

Arntzen, Jan W., Mathieu Denoël, Claude Miaud, Franco Andreone, Milan Vogrin, Paul Edgar, et al. 2009. "*Proteus anguinus*." *The IUCN Red List of Threatened Species* 2009:e.T18377A8173419. Accessed July 1, 2019. https://www.iucnredlist.org/species/18377/8173419.

Baird, Amy B., Jean K. Krejca, James R. Reddell, Colin E. Peden, Meredith J. Mahoney, and David M. Hillis. 2006. "Phylogeographic Structure and Color Pattern Variation among Populations of *Plethodon albagula* on the Edwards Plateau of Central Texas." *Copeia* 2006:760–768.

Baldwin, Andrew S. 2002. "*Systematics of Salamander Genera Stereochilus, Gyrinophilus, and Pseudotriton (Plethodontidae) and Phylogeography of Pseudotriton ruber*." Unpublished PhD diss., The University of Texas at Arlington.

Beachy, Christopher K. 1997. "Courtship Behavior in the Plethodontid Salamander *Gyrinophilus porphyriticus*." *Herpetologica* 54:288–96.

Behrmann-Godel, Jasminca, Arne W. Nolte, Joachim Kreiselmaier, Roland Berka, and Jörg Freyhof. 2017. "The First European Cave Fish." *Current Biology* 27:R257–R258.

Bendik, Nathan F., and Andrew G. Gluesenkamp. 2013. "Body Length Shrinkage in an Endangered Amphibian is Associated with Drought." *Journal of Zoology* 290:35–41.

Bendik, Nathan F., Jesse M. Meik, Andrew G. Gluesenkamp, Corey E. Roelke, and Paul T. Chippindale. 2013. "Biogeography, Phylogeny, and Morphological Evolution of Central Texas Cave and Spring Salamanders." *BMC Evolutionary Biology* 13:201.

Besharse, Joseph C., and Ronald A. Brandon. 1974. "Size and Growth of the Eyes of the Troglobitic Salamander *Typhlotriton spelaeus*." *International Journal of Speleology* 6:255–264.

Besharse, Joseph C., and Ronald A. Brandon. 1976. "Effects of Continuous Light and Darkness on the Eyes of the Troglobitic Salamander *Typhlotriton spelaeus*." *Journal of Morphology* 149:527–546.

Besharse, Joseph C., and John R. Holsinger. 1977. "*Gyrinophilus subterraneus*, a New Troglobitic Salamander from Southern West Virginia." *Copeia* 1977:624–634.

Bishop, Sherman C. 1944. "A New Neotenic Plethodont Salamander, with Notes on Related Species." *Copeia* 1944:1–5.

Blaney, Richard M., and Patricia K. Blaney. 1978. "Significance of Extreme Variation in a Cave Population of the Salamander, *Gyrinophilus porphyriticus*." *Proceedings of the West Virginia Academy of Science* 50:23.

Bonett, Ronald M., and Paul T. Chippindale. 2004. "Speciation, Phylogeography and Evolution of Life History and Morphology in the Salamanders of the *Eurycea multiplicata* Complex." *Molecular Ecology* 13:1189–1203.

Bonett, Ronald M., and Paul T. Chippindale. 2006. "Streambed Microstructure Predicts Evolution of Development and Life History Mode in the Plethodontid Salamander *Eurycea tynerensis*." *BMC Biology* 4:6.

Bonett, Ronald M., Michael A. Steffen, Shea M. Lambert, John J. Wiens, and Paul T. Chippindale. 2014. "Evolution of Paedomorphosis in Plethodontid Salamanders: Ecological Correlates and Re-evolution of Metamorphosis." *Evolution* 68:466–482.

Bonett, Ronald M., Michael A. Steffen, and Grant A. Robison. 2014. "Heterochrony Repolarized: A Phylogenetic Analysis of Developmental Timing in Plethodontid Salamanders." *EvoDevo* 5:27.

Bonett, Ronald M., Ana Lilia Trujano-Alvarez, Michael J. Williams, and Elizabeth K. Timpe. 2013. "Biogeography and Body Size Shuffling of Aquatic Salamander Communities on a Shifting Refuge." *Proceedings of the Royal Society B: Biological Sciences* 280:20130200.

Brandon, Ronald A. 1965. "A New Race of the Neotenic Salamander *Gyrinophilus palleucus*." *Copeia* 1965:346–352.

Brandon, Ronald A. 1966. "Systematics of the Salamander Genus *Gyrinophilus*." *Illinois Biology Monographs* 35:1–85.

Brandon, Ronald A. 1968. "Structure of the Eye of *Haideotriton wallacei*, a North American Troglodytic Salamander." *Journal of Morphology* 124:345–351.

Brandon, Ronald A. 1971. "North American Troglobitic Salamanders: Some Aspects of Modification in Cave Habitats, with Special Reference to *Gyrinophilus palleucus*." *Bulletin of the National Speleological Society* 33:1–21.

Brandon, Ronald A., and Jeffery H. Black. 1970. "The Taxonomic Status of *Typhlotriton braggi* (Caudata, Plethodontidae)." *Copeia* 1970:388–391.

Brandon, Ronald A., Jeremy Jacobs, Addison Wynn, and David M. Sever. 1986. "A Naturally Metamorphosed Tennessee Cave Salamander (*Gyrinophilus palleucus*)." *Journal of the Tennessee Academy of Science* 61:1–2.

Bruce, Richard C. 1979. "Evolution of Paedomorphosis in Salamanders of the Genus *Gyrinophilus*." *Evolution* 33:998–1000.

Burton, Tim, Shaun Killen, John D. Armstrong, and Neil Metcalfe. 2011. "What Causes Intraspecific Variation in Resting Metabolic Rate and What Are Its Ecological Consequences?" *Proceedings of the Royal Society of London B: Biological Sciences* 278:3465–3473.

Calatayud, Natalie E., Monica Stoops, and Barbara S. Durrant. 2018. "Ovarian Control and Monitoring in Amphibians." *Theriogenology* 109:70–81.

Camp, Carlos D., and John B. Jensen. 2007. "Use of Twilight Zones of Caves by Plethodontid Salamanders." *Copeia* 2007:594–604.

Chippindale, Paul T., Andrew H. Price, John J. Wiens, and David M. Hillis. 2000. "Phylogenetic Relationships and Systematic Revision of Central Texas Hemidactyliine Plethodontid Salamanders." *Herpetological Monographs* 14:1–80.

Christiansen, Kenneth 1961. "Convergence and Parallelism in Cave Entomobryinae." *Evolution* 15:288–301.

Clemen, Günter, David M. Sever, and Hartmut Greven. 2009. "Notes on the Cranium of the Paedomorphic *Eurycea rathbuni* (Stejneger, 1896) (Urodela: Plethodontidae) with Special Regard to the Dentition." *Vertebrate Zoology* 59:157–168.

Coyne, Jerry A., and H. Allen Orr. 2004. *Speciation*. Sunderland, Massachusetts: Sinauer Associates.

Cross, Frank B., Richard L. Mayden, and James D. Stewart. 1986. "Fishes in the Western Mississippi Basin (Missouri, Arkansas, and Red Rivers)." In *The Zoogeography of North American Freshwater Fishes,* edited by Charles H. Hocutt and Edward O. Wiley, 363–412. New York: Wiley.

Culver, David C., and Tanja Pipan. 2009. *The Biology of Caves and Other Subterranean Habitats.* Oxford: Oxford University Press.

Devitt, Thomas J., April M. Wright, David C. Cannatella, and David M. Hillis. 2019. "Species Delimitation in Endangered Groundwater Salamanders: Implications for Aquifer Management and Biodiversity Conservation." *Proceedings of the National Academy of Science* 116:2624–2633.

de Waele, Jo. 2017. "Karst Processes and Landforms." In *The International Encyclopedia of Geography: People, the Earth, Environment, and Technology,* edited by Douglas Richardson, 1–13. New York: Wiley.

Dufton, Megan, Brian K. Hall, and Tamara A. Franz-Odendaal. 2012. "Early Lens Ablation Causes Dramatic Long-Term Effects on the Shape of Bones in the Craniofacial Skeleton of *Astyanax mexicanus.*" *PLOS ONE* 7:e50308.

Durand, Jacques P. 1971. "Recherches sur l'Appareil Visuel du Protée, *Proteus anguinus* Laurenti, Urodele Hypogé." *Annales de Spéléologie* 26:497–824.

Durand, Jacques P. 1973. "Développement et Involution Occulaire de *Proteus anguinus* Laurenti Urodéle Cavemicole." *Annales de Spéléologie* 28:193–208.

Durand, Jacques P. 1976. "Ocular Development and Involution in the European Cave Salamander, *Proteus anguinus* Laurenti." *Biological Bulletin* 151:450–466.

Durand, Jacques P., Nicole Keller, Gilles Renard, Robert Thorn, and Yves Pouliquen. 1993. "Residual Cornea and the Degenerate Eye of the Cryptophthalmic *Typhlotriton spelaeus.*" *Cornea* 12:437–447.

Edgington, Hilary A., Colleen M. Ingram, and Douglas R. Taylor. 2016. "Cyto-Nuclear Discordance suggests Complex Evolutionary History in the Cave-Dwelling Salamander, *Eurycea lucifuga.*" *Ecology and Evolution* 6:6121–6138.

Eigenmann, Carl H. 1900. "The Eyes of the Blind Vertebrates of North America, II. The Eyes of *Typhomolge rathbuni* Stejneger." *Transactions of the American Microscopy Society* 21:49–60.

Eigenmann, Carl H. 1909. *Cave Vertebrates of America: A Study in Degenerative Evolution.* Washington, DC: Carnegie Institution of Washington.

Fejervary, Geza Gyula. 1926. "Note sur la Variation du Protée et Description d'Individus Provenants d'un Nouvel Habitat." *Annales Musei Nationalis Hungarici, Budapest* 24:228–236.

Fenolio, Danté B., and Stanley E. Trauth. 2005. "*Eurycea (Typhlotriton) spelaea.*" In *Amphibian Declines in North America,* edited by Michael J. Lannoo, 863–866. Berkeley, California: University of California Press.

Fenolio, Danté B., G. O. Graening, Bret A. Collier, and Jim F. Stout. 2006. "Coprophagy in a Cave-Adapted Salamander: The Importance of Bat Guano Examined Through Stable Isotope and Nutritional Analyses." *Proceedings of the Royal Society B* 273:439–443.

Fenolio, Danté B., Matthew L. Niemiller, Ronald M. Bonett, Gary O. Graening, Bret A. Collier, and Jim F. Stout. 2014. "Life History, Demography, and the Influence of Cave-Roosting Bats on a Population of the Grotto Salamander (*Eurycea spelaea*) from the

Ozark Plateaus of Oklahoma (Caudata: Plethodontidae)." *Herpetological Conservation and Biology* 9:394–405.

Fenolio, Danté B., Matthew L. Niemiller, Michael G. Levy, and Benjamin Martinez. 2013. "Conservation Status of the Georgia Blind Salamander (*Eurycea wallacei*) from the Floridan Aquifer of Florida and Georgia." *IRCF Reptiles & Amphibians: Conservation and Natural History* 20:97–111.

Fitzinger, Leopold J. 1850. "Ueber den *Proteus anguinus* der Autoren." *Sitzungsberichte der Kaiserlichen Akademie der Wissenschaften. Mathematisch-Naturwissenschaftliche Classe* 5:291–303.

Fowler, Henry W., and Emmett R. Dunn. 1917. "Notes on Salamanders." *Proceedings of the Academy of Natural Sciences of Philadelphia* 69:7–28.

Franz-Odendaal, Tamara A., Kerrianne Ryan, and Brian K. Hall. 2007. "Developmental and Morphological Variation in the Teleost Craniofacial Skeleton Reveals an Unusual Mode of Ossification." *Journal of Experimental Zoology B. Molecular Development and Evolution* 308:709–721.

Frost, Daryl R., Taran Grant, Julián Faivovich, Raoul H. Bain, Alexander Haas, Célio F. B. Haddad, et al. 2006. "The Amphibian Tree of Life." *Bulletin of the American Museum of Natural History* 297:1–370.

[GDNR] Georgia Department of Natural Resources. 2015. *Georgia State Wildlife Action Plan.* Social Circle, GA: Georgia Department of Natural Resources.

Glazier, Douglas S. 2014. "Metabolic Scaling in Complex Living Systems." *Systems* 2:451–540.

Gorički, Špela, and Peter Trontelj. 2006. "Structure and Evolution of the Mitochondrial Control Region and Flanking Sequences in the European Cave Salamander *Proteus anguinus*." *Gene* 378:31–41.

Gorički, Špela, Matthew L. Niemiller, and Danté B. Fenolio. 2012. "Salamanders." In *Encyclopedia of Caves*, 2nd ed., edited by David C. Culver and William B. White, 665–676. London: Elsevier, Academic Press.

Gorički, Špela, Matthew L. Niemiller, Danté B. Fenolio, and Andrew G. Gluesenkamp. 2019. "Salamanders." In *Encyclopedia of Caves*, 3rd ed., edited by William B. White, David C. Culver, and Tanja Pipan, 871–884. Amsterdam: Elsevier, Academic Press.

Gorički, Špela, David Stanković, Aleš Snoj, Matjaz Kuntner, William R. Jeffery, Peter Trontelj, et al. 2017. "Environmental DNA in Subterranean Biology: Range Extension and Taxonomic Implications for *Proteus*." *Scientific Reports* 7:45054.

Grigsby, Melanie M. H. 2009. "Feeding Kinematics of the Grotto Salamander, *Eurycea spelaea*." Master's thesis, Wake Forest University.

Gross, Joshua B., and Horst Wilkens. 2013. "Albinism in Phylogenetically and Geographically Distinct Populations of *Astyanax* Cavefish Arises through the Same Loss-of-Function *Oca2* Allele." *Heredity* 111:122–130.

Guillaume, Oliver. 2000. "Role of Chemical Communication and Behavioural Interactions among Conspecifics in the Choice of Shelters by the Cave-Dwelling Salamander *Proteus anguinus* (Caudata, Proteidae)." *Canadian Journal of Zoology* 78:167–173.

Hammerson, Geoffrey. 2004. "*Gyrinophilus gulolineatus*." *The IUCN Red List of Threatened Species* 2004:e.T59280A11896479. Accessed December 20, 2019. https://www.iucnredlist.org/species/59280/11896479.

Hammerson, Geoffrey, and Christopher K. Beachy. 2004a. "*Gyrinophilus palleucus*." *The IUCN Red List of Threatened Species* 2004:e.T59281A1189670. Accessed December 20, 2019. https://www.iucnredlist.org/species/59281/11896704.

Hammerson, Geoffrey, and Christopher K. Beachy. 2004b. "*Gyrinophilus subterraneus*." *The IUCN Red List of Threatened Species* 2004:e.T59283A11897278. Accessed December 20, 2019. https://www.iucnredlist.org/species/59283/11897278.

Hecht, Max K. 1957. "A Case of Parallel Evolution in Salamanders." *Proceedings of the Zoological Society of Calcutta, Mookerjee Mem* 283:292.

Hecht, Max K., and Jamie L. Edwards. 1976. "The Determination of Parallel or Monophyletic Relationships: The Proteid Salamanders–a Test Case." *The American Naturalist* 110:653–677.

Hervant, Frédéric, Jacques Mathieu, and Jacques P. Durand. 2000. "Metabolism and Circadian Rhythms of the European Blind Cave Salamander *Proteus anguinus* and a Facultative Cave Dweller, the Pyrenean Newt (*Euproctus asper*)." *Canadian Journal of Zoology* 78:1427–1432.

Hillis, David M., Dee A. Chamberlain, Thomas P. Wilcox, and Paul T. Chippindale. 2001. "A New Species of Subterranean Blind Salamander (Plethodontidae: Hemidactyliini: *Eurycea*: *Typhlomolge*) from Austin, Texas, and a Systematic Revision of Central Texas Paedomorphic Salamanders." *Herpetologica* 57:266–280.

Holsinger, John R. 2000. "Ecological Derivation, Colonization, and Speciation." In *Ecosystems of the World: Subterranean Ecosystems*, edited by Horst Wilkens, David C. Culver, and William F. Humphreys, 399–415. Amsterdam: Elsevier.

Houck, Lynne D., and Nancy L. Reagan. 1990. "Male Courtship Pheromones Increase Female Receptivity in a Plethodontid Salamander." *Animal Behaviour* 39:729–734.

Houck, Lynne D., Alison M. Bell, Nancy L. Reagan-Wallin, and Richard C. Feldhoff. 1998. "Effects of Experimental Delivery of Male Courtship Pheromones on the Timing of Courtship in a Terrestrial Salamander, *Plethodon jordani* (Caudata: Plethodontidae)." *Copeia* 1998:214–219.

Howard, James H., Richard L. Raesly, and Edward L. Thompson. 1984. "Management of Nongame Species and Ecological Communities." Material prepared for workshop, Univ. of Kentucky, Lexington.

Howarth, Francis G. 1973. "The Cavernicolous Fauna of Hawaiian Lava Tubes. 1. Introduction." *Pacific Insects* 15:139–51.

Howarth, Francis G. 2019. "Adaptive Shifts." In *Encyclopedia of Caves*, 3rd ed., edited by William B. White, David C. Culver, and Tanja Pipan, 47–55. London: Elsevier, Academic Press.

Huntsman, Brock M., Michael P. Venarsky, Jonathan P. Benstead, and Alexander D. Huryn. 2011. "Effects of Organic Matter Availability on the Life History and Production of a Top Vertebrate Predator (Plethodontidae: *Gyrinophilus palleucus*) in Two Cave Streams." *Freshwater Biology* 56:1746–1760.

Istenic, Lili, and Boris Bulog. 1984. "Some Evidence for the Ampullary Organs in the European Cave Salamander *Proteus anguinus* (Urodela, Amphibia)." *Cell Tissue Research* 235:393–402.

[IUCN] International Union for Conservation of Nature. 2019. "Guidelines for Using the IUCN Red List Categories and Criteria, Version 14 (August 2019)." Prepared by the Standards and Petitions Committee of the IUCN Species Survival Commission. Gland, Switzerland: IUCN.

[IUCN] International Union for Conservation of Nature. 2021. *The IUCN Red List of Threatened Species*. Accessed February 2, 2021. https://www.iucnredlist.org/.

Ivanović, Ana, Gregor Aljančič, and Jan W. Artzen. 2013. "Skull Shape Differentiation of Black and White Olms (*Proteus anguinus anguinus* and *Proteus a. parkelj*): An Exploratory Analysis with Micro-CT Scanning." *Contributions to Zoology* 82:107–114.

Jeffery, William R. 2008. "Emerging Model Systems in Evo-Devo: Cavefish and Microevolution of Development." *Evolution and Development* 10:265–272.

Jockusch, Elizabeth L. 1997. "An Evolutionary Correlate of Genome Size Change in Plethodontid Salamanders." *Proceedings of the Royal Society B: Biological Sciences* 264:597–604.

Kapralova, Kalina H., Zophonías O. Jonsson, Arnar Palsson, Sigrídur Rut Franzdottir, Soizic le Deuff, Bjarni K. Kristjansson, and Sigurour S. Snorrason. 2015. "Bones in Motion: Ontogeny of Craniofacial Development in Sympatric Arctic Charr Morphs." *Developmental Dynamics* 244:1168–1178.

[KDWP] Kansas Department of Wildlife & Parks. 2021. "Kansas Threatened & Endangered Species." Accessed January 24, 2021. https://ksoutdoors.com/Services/Threatened-and-Endangered-Wildlife/Kansas-Threatened-and-Endangered-Species-Statewide.

Keinath, Melissa C., Vladimir A. Timoshevskiy, Nataliya Y. Timoshevskaya, Panagiotis A. Tsonis, S. Randal Voss, and Jeramiah R. Smith. 2015. "Initial Characterization of the Large Genome of the Salamander *Ambystoma mexicanum* Using Shotgun and Laser Capture Chromosome Sequencing." *Scientific Reports* 5:16413.

Kezer, James, Takeshi Seto, and Charles M. Pomerat. 1965. "Cytological Evidence against Parallel Evolution of *Necturus* and *Proteus*." *The American Naturalist* 99:153–158.

Kletečki, Eduard, Branko Jalžić, and Tonci Rađa, 1996. "Distribution of the Olm (*Proteus anguinus*, Laur.) in Croatia." *Memoires de Biospeleologie* 23:227–231.

Kos, Marjanca, and Boris Bulog. 1996a. "Functional Morphology of the Pineal Organ of *Proteus anguinus* (Amphibia: Proteidae)." *Memoires de Biospeologie* 23:1–4.

Kos, Marjanca, and Boris Bulog. 1996b. "Pineal and Retinal Photoreceptors of *Proteus anguinus* (Amphibia: Proteidae)." *Journal of Computer Assisted Microscopy* 8:239–240.

Kos, Marjanca, and Boris Bulog. 2000. "The Ultrastructure of Photoreceptor Cells in the Pineal Organ of the Blind Cave Salamander, *Proteus anguinus* (Amphibia, Urodela)." *Pflügers Archive—European Journal of Physiology* 439:R175–R177.

Kos, Marjanca, Boris Bulog, Ágoston Szel, and Pál Rohlich. 2001. "Immunocytochemical Demonstration of Visual Pigments in the Degenerate Retinal and Pineal Photoreceptors of the Blind Cave Salamander (*Proteus anguinus*)." *Cell Tissue Research* 303:15–25.

Kuchta, Shawn R., Michael Haughey, Addison H. Wynn, Jeremy F. Jacobs, and Richard Highton. 2016. "Ancient River Systems and Phylogeographic Structure in the Spring Salamander, *Gyrinophilus porphyriticus*." *Journal of Biogeography* 43:638–652.

Larsen, John H., Jr., and Dan J. Guthrie. 1974. "Parallelism in the Proteidae Reconsidered." *Copeia*, 635–643.

Laurenti, Josephus N. 1768. *Specimen Medicum, Exhibens Synopsin Reptilium Emendatam Cum Experimentis Circa Venenaet Antidota Reptilium Austriacorum*. Wien: Joan Thomae.

Lazell, James D., and Ronald A. Brandon. 1962. "A New Stygian Salamander from the Southern Cumberland Plateau." *Copeia* 1962:300–306.

Martin, Samuel D., Donald B. Shepard, Michael A. Steffen, John G. Phillips, and Ronald M. Bonett. 2016. "Biogeography and Colonization History of Plethodontid Salamanders from the Interior Highlands of Eastern North America." *Journal of Biogeography*, 43:410–422.

McCrady, Edward. 1954. "A New Species of *Gyrinophilus* (Plethodontidae) from Tennessee Caves." *Copeia* 1954:200–206.

Mertens, Robert, and Lorenz Müller. 1940. "Die Amphibien und Reptilien Europas (Zweite Liste, nach dem Stand vom 1. Januar 1940)." *Abhandlungen der Senckenbergischen Naturforschenden Gesellschaft, Frankfurt am Main* 451:1–56.

Miller, Brian T. 1995. "Geographic Distribution: *Gyrinophilus palleucus*." *Herpetological Review* 26:103.

Miller, Brian T., and Matthew L. Niemiller. 2008. "Distribution and Relative Abundance of Tennessee Cave Salamanders (*Gyrinophilus palleucus* and *Gyrinophilus gulolineatus*) with an Emphasis on Tennessee Populations." *Herpetological Conservation and Biology* 3:1–20.

Miller, Brian T., Matthew L. Niemiller, and R. Graham Reynolds. 2008. "Observations on Egg-Laying Behavior and Interactions among Attending Female Red Salamanders (*Pseudotriton ruber*) with Comments on the Use of Caves by This Species." *Herpetological Conservation and Biology* 3:203–210.

Miller, James A. 1986. "Hydrogeologic Framework of the Floridan Aquifer System in Florida and Parts of Georgia, Alabama, and South Carolina: Regional Aquifer-System Analysis." *U.S. Geological Survey Professional Paper 1403-B*. Washington, DC: US Government Printing Office.

Miller, Robert V. 1968. "A Systematic Study of the Greenside Darter, *Etheostoma blennioides* Rafinesque (Pisces: Percidae)." *Copeia* 1968:1–40.

Mitchell, Robert W., and James R. Reddell. 1965. "*Eurycea tridentifera*, a New Species of Troglobitic Salamander from Texas and a Reclassification of *Typhlomolge rathbuni*." *Texas Journal of Science* 17:12–27.

Mitchell, Robert W., and Richard E. Smith. 1965. "Some Aspects of the Osteology and Evolution of the Neotenic Spring and Cave Salamanders (*Eurycea*, Plethodontidae) of Central Texas." *The Texas Journal of Science* 23:343–362.

[MNHP] Missouri Natural Heritage Program. 2019. *Missouri Species and Communities of Conservation Concern Checklist*. Jefferson City: Missouri Department of Conservation.

NatureServe. 2019. "NatureServe Explorer: An Online Encyclopedia of Life." Accessed June 21, 2019. http://explorer.natureserve.org.

Newman, Catherine E., and Christopher C. Austin. 2016. "Sequence Capture and Next-Generation Sequencing of Ultraconserved Elements in a Large-Genome Salamander." *Molecular Ecology* 25:6162–6174.

Niemiller, Matthew L., and Brian T. Miller. 2009. "A Survey of the Cave-Associated Amphibians of the Eastern United States with an Emphasis on Salamanders." *Proceedings of the 15th International Congress of Speleology* 15:249–256.

Niemiller, Matthew L., and Brian T. Miller. 2010. "*Gyrinophilus gulolineatus*, 862.1." *Catalogue of American Amphibians and Reptiles*. Society for the Study of Amphibians and Reptiles.

Niemiller, Matthew L., and K. Denise Kendall Niemiller. 2020. "Species Status Assessment for the Tennessee Cave Salamander (*Gyrinophilus palleucus*) McCrady, 1954." Unpublished technical report submitted to Tennessee Wildlife Resources Agency, Nashville, Tennessee.

Niemiller, Matthew L., and Thomas L. Poulson. 2010. "Subterranean Fishes of North America: Amblyopsidae." In *Biology of Subterranean Fishes*, edited by Elena Trajano, Maria Elina Bichuette, and B. G. Kapoor, 169–280. Enfield, New Hampshire: Science Publishers.

Niemiller, Matthew L., and Daphne Soares. 2015. "Cave Environments." In *Extremophile Fishes: Ecology, Evolution, and Physiology of Teleosts in Extreme Environments*, edited by Rüdiger Riesch, Michael Tobler, and Martin Plath, 161–191. Switzerland: Springer International Publishing.

Niemiller, Matthew L., Evin T. Carter, Nicholas S. Gladstone, Lindsey E. Hayter, and Annette S. Engel. 2018. "New Surveys and Reassessment of the Conservation Status of the Berry Cave Salamander (*Gyrinophilus gulolineatus*)." Unpublished technical report submitted to US Fish and Wildlife Service, Cookeville, Tennessee.

Niemiller, Matthew L., Brad M. Glorioso, Dante B. Fenolio, R. Graham Reynolds, Steven J. Taylor, and Brian T. Miller. 2016. "Growth, Survival, Longevity, and Population Size of the Big Mouth Cave Salamander (*Gyrinophilus palleucus necturoides*) from the Type Locality in Grundy County, Tennessee, USA." *Copeia* 104:35–41.

Niemiller, Matthew L., Benjamin M. Fitzpatrick, and Brian T. Miller. 2008. "Recent Divergence with Gene Flow in Tennessee Cave Salamanders (Plethodontidae: *Gyrinophilus*) Inferred from Gene Genealogies." *Molecular Ecology* 17:2258–75.

Niemiller, Matthew L., Brian T. Miller, and Benjamin M. Fitzpatrick. 2009. "Systematics and Evolutionary History of Subterranean *Gyrinophilus* Salamanders." *Proceedings of the 15th International Congress of Speleology* 1:242–48.

Niemiller, Matthew L., Brian T. Miller, and Benjamin M. Fitzpatrick. 2010. "Review of the Scientific Literature and Research for the USFWS Review for Potential Listing of the Berry Cave Salamander (*Gyrinophilus gulolineatus*)." Unpublished technical report on file with US Fish and Wildlife Service, Cookeville, Tennessee.

Niemiller, Matthew L., Michael S. Osbourn, Danté B. Fenolio, Thomas K. Pauley, Brian T. Miller, and John R. Holsinger. 2010. "Conservation Status and Habitat Use of the West Virginia Spring Salamander (*Gyrinophilus subterraneus*) and Spring Salamander (*G. porphyriticus*) in General Davis Cave, Greenbrier Co., West Virginia." *Herpetological Conservation and Biology* 5:32–43.

Noble, Gladwyn K. 1931. *Biology of the Amphibia*. New York: McGraw-Hill.

Noble, Gladwyn K., and Sarah H. Pope. 1928. "The Effect of Light on the Eyes, Pigmentation, and Behavior of the Cave Salamander, *Typhlotrition*." *The Anatomical Record* 41:21.

Nowoshilow, Sergej, Siegfried Schloissnig, Ji-Feng Fei, Andreas Dahl, Andy W. C. Pang, Martin Pippel, et al. 2018. "The Axolotl Genome and the Evolution of Key Tissue Formation Regulators." *Nature* 554:50–55.

Nunziata, Skyler O., and David W. Weisrock. 2018. "Estimation of Contemporary Effective Population Size and Population Declines Using RAD Sequence Data." *Heredity* 120:196–207.

Nunziata, Skyler O., Stacey L. Lance, David E. Scott, Emily Moriarty Lemmon, and David W. Weisrock. 2017. "Genomic Data Detect Corresponding Signatures of Population Size Change on an Ecological Time Scale in Two Salamander Species." *Molecular Ecology* 26:1060–1074.

Nussbaum, Ronald A. 1987. "Parental Care and Egg Size in Salamanders: An Examination of the Safe Harbor Hypothesis." *Researches on Population Ecology* 29:2–44.

[ODWC] Oklahoma Department of Wildlife Conservation. 2016. *Oklahoma Comprehensive Wildlife Conservation Strategy: A Strategic Plan for Oklahoma's Rare and Declining Wildlife*. Oklahoma City: Oklahoma Department of Wildlife Conservation.

O'Neill, Eric M., Rachel Schwartz, C. Thomas Bullock, Joshua S. Williams, H. Bradley Shaffer, Xochitl Aguilar-Miguel, et al. 2013. "Parallel Tagged Amplicon Sequencing Reveals Major Lineages and Phylogenetic Structure in the North American Tiger Salamander (*Ambystoma tigrinum*) Species Complex." *Molecular Ecology* 22:111–129.

Palmer, Arthur N. 2007. *Cave Geology*. Dayton, OH: Cave Books.

Parra-Olea, Gabriela, David B. Wake, and James Hanken. 2008. "*Chiropterotriton magnipes*." *The IUCN Red List of Threatened Species*. Accessed December 20, 2019. https://www.iucnredlist.org/species/59227/3077636.

Parzefall, Jakob, Jacques P. Durand, and Bernard Richard. 1980. "Chemical Communication in *Necturus maculosus* and His Cave-Living Relative *Proteus anguinus* (Proteidae, Urodela)." *Zeitschrift für Tierpsychologie* 53:133–138.

Peck, Stewart B. 1970. "The Terrestrial Arthropod Fauna of Florida Caves." *Florida Entomologist* 53:203–207.

Petranka, James W. 1998. *Salamanders of the United States and Canada*. Washington, DC: Smithsonian Institute Press.

Pettersen, Amanda K., Craig R. White, and Dustin J. Marshall. 2016. "Metabolic Rate Covaries with Fitness and the Pace of the Life History in the Field." *Proceedings of the Royal Society of London B: Biological Sciences* 283:20160323.

Phillips, John B. 1977. "Use of the Earth's Magnetic Field by Orienting Cave Salamanders (*Eurycea lucifuga*)." *Journal of Comparative Physiology* 121:273–288.

Phillips, John B. 1986. "Magnetic Compass Orientation in the Eastern Red-Spotted Newt (*Notophthalmus viridescens*)." *Journal of Comparative Physiology A* 158:103–109.

Phillips, John B. 1987. "Laboratory Studies of Homing Orientation in the Eastern Red-Spotted Newt, *Notophthalmus viridescens*." *Journal of Experimental Biology* 131: 215–229.

Phillips, John G., Danté B. Fenolio, Sarah L. Emel, and Ronald M. Bonett. 2017. "Hydrologic and Geologic History of the Ozark Plateau Drive Phylogenomic Patterns in a Cave-Obligate Salamander." *Journal of Biogeography* 44:2463–2474.

[POGP] Proteus Olm Genome Project. 2021. Accessed December 10, 2021. https://www.proteusgenome.com.

Potter, Floyd E., Jr., and Samuel S. Sweet. 1981. "Generic Boundaries in Texas Cave Salamanders, and a Redescription of *Typhlomolge robusta* (Amphibia: Plethodontidae)." *Copeia* 1981:64–75.

Poulson, Thomas L. 1963. "Cave Adaptation in Amblyopsid Fishes." *American Midland Naturalist* 70:257–290.

Prelovšek, Petra-Maja, Lilijana B. Mali, and Boris Bulog. 2008. "Hepatic Pigment Cells of Proteidae (Amphibia, Urodela): A Comparative Histochemical and Ultrastructural Study." *Animal Biology* 58:245–256.

Protas, Meredith E., Candace Hersey, Dawn Kochanek, Yi Zhou, Horst Wilkens, William R. Jeffery, et al. 2006. "Genetic Analysis of Cavefish Reveals Molecular Convergence in the Evolution of Albinism." *Nature Genetics* 38:107–111.

Proudlove, Graham S. 2006. *Subterranean Fishes of the World: An Account of the Subterranean (Hypogean) Fishes Described to 2003 with a Bibliography 1541–2004*. Moulis, France: International Society for Subterranean Biology.

Proudlove, Graham S. 2010. "Biodiversity and Distribution of the Subterranean Fishes of the World." In *The Biology of Subterranean Fishes*, edited by Eleonora Trajano, Maria Elina Bichuette and B. G. Kapoor, 41–63. Enfield, New Hampshire: Science Publishers.

Pyron, R. Alexander, and John J. Wiens. 2011. "A Large-Scale Phylogeny of Amphibia including over 2800 Species, and a Revised Classification of Extant Frogs, Salamanders, and Caecilians." *Molecular Phylogenetics and Evolution* 61:543–583.

Resetarits, William J., Jr. 1991. "Ecological Interactions among Predators in Experimental Stream Communities." *Ecology* 72:1782–1793.

Rétaux, Sylvie, and Didier Casane. 2013. "Evolution of Eye Development in the Darkness of Caves: Adaptation, Drift, or Both?" *EvoDevo* 4:26.

Rivera, Malia Ana J., Francis G. Howarth, Stefano Taiti, and George K. Roderick. 2002. "Evolution in Hawaiian Cave-Adapted Isopods (Oniscidea: Philosciidae): Vicariant Speciation or Adaptive Shifts?" *Molecular Phylogenetics and Evolution* 25:1–9.

Rollmann, Stephanie M., Lynne D. Houck, and Richard C. Feldhoff. 1999. "Proteinaceous Pheromone Affecting Female Receptivity in a Terrestrial Salamander." *Science* 285:1907–1909.

Roth, Anton, and Peter A. Schlegel. 1988. "Behavioral Evidence and Supporting Electrophysiological Observations for Electroreception in the Blind Cave Salamander, *Proteus anguinus* (Urodela)." *Brain, Behavior and Evolution* 32:277–280.

Roth, Gerhard. 1987. *Visual Behavior in Salamanders*. Berlin: Springer-Verlag.

Rovito, Sean M., García Parra-Olea, Ernesto Recuero, and David B. Wake. 2015. "Diversification and Biogeographical History of Neotropical Plethodontid Salamanders." *Zoological Journal of the Linnean Society* 175:167–188.

Ryan, Travis J., and Richard C. Bruce. 2000. "Life History Evolution and Adaptive Radiation of Hemidactyliine Salamanders." In *The Biology of Plethodontid Salamanders*, edited by Richard C. Bruce, Robert G. Jaeger, and Lynne D. Houck, 303–325. New York: Kluwer Academic, Plenum Publishers.

Safi, Rachid, Virginie Vlaeminck-Guillem, Marilyne Duffraisse, Isabelle Seugnet, Michelina Plateroti, Alain Margotat, et al. 2006. "Pedomorphosis Revisited: Thyroid Hormone Receptors Are Functional in *Necturus maculosus*." *Evolution & Development* 8:284–92.

Šarić, Katarina K., and Petra K. Konrad. 2017. "The Deepest Finding of an Olm (*Proteus anguinus*): Zagorska peć, Ogulin, Croatia." *Acta Carsologica* 46:347–351.

Schlegel, Peter A. 1997. "Behavioral Sensitivity of the European Blind Cave Salamander, *Proteus anguinus*, and a Pyrenean newt, *Euproctus asper*, to Electrical Fields in Water." *Brain, Behavior and Evolution* 49:121–131.

Schlegel, Peter A., and A. Roth. 1997. "Tuning of Electroreceptors in the Blind Cave Salamander, *Proteus anguinus* L." *Brain, Behavior and Evolution* 49:132–136.

Schlegel, Peter A., Wolfgang Briegleb, Boris Bulog, and Sebastian Steinfartz. 2006. "Revue et Nouvelles données sur la Sensitivité à la Lumière et Orientation Non-Visuelle chez *Proteus anguinus*, Calotriton asper et *Desmognathus ochrophaeus* (Amphibiens Urodèles Hypogés)." *Bulletin de la Société Herpétologique de France* 118:1–31.

Schlegel, Peter A., Sebastian Steinfartz, and Boris Bulog. 2009. "Non-visual Sensory Physiology and Magnetic Orientation in the Blind Cave Salamander, *Proteus anguinus* (and Some Other Cave-Dwelling Urodele Species). Review and New Results on Light-Sensitivity and Non-visual Orientation in Subterranean Urodeles (Amphibia)." *Animal Biology* 59:351–384.

Semlitsch, Raymond D., and Henry M. Wilbur. 1989. "Artificial Selection for Paedomorphosis in the Salamander *Ambystoma talpoideum*." *Evolution* 43:105–12.

Sever, David M., and Henry L. Bart Jr. 1996. "Ultrastructure of the Spermathecae of *Necturus beyeri* (Amphibia: Proteidae) in Relation to Its Breeding Season." *Copeia* 1996:927–937.

Shaw, Trevor R. 1999. "*Proteus* for Sale and for Science in the 19th Century." *Acta Carsologica* 28:229–304.

Shen, Xing X., Dan Liang, Yan J. Feng, Meng Y. Chen, and Peng Zhang. 2013. "A Versatile and Highly Efficient Toolkit including 102 Nuclear Markers for Vertebrate Phylogenomics, Tested by Resolving Higher Level Relationships of the Caudata." *Molecular Biology and Evolution* 30:2235–2248.

Shoop, C. Robert 1965. "Aspects of Reproduction in Louisiana *Necturus* Populations." *The American Midland Naturalist* 74:357–367.

Simmons, Douglas D. 1976. "A Naturally Metamorphosed *Gyrinophilus palleucus* (Amphibia, Urodela, Plethodontidae)." *Journal of Herpetology* 10:255–257.

Sket, Boris. 1997. "Distribution of *Proteus* (Amphibia: Urodela: Proteidae) and Its Possible Explanation." *Journal of Biogeography* 24:263–280.

Sket, Boris. 2017. "Discovering the Black Proteus *Proteus anguinus parkelj* (Amphibia: Caudata)." *Natura Sloveniae* 19:27–28.

Sket, Boris, and Jan W. Arntzen. 1994. "A Black, Non-troglomorphic Amphibian from the Karst of Slovenia: *Proteus anguinus parkelj* n. ssp. (Urodela: Proteidae)." *Bijdragen tot de Dierkunde* 64:33–53.

Smith, Charles C. 1968. "A New *Typhlotriton* from Arkansas (Amphibia Caudata)." *The Wasmann Journal of Biology* 26:155–159.

Smith, Philip W. 1948. "Food Habits of Cave Dwelling Amphibians." *Herpetologica* 4:205–208.

Soares, Daphne, and Matthew L. Niemiller. 2013. "Sensory Adaptations of Fishes to Subterranean Environments." *BioScience* 63:274–283.

Soares, Daphne, and Matthew L. Niemiller. 2020. "Extreme Adaptations in Caves." *The Anatomical Record* 303:15–23.

Soares, Daphne, Yoshiyuki Yamamoto, Allen G. Strickler, and William R. Jeffery. 2004. "The Lens Has a Specific Influence on Optic Nerve and Tectum Development in the Blind Cavefish *Astyanax*." *Developmental Neuroscience* 26:308–317.

Stamm, John F., Mary F. Poteet, Amy J. Symstad, MaryLynn Musgrove, Andrew J. Long, Barbara J. Mahler, and Parker A. Norton. 2015. "Historical and Projected Climate (1901–2050) and Hydrologic Response of Karst Aquifers, and Species Vulnerability in South-Central Texas and Western South Dakota." Scientific Investigations Report 2014–5089. Reston, VA: US Geological Survey.

Steffen, Michael A., Kelly J. Irwin, Andrea L. Blair, and Ronald M. Bonett. 2014. "Larval Masquerade: A New Species of Paedomorphic Salamander (Caudata: Plethodontidae: *Eurycea*) from the Ouachita Mountains of North America." *Zootaxa* 3786:423–442.

Stejneger, Leonard H. 1892. "Preliminary Description of a New Genus and Species of Blind Cave Salamander from North America." *Proceedings of the United States National Museum* 15:115–117.

Stone, Leon S. 1964. "The Structure and Visual Function of the Eye of Larval and Adult Cave Salamanders *Typhlotriton spelaeus*." *The Journal of Experimental Zoology* 156:201–218.

Sun, Cheng, and Rachel L. Mueller. 2014. "Hellbender Genome Sequences Shed Light on Genomic Expansion at the Base of Crown Salamanders." *Genome Biology and Evolution* 6:1818–1829.

Sun, Cheng, Donald B. Shepard, Rebecca A. Chong, José López Arriaza, Kathryn Hall, Todd A. Castoe, et al. 2012. "LTR Retrotransposons Contribute to Genomic Gigantism in Plethodontid Salamanders." *Genome Biology and Evolution* 4:168–183.

Sweet, Samuel S. 1977. "Natural Metamorphosis in *Eurycea neotenes*, and the Generic Allocation of the Texas *Eurycea* (Amphibia: Plethodontidae)." *Herpetologica* 33:364–75.

[TPWD] Texas Parks and Wildlife Department. 2019. "Species of Greatest Conservation Need in Texas." Accessed December 20, 2019. https://tpwd.texas.gov/huntwild/wild/wildlife_diversity/nongame/tcap/sgcn.phtml.

Trontelj, Peter, and Špela Gorički. 2003. "Monophyly of the Family Proteidae (Amphibia: Caudata) Tested by Phylogenetic Analysis of Mitochondrial 12S rDNA Sequences." *Natura Croatica: Periodicum Musei Historiae Naturalis Croatici* 12:113–120.

Trontelj, Peter, Christophe J. Douady, Cene Fišer, Janine Gibert, Špela Gorički, Tristan Lefébure, et al. 2009. "A Molecular Test for Cryptic Diversity in Ground Water: How Large Are the Ranges of Macro-Stygobionts?" *Freshwater Biology* 54:727–744.

Trontelj, Peter, Špela Gorički, Slavko Polak, Rudi Verovnik, Valerija Zakšek, and Boris Sket. 2007. "Age Estimates for Some Subterranean Taxa and Lineages in the Dinaric Karst." *Acta Carsologica* 36:183–189.

Uiblein, Franz, and Jakob Parzefall. 1993. "Does the Cave Salamander *Proteus anguinus* Detect Mobile Prey by Mechanical Cues?" *Mémoires de Biospéologie* 20:261–264.

Uiblein, Franz, Jacques P. Durand, Christian Juberthie, and Jakob Parzefall. 1992. "Predation in Caves: The Effects of Prey Immobility and Darkness on the Foraging Behaviour of Two Salamanders, *Euproctus asper* and *Proteus anguinus*." *Behavioural Processes* 28:33–40.

Valentine, Barry D. 1964. "The External Morphology of the Plethodontid Salamander, *Haideotriton*." *Journal of the Ohio Herpetological Society* 4:99–102.

Valvasor, Johann W. 1689. "Die Ehre dess Hertzogthums Crain." *Laybach & Nhrnberg, WJ Endter* 1:275–276.

Vandel, Albert. 1964. *Biospéologie. La Biologie des Animaux Cavernicoles*. Paris: Gauthier -Villars.

Verdeny-Vilalta, Oriol, Charles W. Fox, David H. Wise, and Jordi Moya-Laraño. 2015. "Foraging Mode Affects the Evolution of Egg Size in Generalist Predators Embedded in Complex Food Webs." *Journal of Evolutionary Biology* 28:1225–1233.

Voituron, Yann, Michelle de Fraipont, Julien Issartel, Olivier Guillaume, and Jean Clobert. 2011. "Extreme Lifespan of the Human Fish (*Proteus anguinus*): A Challenge for Ageing Mechanisms." *Biology Letters* 7:105–107.

Wake, David B. 1966. "Comparative Osteology and Evolution of the Lungless Salamanders, Family Plethodontidae." *Memoirs of the Southern California Academy of Sciences* 4:1–111.

Wake, David B. 1993. "Phylogenetic and Taxonomic Issues Relating to Salamander of the Family Plethodontidae." *Herpetologica* 49:229–237.

Weisrock, David W., Paul M. Hime, Skyler O. Nunziata, Kara S. Jones, Mason O. Murphy, Scott Hotaling, and Justin D. Kratovil. 2018. "Surmounting the Large-Fenome 'Problem' for Genomic Data Generation in Salamanders." In *Population Genomics: Wildlife*, edited by Paul Hohenlohe and Om P. Rajora, 1–28. London: Springer.

Whiteman, Howard H., John J. Gutrich, and Randall S. Moorman. 1999. "Courtship Behavior in a Polymorphic Population of the Tiger Salamander, *Ambystoma tigrinum nebulosum*." *Journal of Herpetology* 33:348–51.

Wielstra, Ben, Elza Duijm, Patricia Lagler, Youri Lammers, Willem R. M. Meilink, Janine M. Ziermann, and Jan W. Arntzen. 2014. "Parallel Tagged Amplicon Sequencing of Transcriptome-Based Genetic Markers for *Triturus* Newts with the Ion Torrent Next-Generation Sequencing Platform." *Molecular Ecology Resources* 14:1080–1089.

Wiens, John J., Ronald M. Bonett, and Paul T. Chippindale. 2005. "Ontogeny Discombobulates Phylogeny: Paedomorphosis and Higher-Level Salamander Relationships." *Systematic Biology* 54:91–110.

Wiens, John J., Paul T. Chippindale, and David M. Hillis. 2003. "When Are Phylogenetic Analyses Misled by Convergence? A Case Study in Texas Cave Salamanders." *Systematic Biology* 52:501–514.

Wiens, John J., Gabriela Parra-Olea, Mario Garcia-Paris, and David B. Wake, 2007. "Phylogenetic History Underlies Elevational Biodiversity Patterns in Tropical Salamanders." *Proceedings of the Royal Society B: Biological Sciences* 274:919–928.

Wilbur, Henry M., and James P. Collins. 1973. "Ecological Aspects of Amphibian Metamorphosis." *Science* 182:1305–1314.

Withers, David I. 2016. *A Guide to the Rare Animals of Tennessee*. Nashville: Division of Natural Areas, Tennessee Department of Environment and Conservation.

Wray, Kenneth P., and Scott J. Steppan. 2017. "Ecological Opportunity, Historical Biogeography and Diversification in a Major Lineage of Salamanders." *Journal of Biogeography* 44:797–809.

Yamamoto, Yoshiyuki, Luis Espinasa, David W. Stock, and William R. Jeffery. 2003. "Development and Evolution of Craniofacial Patterning is Mediated by Eye-Dependent and -Independent Processes in the Cavefish *Astyanax*." *Evolution & Development* 5:435–446.

Yoshizawa, Masato, Spela Gorički, Daphne Soares, and William R. Jeffery. 2010. "Evolution of a Behavioral Shift Mediated by Superficial Neuromasts Helps Cavefish Find Food in Darkness." *Current Biology* 20:1631–1636.

Zakšek, Valerija, and Peter Trontelj. 2017. "Conservation Genetics of Proteus in the Postojna-Planina Cave System." *Natura Sloveniae* 19:33–34.

Zakšek, Valerija, Marjeta Konec, and Peter Trontelj. 2018. "First Microsatellite Data on *Proteus anguinus* Reveal Weak Genetic Structure between the Caves of Postojna and Planina." *Aquatic Conservation: Marine and Freshwater Ecosystems* 28:241–246.

Zielinski, Piotr, Michal T. Stuglik, Katarzyna Dudek, Mateusz Konczal, and Wieslaw Babik. 2014. "Development, Validation and High-Throughput Analysis of Sequence Markers in Nonmodel Species." Molecular Ecology Resources 14:352–360.

7

Diversity, Distribution, and Conservation of Cavefishes in China

Yahui Zhao, Andrew G. Gluesenkamp, J. Judson Wynne, Danté B. Fenolio, Daphne Soares, Matthew L. Niemiller, Maria E. Bichuette, and Prosanta Chakrabarty

Introduction

Karst landscapes are widespread and diverse in China. The exposed and near-surface soluble rocks cover 1.3 million km^2, which is approximately one-seventh of the country's territory (Z. Zhang 1980). Importantly, about one-quarter of the world's total terrestrial carbonate rocks occur in China, and almost every geomorphological type of karst may also be found in China (Sweeting 1995). Moreover, almost one-quarter of the total water resources of China is karstic in origin (Yuan et al. 1991). In southern China, limestone deposits have been subjected to intense fluvial dissection by large rivers in a wet climate since the Neogene period (between 23.03 and 2.58 million years ago; Sweeting 1995).

The South China Karst (SCK), the largest continuous karst area in the world (Yuan et al. 2016), is situated within Guizhou Province, Yunnan Province, Guangxi Zhuang Autonomous Region, and Chongqing Municipality (Fig. 7.1). The karst terrain displays a geomorphic transition as the terrain gradually ascends ~2,000 m over 700 km from the eastern Guangxi Basin to the western Yunnan-Guizhou Plateau—with an average elevation of 110 m. This region supports several types of karst landforms, including tower karst, pinnacle karst, cone karst, natural bridges, gorges, and large cave systems (UNESCO 2018).

Caves are one of the dominant habitats in the SCK, supporting a substantial diversity of subterranean-adapted organisms. With rich aquifer resources and complex subterranean water systems, the SCK is also home

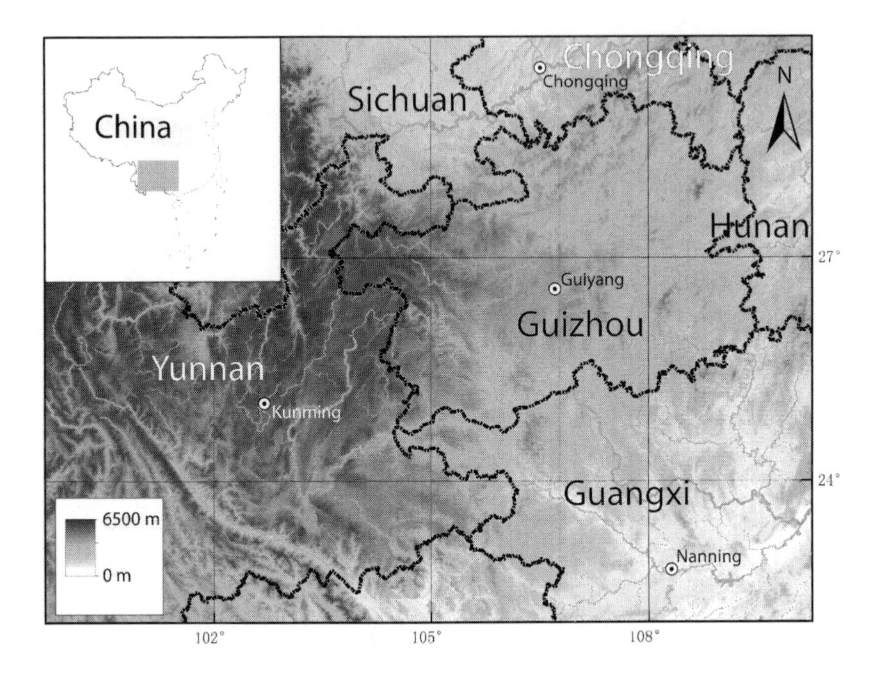

Figure 7.1. The geographic extent of the 1,762 km² South China Karst. Legend (*bottom left*) represents elevational gradient.

to a great diversity of aquatic cave animals—of which, cavefishes are the most diverse and distinctive group. Incidentally, cavefishes represent the only subterranean-adapted vertebrate species in China—and may arguably be considered to retain the highest conservation value.

The earliest written records of cavefishes are from China (Y. Zhao et al. 2011). In 1436, Mao Lan (兰茂), in his book *Materia Medica of South Yunnan*, described a "golden lined fish" (M. Zhang et al. 2016). Nearly 500 years later, the fish was formally described as *Sinocyclocheilus grahami* (Regan, 1904). In 1541, the governor of Luxi County, Yunnan Province, Yijing Xie (解一经), mentioned an eyeless cavefish in an article entitled "Records on Alu Cave" (Y. Zhao and Zhang 2009a). He wrote: "I heard there was a kind of transparent fish coming out when the subterranean river flooded." This fish was later described as *S. hyalinus* Chen & Yang, 1993 (Y. Zhao et al. 2011).

Although Chinese cavefishes are discussed in ancient texts, their descriptions and relationship to the subterranean environment were not addressed until much later. The discovery of several eyeless cavefish species in the late 1970s and early 1980s heralded modern cavefish research in China

(Y. Zhao and Zhang 2006). Over the past 40 years, 148 cavefish species have been formally described. Unfortunately, their habitats are threatened by rapid economic development, and the persistence of several species remains uncertain (Y. Zhao et al. 2011). In this chapter, we discuss the diversity, distribution, and conservation status of Chinese cavefishes. We also identify future research and conservation needs for these amazing animals.

Cavefish Diversity in China

Species Diversity

China boasts the highest diversity of cavefishes globally, with 148 species accounting for more than 11% of the total known freshwater fish diversity (Xing et al. 2016; L. Ma et al. 2019). All of these species are endemic to China. Chinese cavefishes belong to two orders, four families, and 20 genera. Families with the highest diversity include Cyprinidae (77 species) and Nemacheilidae (65 species; Fig. 7.2). Most Chinese cavefishes (99%) belong to the order Cypriniformes (147 species), while Siluriformes is represented by a single species, *Xiurenbagrus dorsalis* Xiu, Yang & Zheng, 2014, which is the only stygobiontic catfish found in China and in eastern Asia (Xiu et al.

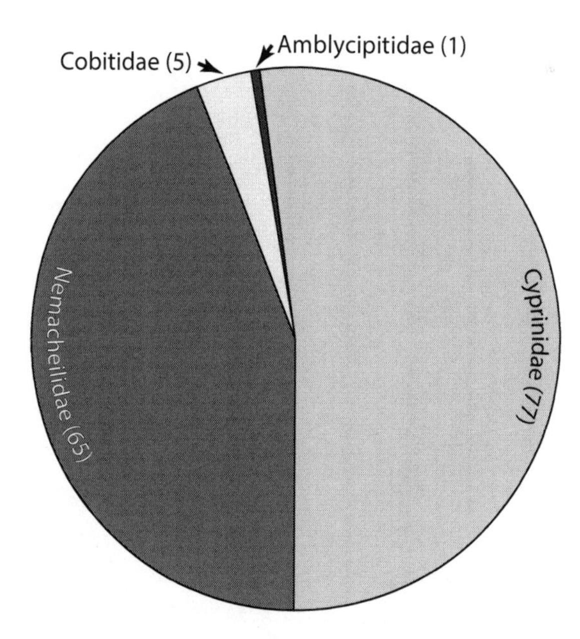

Figure 7.2. Species diversity per family of Chinese cavefishes.

2014). More than half of Chinese cavefishes (78 species) are considered stygobionts (aquatic subterranean-adapted organisms), which accounts for 39% of the global diversity of stygobiontic fishes.

Cyprinidae

Cyprinidae is the most species-rich freshwater fish family in China with over 660 species and subspecies (C. Zhang and Zhao 2016). Eight genera comprise 77 species of cave dwellers, accounting for 52% of the known cavefishes in China. The first stygobiontic fish species described from China, *Typhlobarbus nudiventris* Chu & Chen, 1982, was discovered in 1976 and was based on a single specimen (Chu and Chen 1982). To date, it is the only species known in this genus (Romero et al. 2009). Unfortunately, this species has not been observed in over 40 years, and it may now be extinct.

With 69 species, *Sinocyclocheilus* supports the highest species richness of any cavefish genus in the world. It is the most diverse cyprinid genus in China and the second-most diverse genus among Chinese freshwater fishes (Xing et al. 2016). This genus includes four clades: "*jii*," "*angularis*," "*cyphotergous*," and "*tingi*" (Romero et al. 2009; Y. Zhao and Zhang 2009a). However, based upon molecular analysis, there is potentially a fifth clade (Y. Zhao, unpublished data). Rapid speciation within this genus is reflected by considerable morphological variation (Y. Zhao and Zhang 2009a; Plate 7.1) in which each clade has a unique morphology and largely nonoverlapping distributions.

In addition to stygobiontic traits widely observed in cavefishes—for example, reduction or loss of eyes, depigmentation, loss of scales and swim bladders, enhancement of mechano- and chemosensation, and the elongation of fin rays (Romero and Green 2005; Y. Zhao et al. 2011; Soares and Niemiller 2013; Niemiller and Soares 2015; Niemiller et al. 2019)—some members of the genus *Sinocyclocheilus* also have a horn-like structure on the nape (Plate 7.1B) or a humpback profile (Plate 7.1C). Both characters are considered traits of subterranean adaptation (Romero et al. 2009; Y. Zhao and Zhang 2009a; Y. Zhao et al. 2011; Niemiller et al. 2019).

More than 10 *Sinocyclocheilus* species (all in the "*angularis*" clade) have a distinct horn on the nape. With considerable interspecific variation in horn structure, the horn of *S. angularis* Zheng & Wang, 1990 is relatively short (Plate 7.2A), whereas the horn in both *S. bicornatus* Wang & Liao, 1997 and *S. furcodorsalis* Chen, Yang & Lan, 1997 is more prominent and forked (Plate

7.2B and C). In *S. rhinocerous* Li & Tao, 1994 and *S. aquihornes* Li & Yang, 2007, the horn is rod-like (Plate 7.2D and E), whereas the horn of *S. tileihornes* Mao, Lu & Li, 2003 has a flattened, curved tile shape (Plate 7.2F). Microcomputed tomography scans of *S. furcodorsalis* revealed the horn is composed of the supraoccipital and parietal bones fused in a complex structure (Soares et al. 2019).

A similar cranial morphology is present in *Kurtus gulliveri* Castelnau, 1878 (Perciformes: Kurtidae) and represents the only known modification of the supraoccipital crest. This species is a surface-dwelling brackish water fish from New Guinea and northern Australia (Froese and Pauly 2019). Nape horns occur only on males and are used to carry fertilized eggs (Berra and Humphrey 2002). In contrast, the head horns of *Sinocyclocheilus* species occur in both sexes. Several hypotheses suggest this feature may have evolved as adaptations for fat storage, a mechanical advantage in competing for mates, improved hydrodynamics, or predator defense (Y. Chen et al. 1994; W. Li and Tao 2002; Y. Zhao and Zhang 2009a; Soares et al. 2019). However, none of these hypotheses presently have empirical support.

A humpback is more common than horns in most species of *Sinocyclocheilus*—as it occurs in every clade outside of the "*angularis*" group. Humpback shape varies from a slight humpback observed in *S. brevis* Lan & Chen, 1992 to the extreme bulge of *S. altishoulderus* (Li & Lan, 1992). Interestingly, a horn-like humpback located on the dorsum is present on *S. cyphotergous* (Dai, 1988) (Huang et al. 2017); microcomputed tomography revealed it was composed entirely of adipose tissue (Plate 7.3).

Cobitidae

The family Cobitidae is a group of elongate and thin loaches. Widely distributed throughout Eurasia and Morocco, there are about 195 described species (Nelson et al. 2016). With 56 species, China contains nearly 30% of cobitid diversity (C. Zhang and Zhao 2016). Of these, five species across three genera are considered stygobionts.

The Chinese genus *Protocobitis* was described in 1993 with the type species *P. typhlops* Yang, Chen & Lan, 1993 and includes three species. More than 20 years later, *Protocobitis typhlops* Yang & Chen, 1993 (Fig. 7.3) and *P. anteroventris* Lan, 2013 were described (Zhu et al. 2008; Lan et al. 2013). Pectoral and pelvic fins of this group have evolved in a manner that enable it to "crawl" along the bottom of cave pools (Fenolio et al. 2013). The first

Figure 7.3. *Protocobitis typhlops* Yang & Chen, 1993 from Guangxi Zhuang Autonomous Region. Image courtesy of Danté B. Fenolio.

branched ray of the pectoral fin in males is longer than in females (J. Yang et al. 1994). All *Protocobitis* species are eyeless and depigmented; they possess short barbels and are considered single-cave endemics with small population sizes (Romero et al. 2009).

The remaining two species in this family, *Bibarba parvoculus* Wu, Yang & Xiu, 2015 and *Paralepidocephalus translucens* Liu, Yang & Chen, 2016 are also single-cave endemic stygobionts. *Bibarba parvoculus* was described based on two specimens (T. Wu et al. 2015), while the description of *Paralepidocephalus translucens* was based on one specimen (Liu et al. 2016).

Nemacheilidae

Nemacheilidae (stone loaches) is found throughout Eurasia, with the greatest diversity in the Indian subcontinent, Indochina, and China. This family also includes one species from Ethiopia (Nelson et al. 2016). Cave-dwelling species are known from Iran, India, Thailand, Malaysia, and China (Niemiller et al. 2019). In China, Nemacheilidae is both the second-most diverse freshwater fish family with 217 species (Xing et al. 2016) and the second-most diverse Chinese cavefish family with 65 species (or 44% of the known cavefishes; C. Zhang and Zhao 2016). Most of these cave-dwelling species occur across two genera, *Troglonectes* and *Triplophysa* (L. Ma et al. 2019).

All 10 known species within the genus *Troglonectes* are stygobionts and endemic to China (C. Zhang and Zhao 2016). The single-cave endemic, *Troglonectes macrolepis* (Huang, Chen & Yang, 2009) (Plate 7.4A), is sympatric with an undescribed species of *Troglonectes*. Interestingly, *T. macrolepis* displays a burying behavior when disturbed (Fenolio et al. 2013). While exhibited in several epigean loach species, this burying behavior is known to occur in only one other subterranean species from India, *Indoreonectes evezardi* (Day, 1872) (Fenolio et al. 2013).

One of largest genera in the family Nemacheilidae, *Triplophysa* contains more than 100 species and is found primarily in and adjacent to the Qinghai-Xizang (Tibet) Plateau (Nelson et al. 2016). In China, this genus consists of at least 27 cave-dwelling fishes (C. Zhang and Zhao 2016). Several species have elongated fin rays; this trait is believed to be indicative of troglomorphism (Trajano 2001). For example, two subterranean-adapted *Triplophysa* exhibit similar fin ray morphology. *Triplophysa rosa* Chen & Yang, 2005 (Plate 7.4B) has long pectoral and pelvic fins compared to epigean relatives (X. Chen and Yang 2005), while *Triplophysa xiangxiensis* (Yang, Yuan & Liao, 1986) has similarly elongated fin rays (He et al. 2006).

Amblycipidae

Amblycipidae is a small family of catfishes (Siluriformes) consisting of 39 species known to inhabit swift streams from Pakistan and Malaysia to southern Japan (Nelson et al. 2016; Froese and Pauly 2019). Fifteen species are known from China (C. Zhang and Zhao 2016), including one stygobiont, *Xiurenbagrus dorsalis* Xiu, Yang & Zheng, 2014 (Fig. 7.4). This species was initially described from a single individual collected from a residential well in Guangxi Zhuang Autonomous Region; however, the lead author collected a second specimen in 2016. Most stygobiontic catfishes are known from South America. This species represents the only stygobiontic catfish found in eastern Asia to date. Unfortunately, nothing is known concerning this animal's life history, habitat requirements, or ecology—as only two individuals have been examined.

Figure 7.4. The stygobiontic *Xiurenbagrus dorsalis* Xiu, Yang & Zheng, 2014, from Guangxi Zhuang Autonomous Region. Image courtesy of Danté B. Fenolio.

Cavefish Zoogeography and Habitat

Distribution of Stygobiontic Cavefish

Cavefishes have been documented on every continent except Antarctica (Niemiller et al. 2019). Most cavefishes are distributed latitudinally in the tropics and subtropics between latitude 40° north and the Tropic of Capricorn (Romero and Paulson 2001; Y. Zhao and Zhang 2006). The highest species diversity of stygobiontic cavefishes is known from the karst habitats of Asia with 124 species, followed by South America and North America (Fig. 7.5). Cavefishes are also known from Africa (including four from Madagascar), Oceania (specifically, Australia and Papua New Guinea), and Europe (one undescribed loach from Germany; Niemiller et al. 2019).

Of the 148 known Chinese cavefishes, only one species occurs north of the SCK. The provinces of Yunnan, Guizhou, and Guangxi support the greatest diversity. Guangxi Zhuang Autonomous Region boasts the highest diversity with 75 species, which include 46 stygobionts. Yunnan and Guizhou have 53 and 21 species, respectively. On the edge of the Yunnan-Guizhou Plateau, Chongqing Municipality and Hunan Province each have one stygobiontic species (Fig. 7.6; Table 7.1).

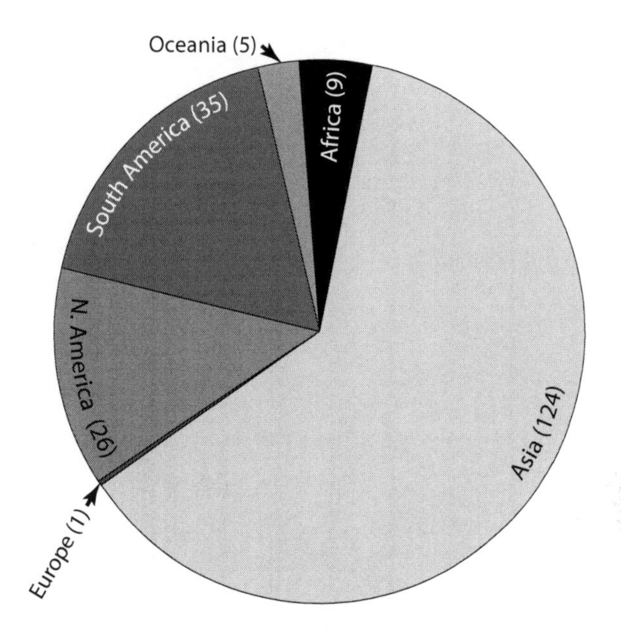

Figure 7.5. Diversity of stygobiontic cavefish per geographic region.

One species, *Onychostoma macrolepis* (Bleeker, 1871), a stygoxenic cyprinid cavefish occurs in several caves north of the Changjiang (Yangtze) River (Fig. 7.6) and has the largest distribution of any cavefish in China (Y. Zhao et al. 2011). This fish inhabits caves and springs during the winter (from November through April) and returns to the surface from late April through October to feed and reproduce (C. Zhang 1986). *Onychostoma macrolepis* represents the only cavefish that occurs outside of the SCK.

Single-Site Endemism and Syntopy

Of the 148 known cavefish species, at least 85 are likely single-site endemics, known from a single cave system or subterranean stream. For example, *Sinocyclocheilus anophthalmus* Chen & Chu, 1988, a stygobiont, is known only from Xiaogou Cave in Yiliang County, Yunnan (Y. Zhao and Zhang 2009b). While there are several examples of short-range endemics among cavefishes, such as *Speoplatyrhinus poulsoni* Cooper & Kuehne, 1974 in the United States and *Rhamdia enfurnada* Bichuette & Trajano, 2005 in Brazil, we recognize most cave-bearing regions have not been thoroughly surveyed for cavefishes, and that most subterranean waters are difficult to impossible

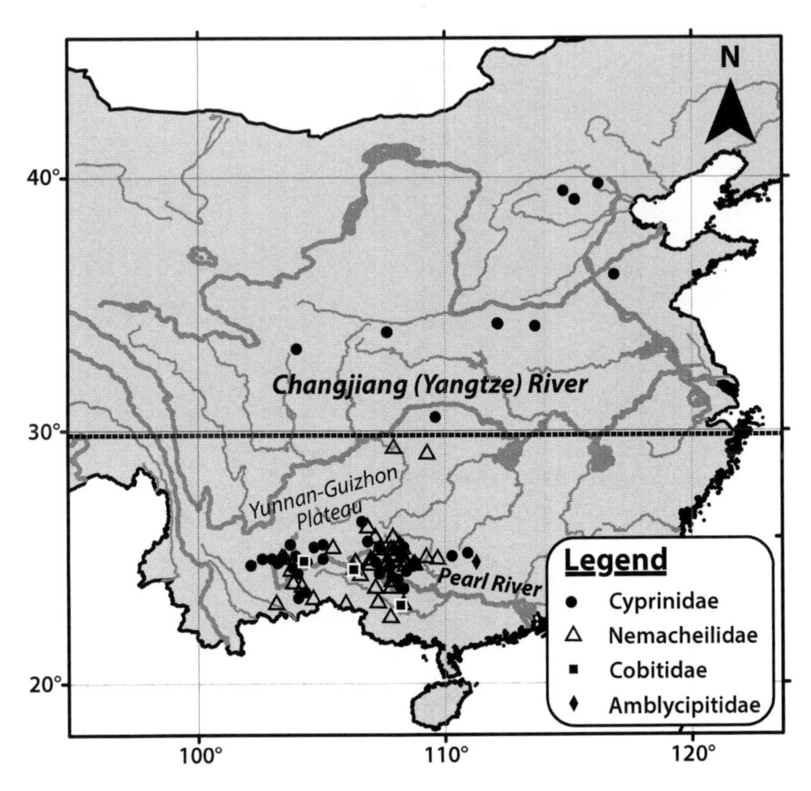

Figure 7.6. Distribution of cavefishes in China. Black dots located north of the Changjiang (Yangtze) River represent known locations for stygoxenic *Onychostoma macrolepis* (Bleeker, 1871).

for humans to access (barring accessible wellheads). Given this, we suggest some of the species presently considered short-range endemic species may represent regional endemics. Due to their high endemicity (either from a single cave, stream or region), we suggest these Chinese cavefish species are likely of conservation concern.

Globally, multiple cavefish species occurring syntopically is not rare. Nearly 40% of all stygobiontic fishes coexist with at least one other cavefish species (Niemiller et al. 2019). For example, a cave in Iran supports three cavefish species, *Eidinemacheilus smithi* (Greenwood, 1976), *Garra typhlops* (Bruun & Kaiser, 1944), and *G. lorestanensis* Mousavi-Sabet & Eagderi, 2016 (Mousavi-Sabet and Eagderi 2016). Another example is from Passa Três, São Vicente, in central Brazil, where two stygobiontic catfishes, *Ancistrus*

Table 7.1. Cavefish species diversity in South China Karst by administrative unit (Chinese provinces, autonomous regions, and municipalities). "Karst (km²)" denotes the extent of karst area, "Karst (%)" represents the total amount of karst, and "River" is the total number of subterranean river systems per administrative unit. The total numbers of stygobiontic ("SB") and stygophilic ("SP") species and "Richness" for each administrative unit are also provided.

Administrative unit	Karst (km^2)[a]	Karst (%)	River[b]	SB	SP	Richness[c]
Guangxi	98,700	23.6	347	46	29	75
Yunnan	110,876	26.5	96	16	37	53
Guizhou	109,085	26.1	219	12	9	21
Hunan	67,200	16.1	206	1	0	1
Chongqing	32,038	7.7	85	1	0	1

[a] Karst area estimates are from Zongfa Li (1994), Zhibin Li (2011), Shao et al. (2006), Y. Wang et al. (2005), and Wei et al. (2018).

[b] Total number of subterranean river systems are from Pei et al. (2008).

[c] Four species occur in more than one administrative unit; thus, richness values per administrative unit total more than 148 species.

cryptophthalmus Reis, 1987 and *Ituglanis passensis* Fernandéz & Bichuette, 2002, co-occur (Bichuette and Trajano 2003).

In the United States, *Amblyopsis spelaea* DeKay, 1842 and *Typhlichthys subterraneus* Giard, 1859 occur in the Mammoth Cave system; however, this cave system is extensive, and these species appear to be filling different niches. *Amblyopsis spelaea* inhabits deeper water further downstream, whereas *T. subterraneus* is found primarily within the plunge pools at the base of vertical shafts (Niemiller and Poulson 2010). Because cavefishes are typically apex predators in subterranean aquatic ecosystems, habitat partitioning may promote the coexistence of multiple cavefish species in a given subterranean stream, thus reducing interspecific competition (Trajano and Bichuette 2010).

For China, there are several examples of syntopy in cavefishes. In Xiaogou Cave, Yunnan Province, the stygobiontic *Sinocyclocheilus anophthalmus* occurs with the stygophilic *S. maitianheensis* Li, 1992, which is always found near the entrance of the cave (Y. Zhao and Zhang 2009a). Additionally, *S. rhinocerous*, *S. angustiporus* Zheng & Xie, 1985, and *S. malacopterus* Chu & Cui, 1985 are found in a single dragon pool (i.e., a small pool occurring at the entrance of a cave, representing the terminus of a subterranean stream or river) near Xinzhai Village, Luoping County, Yunnan (W. Li and Tao 1994). Moreover, three stygobiontic cavefishes, *S. microphthalmus* Li, 1989, *S. lingyunensis* Li, Xiao & Luo, 2000, and *Triplophysa lingyunensis* occur in the same subterranean stream of Shadong Cave, Lingyun County, Guangxi (W. Li et al. 2000).

Regarding syntopy and the use of different niches by different species, most species of *Sinocyclocheilus* occur in habitats with slow-flowing water, whereas nemacheilid species occur in fast-flowing waters. Additionally, stygophiles frequently inhabit areas near cave entrances or exits of subterranean rivers, while stygobionts are found in deeper subterranean habitats. For example, *S. hyalinus* Chen & Yang, 1993, *Triplophysa aluensis* Li & Zhu, 2000, and *S. aluensis* Li & Xiao, 2005 occur in one subterranean river system. The former two species inhabit the deeper reaches of the river, whereas *S. aluensis* is known to occur near the surface where the river exits the cave (Y. Zhao and Zhang 2013).

Habitat

Cavefish habitat in China may be best described at three scales. The landscape scale, which is subdivided into three geologic regions: high plateau, fengcong (cone karst), and fenglin (tower karst). The second scale is the subterranean river or stream level, while the finest scale consists of the individual niches within the aquatic habitat.

As with the evolution and diversification of most subterranean-adapted organisms, the distribution of stygobiontic cavefish species in southern China is due to geologic age. Thus, generally speaking, the landscape scale appears to drive the distribution of cavefish species.

The Yunnan-Guizhou karst, which is strongly influenced by the uplift of Qinghai-Xizang (Tibet) Plateau from the west to the east (L. Zhang et al. 2012), can be viewed as three sections. The western section, located in eastern Yunnan, is characterized by a relatively flat plateau (ranging 1,500 m to 2,000 m above sea level) and contains some fenglin landscape. Comprised primarily of rugged and steep fengcong landscape (ranging in elevation from 200 to 1,500 m), the middle section occurs within southern Guizhou and northwestern Guangxi. Finally, the easternmost section, located in northeastern Guangxi, is a relatively flat fenglin terrain (with an average elevation ~200 m) and is considered the oldest geologically (Y. Zhao et al. 2011).

Within these three karst sections, nine distinct stages of karst terrain formation are underway. For this chapter, however, we are providing a succinct overview concerning the evolution of karst terrain over time (refer to Waltham 2008 for specific details). The first stage is a karst plain. Dolines slowly form within the plain over time. As the doline landscape continues to erode, it transitions to a fengcong terrain. As the fengcong matures and regional uplift occurs, it is transformed to fenglin. More uplift occurs and, over time, the fenglin erodes and ultimately returns to karst plain.

The fengcong terrain represents younger formations and subterranean habitats that are still developing, while fenglin is older with more developed subterranean habitats (Che and Yu 1985). This may be somewhat counterintuitive, as the development of subterranean-adapted forms are typically associated with age—the older the geologic unit has a longer evolutionary time span between colonization from the surface and the evolution of troglobiontic forms (e.g., Culver 1976; Engel 2007). However, in the SCK, karst terrain of an intermediate age disproportionately supports stygobiontic cavefishes, while the older formations support primarily stygophilic cavefishes (Y. Zhao and Zhang 2009a; Y. Zhao et al. 2021).

The fengcong landscape contains the most well-developed subterranean rivers and streams. Subsequently, we suggest a greater diversity of habitats and niches occur within this substrate. This terrain supports most of the stygobionts that occur in China. Typical inhabitants of the fengcong include *Sinocyclocheilus furcodorsalis* Chen, Yang & Lan, 1997, *S. tianlinensis* Zhou, Zhang & He, 2003, *Protocobitis anteroventris* Lan, 2013, and *Triplophysa tianeensis* Chen, Cui & Yang, 2004.

Fenglin terrain occurs in eastern Yunnan and northeastern Guangxi and supports a large number of caves, but these features are typically shallow in depth. Subterranean rivers generally flow shorter distances and often connect to surface rivers or drain into dragon pools. In this region, stygophilic fishes are more abundant (Y. Zhao and Zhang 2009a; Y. Zhao et al. 2011). For example, *Sinocyclocheilus luopingensis* Li & Tao, 2003 occurs in Yunnan Province, while both *S. guilinensis* Ji, 1982 and *Oreonectes polystigmus* Du, Chen & Yang, 2008 are known from Guangxi.

The subterranean river/stream scale includes (1) the entire reach of the river/stream, (2) intermittent openings to the surface, and (3) the entrance and exit of the subterranean river. Regarding intermittent openings, subterranean rivers are often intersected by sinkholes and tiankeng (large dolines formed in carbonate rocks greater than 100 m deep and wide), which permit the allochthonous transport of nutrients underground. These openings often support larger and more diverse cavefish communities (Y. Zhao et al. 2011). Subsequently, stygobionts and stygophiles often co-occur in these habitats. For example, *Sinocyclocheilus tileihornes* (stygobiont) and *S. angustiporus* (stygophile) are found in a small tiankeng in Yunnan Province (Y. Zhao and Zhang 2009a). A similar phenomenon has been observed in Caballo Moro Cave in northern Mexico; two clades of *Astyanax* cavefish, one with highly reduced eyes and the other with fully developed eyes, co-occur in a pool beneath a skylight (Strecker et al. 2012).

As most subterranean aquatic habitats extend beyond what is observable within a cave or its surface expression (e.g., sinking streams and springs), quantifying the habitat requirements of any given aquatic species remains challenging. Subsequently, most information on habitat requirements of Chinese cavefishes is largely anecdotal. As most species occur within subterranean streams or rivers, which are often connected to larger groundwater systems, many Chinese cavefish populations may occur principally within aquifer habitats and secondarily within accessible subterranean watercourses (where they are detected by researchers). Additionally, some subterranean streams and rivers drain into surface lakes or dragon pools. Both fengcong and fenglin of southwestern China contain dragon pool and subterranean stream/river habitats (Sweeting 1995).

Subterranean streams, like surface streams, provide several niches (e.g., riffles, runs, still water, flowing water, and pools), which species likely had to adapt to exploit. Presently, information is lacking regarding niche specialization of these microhabitats within subterranean streams. Therefore, we provide the following examples based upon direct observations of cavefishes. Stygophiles, including *Sinocyclocheilus angustiporus*, *S. guishanensis* Li, 2003, and *S. qiubeiensis* Li, 2002, have been observed foraging near cave entrances (Y. Zhao et al. 2011), while other stygophilic species, *S. grahami*, *S. yangzongensis* Tsu & Chen, 1977, and *S. tingi* Fang, 1936, occur in Dianchi, Yangzonghai, and Fuxian Lakes, respectively; these surface lakes are connected to subterranean streams and/or rivers. "Aquifer dwelling" stygobiontic species include *Protocobitis polylepis* Zhu, Lü, Yang & Zhang, 2008 and *Xiurenbagrus dorsalis*, which have only been detected from human-made wellheads (Lan et al. 2013). However, it is possible they may ultimately be found in caves where conditions are appropriate.

Anthropogenic Threats

In the SCK, intensive human activity and development typically proceed without consideration of the impacts on sensitive and endemic subterranean animal populations (Whitten 2009). The most pressing human activities affecting SCK subterranean resources and cavefish populations, include groundwater pollution and withdrawal, habitat degradation and loss, commercial development of caves, and local harvesting of cavefishes for human consumption.

Groundwater Pollution

Groundwater pollution in subterranean ecosystems is a significant concern given the highly porous nature of karst substrates (Ford and Williams 2007). Both point and nonpoint source pollution are typically widespread. Point source pollution includes factories, waste disposal, wastewater discharge, and agricultural runoff (Y. Wu and Chen 2013). These activities may introduce high levels of chemicals, nutrients, and heavy metals (such as zinc, aluminum, and manganese) into cave ecosystems (Du et al. 2015).

In 2012, several industrial plants and smelters illegally discharged cadmium wastewater at proximity to Longjiang River, Guangxi Province, which resulted in short-term widespread contamination events in the region (Zhou 2012; J. Wang et al. 2019). Following these discharges, J. Wang et al. (2019) reported cadmium concentrations from 0.3 to 8 times higher than the threshold value of national drinking water quality standards for China at several sample sites along the Longjiang River. A series of emergency chemical treatments were later administered, and cadmium levels fell to safe levels for drinking water (J. Wang et al. 2019). The Longjiang River system is considered a cavefish hotspot (Y. Zhao and Zhang 2009a; Lan et al. 2013); unfortunately, it is unknown how both the cadmium spill and the subsequent emergency chemical treatments affected cavefish populations.

Nonpoint sources consist primarily of erosion from road construction and agriculture, which can result in siltation and agrochemical pollution (Jin et al. 2016). Current agricultural practices introduce nutrients via fertilizers (e.g., nitrogen, phosphorus, and potassium) and other agrochemicals (e.g., fungicides, herbicides, and insecticides) into groundwater. In intensive agricultural karstic areas, nitrate is possibly the most widespread contaminant in groundwater and surface water systems due to pervasive fertilizer application (Yue et al. 2018). High nitrate concentrations were detected in the mixed-land-use karst catchment in Guizhou Province—a region particularly vulnerable to groundwater pollution. Yue et al. (2018) indicated nitrate flux was strongly correlated with agricultural activities and exacerbated by precipitation intensity, which accelerated transport from surface to subterranean streams and rivers, pools, and aquifers. This impact is of grave concern and warrants further study as most caves and subterranean water bodies are directly linked to agricultural fields.

Groundwater Withdrawal

Overexploitation of groundwater in the SCK is considered one of the greatest threats to Chinese cavefishes. Although precipitation in the SCK is relatively high, droughts associated with climate change are occurring more frequently (Shu et al. 2013; Q. Zhang et al. 2019; J. Zhao et al. 2019). As surface water sources are growing more scarce, groundwater is becoming the primary water source in many areas; as such, extraction for industry, agriculture, and residential needs is rapidly increasing. Subsequently, balancing groundwater recharge with increased drought frequency and increased withdrawals has become a pressing environmental issue (Xiong and Chi 2015; T. Ma et al. 2020).

Large-scale reservoirs have been built, including the Maguan Reservoir in Guizhou, Xiaopingyang Reservoir in Guangxi, Baiyanhe Reservoir in Hunan, and Haidigou Reservoir in Chongqing (Gu and Yan 2006; P. Wu and Yan 2006). For Guizhou, a total of 29 reservoirs have been developed, accounting for more than half of the total groundwater reservoirs in China (Z. Deng 1995). These reservoirs change the hydrology of subterranean rivers; for example, water temperature, flow rate, and flow capacity, all of which are important abiotic factors that trigger reproduction in fish. Moreover, the environment will be reshaped from a lotic to a more lentic system due to the presence of these reservoirs.

These changes are expected to adversely affect community composition and may disrupt dispersal of stygobiontic cavefishes. For example, genetic exchange of riverine epigean fish species is likely to be interrupted between populations of upper and lower streams by even a low-head dam (Bunn and Arthington 2002; Smith et al. 2017). To date, the distributions of the stygobionts and stygophiles have been negatively impacted by reservoir development; species affected include *Sinocyclocheilus cyphotergous*, *S. qujingensis* Li, Mao & Lu, 2002, *S. yishanensis* Li & Lan, 1992, *S. oxycephalus* Li, 1985, and *S. macrocephalus* Li, 1985 (Y. Zhao and Zhang 2009a). Conversely, development of large dams may positively influence stygophilic cavefish populations as reservoirs associated with dams could generate—at least in the short term—more nutrients.

Subterranean watercourses represent an important water source for rural people in many regions of the SCK. Pump stations and associated infrastructure are often constructed within or at close proximity to a cave (e.g., Lan et al. 1996, 2013; Y. Zhao et al. 2011; Fenolio et al. 2013; Feng et al. 2020). Namely, the type locality of *Protocobitis typhlops*, listed as "Vulnerable" by the IUCN, is located beneath a major road outside of the town of Du'an in Guangxi Province (J. Yang et al. 1994). There are several other species that occur in this cave within a ~50 m^2 cave pool that contains infrastructure for water extraction (Fenolio et al. 2013). Also, several *Protocobitis polylepis* were collected from a well used to extract subterranean water to supply a fish farm (Lan et al. 2013). Furthermore, water extraction can temporarily reduce habitat extent and nutrient resources, and individual cavefish may be inadvertently removed during pumping operations (Zhu et al. 2008; Lan et al. 2013). Noise and vibrations associated with water pumps can also stress cavefish populations (Fenolio et al. 2013). In particular, the only known locality for the stygobiontic *Heminoemacheilus hyalinus* Lan, Yang & Chen, 1996 is used for water extraction (Fig. 7.7). This cavefish has not been observed in the past 20 years (Y. Zhao, unpublished data) and is believed to either be extirpated or potentially extinct.

Habitat Degradation and Loss

Mining and limestone quarrying for concrete production represents one of the most important economic activities in some provinces of China. Nearly 4,000 mines operate in Guizhou. Of those, over 40% are limestone quarries. These operations generated US$8.4 billion in 2018, accounting for 10.23% of the province's gross domestic product (Statistics Bureau of Guizhou Province, 2020).

These activities can severely degrade or eliminate subterranean habitats (Whitten 2009; F. Zhang et al. 2011; Fig. 7.8). As expected, this activity poses a serious threat to short-range endemic cavefish species. Moreover, mining activities can artificially connect surface habitats and groundwater (Ding et al. 2018), which can further degrade subterranean aquatic habitats by serving as a point source for sedimentation and pollution.

Figure 7.7. Water extraction infrastructure in a cave in Du'an town, Guangxi Zhuang Autonomous Region. This cave is also the type locality of *Heminoemacheilus hyalinus* Lan, Yang & Chen, 1996. Image courtesy of Yahui Zhao.

Figure 7.8. Removal of entire limestone formations for concrete production is common in China. This mine is located approximately 2 km south from the type locality of *Oreonectes luochengensis* Yang, Wu, Wei & Yang, 2011. Image courtesy of Yahui Zhao.

Construction activities may also directly threaten subterranean environments. Infrastructure projects associated with road development can entirely or partially destroy subterranean habitats (Mammola et al. 2019). For instance, a road construction project resulted in the loss of the type locality of *Sinocyclocheilus anatirostris* (Y. Zhao and Zhang 2009a; Y. Zhao et al. 2011).

Rocky desertification, associated with the removal of vegetation and subsequent soil erosion in rocky landscapes, can impede subterranean hydrologic regimes and exacerbate nutrient loading from the surface (Xiong and Chi 2015). This process can threaten the persistence of some subterranean animal populations (Mammola et al. 2019). Specifically, rocky desertification is expected to either reduce populations or potentially result in the extinction of some short-range endemic cavefish species.

Commercial Cave Development

In China, caves slated for tourism are developed without consideration for subterranean resources or the rich biological diversity these systems often support (Whitten 2009). With at least 708 commercial caves and developed scenic areas (X. Cao et al. 2017), cave tourism is rapidly growing throughout the country. Guangxi Zhuang Autonomous Region has the most show caves and scenic areas (totaling 91), while Guizhou and Yunnan Provinces have 78 and 54, respectively (X. Cao et al. 2017). Show caves typically involve the construction of paved walkways, the creation of additional entrances (which alters airflow), alterations of natural water flow, and the use of artificial lighting (Y. Deng et al. 2011; Pellegrini and Ferreira 2012; Cigna 2016; Gao et al. 2018). Additionally, temperature and carbon dioxide levels have been positively correlated with the light intensity and human visitation (Song et al. 2000; X. Yang et al. 2007; Russel and MacLean 2008; Cigna and Forti 2013; Cigna 2016).

In the aggregate, the loss of cave habitats, altered cave climates and water flow, and intensive human use can stress and imperil sensitive cavefish populations. For example, Ancient Alu Cave, known as "the first cave in Yunnan," is a large underground karst cave system, which also includes Luyuan, Yuzhu, and Biyu Caves and the Yusun subterranean river system (F. Zhang 1998). Developed as a show cave in 1988 (F. Zhang 1998), it receives millions of tourists each year. This cave contains at least 3 km of paved concrete trails, artificial lighting, and audio speakers for music (Fig. 7.9).

The Ancient Alu Cave karst system is also the type locality of *Sinocyclocheilus hyalinus* (Y. Chen et al. 1994). About 50 specimens were curated in the park museum in the early 1990s, which suggests the population was once abundant. However, since 2000, several investigations conducted by the first author failed to detect *S. hyalinus*. While five juveniles of *S. hyalinus* were collected in unopened areas of the upper and lower streams of the Yusun subterranean river in 2009 (KIZ-CAS 2009), this species has not been observed since.

Other species, such as *Oreonectes anophthalmus* Zheng, 1981 and *Troglonectes macrolepis*, have been confirmed in other show caves, but they are exposed to varying degrees of human disturbance (Fenolio et al. 2013). Their current status within these cave systems is unknown.

Figure 7.9. Commercially developed portion of Ancient Alu Cave, Yunnan Province, with artificial lighting featured (as depicted by light areas in this image). Image courtesy of Andrew G. Gluesenkamp.

Harvesting Cavefishes

Human consumption and trafficking of cavefishes occurs in many rural areas. Cavefishes are collected for commercial and subsistence consumption, as well as for the aquarium trade (Fenolio et al. 2013). Additionally, there have been increased demands for cavefish in traditional Chinese medicine (M. Zhang et al. 2016).

Some cavefish, also known known as "oil fish," are traditionally harvested as an important food resource in rural communities. Regionally, oil fish soup is considered a delicacy and has recently surged in popularity. This soup is typically made with *Sinocyclocheilus grahami* and *S. tingi*, both of which are cavefish species. Market price can be as much as 50 times higher than other commercially available fish. Moreover, high market prices incentivize locals to collect more cavefish, which only exacerbates the fish harvests (Y. Zhao and Zhang 2009a) and may further threaten the persistence of these populations.

The aquarium trade places additional unsustainable pressures on cavefish populations. Given the ease of human access to some sites, cavefish populations may suffer intense collection pressure (Fenolio et al. 2013). Currently, *Sinocyclocheilus grahami* is the only Chinese cavefish with legal protection (Y. Zhao and Zhang 2009a). However, there are no seasonal or take limits for other cavefish species.

Conservation and Future Research

IUCN Red List Status

Presently, 126 (or 85%) Chinese cavefish species have been assessed using the International Union for Conservation of Nature's *IUCN Red List of Threatened Species* assessment criteria (Jiang et al. 2016). Thirty-four species (or 23%) are listed as species of conservation concern (Cao et al. 2016). These include five species considered "Critically Endangered," seven "Endangered" species, and 22 "Vulnerable" species (Table 7.2). Importantly, more than half of the cavefish species evaluated were identified as "Data Deficient." This lack of knowledge limits the ability of researchers and conservation biologists to effectively manage and protect these species. Moreover, some of these species, which have narrow distributional ranges, may be threatened by human activities. Given our limited knowledge, we are unable to effectively evaluate these species and provide concordant conservation recommendations.

Conservation

There is little doubt the impressive diversity of subterranean animals, including cavefishes, is one of the most important resources of the

Table 7.2. Summary of IUCN Red List status for Chinese cavefishes by family. The Red List categories are Critically Endangered (CR), Endangered (EN), Vulnerable (VU), Near Threatened (NT), Least Concern (LC), Data Deficient (DD), and Not Evaluated (NE).

Family	CR	EN	VU	NT	LC	DD	NE
Cyprinidae	3	4	14	4	17	28	7
Nemacheilidae	2	2	7	2	5	34	13
Cobitidae		1	1			1	2
Amblycipitidae						1	
Total	5	7	22	6	22	64	22

South China Karst. Given the increasing scale and magnitude of anthropogenic threats, some cavefish species may likely go extinct before they are formally described (Y. Zhao et al. 2011; Mammola et al. 2019). Chinese cavefishes and their habitats should be considered biological treasures worthy of celebration and protection.

There are several natural parks within the SCK, which were established to protect endemic animal populations or cavefishes specifically. Unfortunately, some reserves in the region are "paper parks" (e.g., Dudley and Stolton 1999), where human activities continue largely unabated and management policies require revision and/or enforcement. Luliang Natural Reserve of Golden-Line Barbel represents the first protected area for cavefishes in China. It was established to protect *Sinocyclocheilus lateristriatus* Li, 1992 and *S. macroscalus* Li, 1992. Unfortunately, the local municipality lacks the financial support to best manage this reserve. The extent of protective management involves fencing and gating the dragon pools supporting these species to deter local residents from entering the area and potentially harvesting the fish. Another example is the Lingyun Natural Reserve of Cavefish in Guangxi Zhuang Autonomous Region, which was established in 2008. Consisting of 684 hectares, including one subterranean river system and six other scattered cave systems (Guangxi People's Government 2008), the goal of this reserve is to protect five cavefish species (*S. lingyunensis, S. microphthalmus, S. anatirostris, S. anshuiensis* Gan, Wu, Wei & Yang, 2013, and *Triplophysa lingyunensis* (Liao & Luo, 1997)). Guizhou Mao Lan Karst Forest, Guangxi Mulun National Nature Reserve, and Yunnan Jiuxiang National Geopark are other examples of protected areas that could improve their protections of endemic subterranean faunas and their habitats.

Sinocyclocheilus grahami, a national protected animal, occurs in Dianchi Lake, its tributaries, and springs. Due to human activities, populations have precipitously declined (Pan et al. 2009). This species is now known from at least 20 dragon pools and one tributary of Dianchi Lake (Z. Chen et al. 2001). To safeguard this species and increase its numbers, an artificial breeding program was started in 2007 (J. Yang et al. 2018) and a reintroduction effort was undertaken. By the end of 2019, more than 2 million fish fry were released in Dianchi Lake (Y. Zhao, unpublished data). Captive breeding of other stygophiles, such as *S. qujingensis* and *S. tingi*, is also underway. Unfortunately, no Chinese stygobiontic species have been successfully bred in captivity.

Future Research

A national program on Chinese cavefishes will be required to obtain the information required to effectively conserve and manage sensitive cavefish populations and their habitats. In the past, cavefish research in southern China was largely taxonomically focused—due to a lack of specialists studying cavefishes in China. This lacuna in expertise can be addressed by assembling an interdisciplinary team of Chinese and international scientists, who could be tasked with developing a strategy to systematically collect and synthesize data on cavefish life history, population dynamics, habitat requirements, distributions, and reference DNA sequences. Additionally, such a team could assess the vulnerability to human impacts for all subterranean waterways occurring in the SCK. Given this large scope, survey areas should be prioritized based on their proximity to intensive human activities. This first-order vulnerability assessment should include geospatial analyses using cave and aquifer locations for all cavefish morphospecies, subterranean watercourses and aquifers, and human impacts. This comprehensive dataset, related data products, and vulnerability assessments developed from such a project can then be integrated into community conservation and development programs. Such programs should weigh the impacts of status quo human activities against protecting vulnerable cavefish populations and their habitats while promoting sustainable economic needs of local communities (e.g., Berkes 2007).

Given the challenges with captive breeding of stygobiontic fish, data on life history and population dynamics may be insurmountable—at least in the short term. Furthermore, collecting data on species distributional ranges and habitat requirements may be hampered by limited access to subterranean waters. As most accessible water in caves is connected to aquifers and/or larger subterranean watercourses, the actual extent of cavefish habitat may be hundreds of square meters up to tens of square kilometers beyond cave walls. Thus, a systematic approach to remotely identify the occurrence of specific cavefish species will be required.

Environmental DNA (eDNA), a tool which can confirm the presence of an organism by proxy, should be tested as both a survey and monitoring tool. To date, this approach has been used successfully to detect subterranean salamanders (Gorički et al. 2017; Vörös et al. 2017; Lyons 2019; White et al. 2020), amphipods (Niemiller et al. 2018), crayfishes (Mouser 2019; DiStefano et al. 2020), and cavefishes (Lyons 2019; Mouser 2019; White et al.

2020). To do this, reference DNA sequences will be required for all species of concern. Depending on preservation techniques, viable DNA may be extracted from specimens in museum collections. Additionally, DNA samples can be extracted nonlethally via fin clipping, and/or collecting blood or scales in the field (e.g., Mirimin et al. 2011; Komrakova et al. 2018). However, these techniques require specialized training and can still result in incidental takes; when populations are believed to be small, the loss of even one individual may be unacceptable. Alternatively, a similar technique applied to acquiring eDNA from fish larvae (Espinoza et al. 2017) could be developed as a field-based approach for collecting cavefish DNA. Once the best technique(s) have been selected and/or developed, long-term monitoring programs should be established first for *Sinocyclocheilus grahami*, *Oreonectes anophthalmus*, and *Xiurenbagrus dorsalis*.

Importantly, as both surface and groundwater contamination can degrade subterranean habitats (Graening and Brown 2003), we recommend developing a water quality monitoring program in the SCK that would enable researchers to establish baseline conditions for water quality, as well as monitor water quality changes. While the impacts of pollutants on cave ecosystems have been broadly discussed (Poulson 1976; Sket 2005; Stone and Howarth 2007; Whitten 2009; and herein), data on subterranean pollution is largely lacking for most regions globally (Castaño-Sánchez et al. 2020). In a study of aquatic invertebrates in an Arkansas stream cave, Graening and Brown (2003) found that organic pollutant concentrations of heavy metals exceeded state limits for chronic and acute toxicity to aquatic organisms and may have altered ecosystem dynamics. Unfortunately, we are unaware of any similar studies of agrochemical toxicities in caves. However, as most of the SCK is rural agricultural areas, both heavy metals and agrochemicals should be monitored. As such, China has the opportunity to develop the first ecotoxicological monitoring program of subterranean aquatic habitats.

Data collected through a national cavefish and habitat monitoring program should be made available globally to further facilitate collaborations with other researchers and to spur future research. Although the IUCN Red List has been widely accepted in China as a mechanism for effecting conservation and management of sensitive species, some biases and limitations exist (Cardoso et al. 2011; Niemiller and Taylor 2019). Additionally, as we have mentioned, we currently lack a robust standardized methodology for sampling cavefishes; thus, it is possible current IUCN Red List designations do not reflect the actual vulnerability of most Chinese cavefishes.

For all these programs that we propose, we also recommend that resultant datasets (both current and those collected in the future) be archived in a central repository, such as China's National Specimen Information Infrastructure (NSII 2017).

Finally, one rarely considered conservation tool for cavefishes is captive breeding. To date, a few cavefish species have been successfully bred in the lab, which could serve as a roadmap for developing captive-breeding efforts of Chinese cavefishes. As mentioned earlier, stygophilic *S. grahami* has been captively bred in large numbers (J. Yang et al. 2018). Additionally, the Mexican cave tetra *Astyanax mexicanus* has been part of the aquarium trade since the 1940s (Peña-Herrejón et al. 2017) and has become a model laboratory species (e.g., Sadoglu 1979; Peña-Herrejón et al. 2017; Riddle et al. 2018); subsequently, there are now numerous lab procedures for captively breeding cavefishes. Also, two Brazilian cavefish species have been successfully bred in the lab—*Ancistrus cryptophthalmus* (Loricariidae, Ancistrinae; Secutti and Trajano 2009) and *Aspidoras mephisto* (Siluriformes: Callichthyidae; Secutti et al. 2011). Furthermore, many advancements in captively breeding difficult-to-breed species have been made in the aquarium trade. This has resulted in a plethora of skilled aquarists and tested lab procedures.

As such, captive-breeding information on other cavefish species and the skillsets of aquarists could be harnessed to develop captive-breeding protocols for critically imperiled species utilizing aquaculture facilities, zoos and aquariums, and university laboratories in China. For many short-range endemic stygobiontic cavefish species (such as *S. anatirostris*, *S. aquihornes* Li & Yang, 2007, *Typhlobarbus nudiventris*, *Protocobitis anteroventris*, and *Xiurenbagrus dorsalis*), captive-breeding programs (both within and outside of China) may likely be the only mechanism to prevent these species from going extinct.

Conclusions

Most local communities are completely unaware that China boasts the highest diversity of cavefishes on the planet. Furthermore, SCK residents are either unaware of the damage wrought by their activities or are indifferent because their actions are motivated by economic hardship. Developing karst management strategies that promote biodiversity conservation while considering sustainable economic development will be crucial to the long-term protection of cavefishes and other sensitive cave-dwelling species, and to ensuring a good quality of life for rural communities.

As cavefishes represent one of the most difficult cave organisms to study, developing systematic sampling procedures for both cavefishes and their habitat could result in pioneering new techniques that could have significant implications on how researchers study subterranean, as well as surface, ecosystems. Thus, the development of a program to study China's cavefishes could render dividends far beyond the management and protection of the most diverse cavefish region in the world.

The protection of cavefishes and subterranean habitats remains a low priority in China. We are hopeful that as cave ecological studies are expanded and the importance of SCK endemic animal communities and unique subterranean habitats continues to gain recognition globally, natural resource agencies will similarly recognize their importance. As a result, we remain hopeful the endemic cavefishes and other endemic subterranean populations occurring within the SCK will ultimately receive the recognition and protection they warrant.

ACKNOWLEDGMENTS

The authors thank Larry Page and Aldemaro Romero, who reviewed an earlier version of this chapter. This study was conducted under the auspices of the Chinese Cavefish Working Group. The first author (YZ) was supported by grants (NSFC-31972868, 31970382) from the National Natural Science Foundation of China and National Geographic Society grants (GEFC15-13 and GEFC15-16). The San Antonio Zoo supported AG and DF.

REFERENCES

Berkes, Fikret. 2007. "Community-Based Conservation in a Globalized World." *Proceedings of the National Academy of Sciences* 104:15188–15193.

Berra, Tim M., and John D. Humphrey. 2002. "Gross Anatomy and Histology of the Hook and Skin of Forehead Brooding Male Nurseryfish, *Kurtus gulliveri*, from Northern Australia." *Environmental Biology of Fishes* 65:263–270.

Bichuette, Maria E., and Eleonora Trajano. 2003. "Epigean and Subterranean Ichthyofauna from São Domingos Karst Area, Upper Tocantins River Basin, Central Brazil." *Journal of Fish Biology* 63:1100–1121.

Bunn, Stuart E., and Angela H. Arthington. 2002. "Basic Principles and Ecological Consequences of Altered Flow Regimes for Aquatic Biodiversity." *Environmental Management* 30:492–507.

Cao, Liang, E. Zhang, Chunxin Zang, and Wenxuan Cao. 2016. "Evaluating the Status of China's Continental Fish and Analyzing Their Causes of Endangerment through the Red List Assessment." *Biodiversity Science* 24:598–609.

Cao, Xiang, Xiaoxia Yang, Xi Li, Xu Xiang, and Xiaobei Sun. 2017. "Statistical Analysis of Show Cave Scenic Areas (Spots) in China." *Carsologica Sinica* 36:264–274.

Cardoso, Pedro, Paulo A. V. Borges, Kostas A. Triantis, Miguel A. Ferrández, and José L. Martín. 2011. "Adapting the IUCN Red List Criteria for Invertebrates." *Biological Conservation* 144:2432–2440.

Castaño-Sánchez, Andrea, Grant C. Hose, and Ana Sofia P. S. Reboleira. 2020. "Ecotoxicological Effects of Anthropogenic Stressors in Subterranean Organisms: A Review." *Chemosphere* 244:125422.

Che, Yongtai, and Jinzi Yu. 1985. *Karst in China*. Beijing: Science Press.

Chen, Xiaoyong, and Junxing Yang. 2005. "*Triplophysa rosa* sp. nov.: A New Blind Loach from China." *Journal of Fish Biology* 66:599–608.

Chen, Yinrui, Junxing Yang, and Zhigang Zhu. 1994. "A New Fish of the Genus *Sinocyclocheilus* from Yunnan with Comments on Its Characteristic Adaptation (Cypriniformes: Cyprinidae)." *Acta Zootaxonomica Sinica* 19:246–253.

Chen, Ziming, Junxing Yang, Ruifeng Su, and Xiaoyong Chen. 2001. "Present Status of the Indigenous Fishes in Dianchi Lake, Yunnan." *Biodiversity Science* 9:407–413.

Chu, Xinluo, and Yinrui Chen. 1982. "A New Genus and Species of Blind Cyprinid Fish from China with Special Reference to its Relationships." *Acta Zoologica Sinica* 28:383–388.

Cigna, Arrigo A. 2016. "Tourism and Show Caves." *Zeitschrift für Geomorphologie* 60:217–233.

Cigna, Arrigo A., and Paolo Forti. 2013. "Caves: The Most Important Geotouristic Features in the World." *Tourism and Karst Areas* 6:9–26.

Culver, David C. 1976. "The Evolution of Aquatic Cave Communities." *The American Naturalist* 110:945–957.

Deng, Yadong, Weihai Chen, Yuanhai Zhang, Wang Cheng, and Xumin Yi. 2011. "Ecological Environment Influenced by Karst Show Cave Exploitation." *Journal of Guilin University of Technology* 31:412–417.

Deng, Zimin. 1995. "Factors for Building Underground Reservoir in Karst Area." *Guizhou Science* 13:16–22.

Ding, Hanghang, Qiang Wu, Shuai Yu, and Jie Zhang. 2018. "Risk Assessment of Mine Environmental Problems: A Case Study of Karst Collapse." *China Energy and Environmental Protection* 40:46–51.

DiStefano, Robert J., David Ashley, Shannon K. Brewer, Joshua B. Mouser, and Matthew L. Niemiller. 2020. "Preliminary Investigation of the Critically Imperiled Caney Mountain Cave Crayfish *Orconectes stygocaneyi* Hobbs III, 2001 (Decapoda: Cambaridae) in Missouri, USA." *Freshwater Crayfish* 25:47–57.

Du, Yuchao, Jiangguo Pei, Li Lu, Yongsheng Lin, and Lianjie Fan. 2015. "Characteristics of Heavy Metal Element Distribution in the Groundwater System of Typical Karst Regions: A Case Study in Poyue Underground River, Guangxi, China." *Carsologica Sinica* 34:348–353.

Dudley, N., and S Stolton. 1999. "Conversion of Paper Parks to Effective Management: Developing a Target." Report to the WWF–World Bank Alliance from the IUCN/WWF

Forest Innovation Project. Gland, Switzerland: International Union for Conservation of Nature.

Engel, Annette S. 2007. "Observations on the Biodiversity of Sulfidic Karst Habitats." *Journal of Cave and Karst Studies* 69:187–206.

Espinoza, G. Janelle, J. Michael Poland, and Jaime R. Alvarado Bremer. 2017. "Genotyping Live Fish Larvae: Non-lethal and Noninvasive DNA Isolation from 3–5 Day Old Hatchlings." *BioTechniques* 63:181–186.

Feng, Zegang, J. Judson Wynne, and Feng Zhang. 2020. "Cave-Dwelling Pseudoscorpions of China with Descriptions of Four New Hypogean Species of *Parobisium* (Pseudoscorpiones, Neobisiidae) from Guizhou Province." *Subterranean Biology* 34:61–98.

Fenolio, Danté B., Yahui Zhao, Matthew L. Niemiller, and Jim F. Stout. 2013. "In-situ Observations of Seven Enigmatic Cave Loaches and One Cave Barbel from Guangxi, China, with Notes on Conservation Status." *Speleobiology Notes* 5:16–33.

Ford, Derek, and Paul Williams. 2007. *Karst Hydrogeology and Geomorphology*. Chichester: John Wiley & Sons, Ltd.

Froese, R., and D. Pauly. 2019. "FishBase." Accessed December 20, 2019. http://www.fishbase.org.

Gao, Zhizhong, J. Judson Wynne, and Feng Zhang. 2018. "Two New Species of Cave-Adapted Pseudoscorpions (Pseudoscorpiones: Neobisiidae, Chthoniidae) from Guangxi, China." *The Journal of Arachnology* 46:345–354.

Gorički, Špela, David Stanković, Aleš Snoj, Matjaž Kuntner, William R. Jeffery, Peter Trontelj, et al. 2017. "Environmental DNA in Subterranean Biology: Range Extension and Taxonomic Implications for *Proteus*." *Scientific Reports* 7:45054.

Graening, Gary O., and Arthur V. Brown. 2003. "Ecosystem Dynamics and Pollution Effects in an Ozark Cave Stream." *Journal of the American Water Resources Association* 39:1497–1505.

Gu, Shangyi, and Guiquan Yan. 2006. "Karst Underground Reservoir Study in Guizhou Province: Review and Prospect." *Guizhou Science* 24:28–31.

Guangxi People's Government. 2008. "The First Natural Reserve of Cavefish in China Was Established in Guangxi Zhuang Autonomous Region." The Central People's Government of the People's Republic of China. Accessed April 6, 2020. http://www.gov.cn/gzdt/2008-04/06/content_937457.htm.

He, Li, Xueguang Wang, Qingchun Chen, and Junzu Xiang. 2006. "Morphological Description on *Triplophysa xiangxiensis*." *Freshwater Fisheries* 36:56–58.

Huang, Jinqing, Andrew Gluesenkamp, Danté Fenolio, Zhiqiang Wu, and Yahui Zhao. 2017. "Neotype Designation and Redescription of *Sinocyclocheilus cyphotergous* (Dai) 1988, a Rare and Bizarre Cavefish Species Distributed in China (Cypriniformes: Cyprinidae)." *Environmental Biology of Fishes* 100:1483–1488.

Jiang, Zhigang, Jianping Jiang, Yuezhao Wang, E. Zhang, Yanyun Zhang, Lili Li, et al. 2016. "Red List of China's Vertebrates." *Biodiversity Science* 24:500–551.

Jin, Lin, Rongguo Sun, Zhuo Chen, and Pengwu Li. 2016. "Research Progress of Rural Non-point Source Pollution in Karst Area of Guizhou." *Chinese Agricultural Science Bulletin* 32:94–98.

[KIZ-CAS] Kunming Institute of Zoology, Chinese Academy of Sciences. 2009. "Scientists found Juvenile Fish of *Sinocyclocheilus hyalinus* in Ancient Alu Cave of Yunnan." Accessed November 10, 2020. http://news.sciencenet.cn/htmlnews/2009/2/216674.html.

Komrakova, Marina, Christoph Knorr, Bertram Brenig, Gabriele Hoerstgen-Schwark, and Wolfgang Holtz. 2018. "Sex Discrimination in Rainbow Trout (*Oncorhynchus mykiss*) Using Various Sources of DNA and Different Genetic Markers." *Aquaculture* 497:373–379.

Lan, Jiahu, Xi Gan, Tiejun Wu, and Jian Yang. 2013. *Cave Fishes of Guangxi, China*. Beijing: Science Press.

Lan, Jiahu, Junxing Yang, and Yinrui Chen. 1996. "One New Species of Cavefish from Guangxi." *Zoological Research* 17:109–112.

Li, Weixian, and Jinneng Tao. 1994. "A New Species of Cyprinidae from Yunnan—*Sinocyclocheilus rhinocerous* sp. nov." *Journal of Zhanjiang Ocean University* (formerly *Journal of Zhanjiang Fisheries College*) 14:1–3.

Li, Weixian, and Jinneng Tao. 2002. "Local Dissection of Body of the Fishes *Sinocyclocheilus rhinocerous*." *Journal of Yunnan Agricultural University* 17:207–209.

Li, Weixian, Heng Xiao, Ruiguang Zan, Zhifa Luo, and Hengmeng Li. 2000. "A New Species of *Sinocyclocheilus* from Guangxi, China." *Zoological Research* 21:155–157.

Li, Zhibin. 1994. "Characteristics and Exploitation of Karst Water Resources in Hunan Province." *Tropical Geography* 12:364–371.

Li, Zongfa. 2011. "Division of Karst Landform in Guizhou." *Guuizhou Geology* 28:177–181.

Liu, Shuwei, Junxing Yang, and Xiaoyong Chen. 2016. "*Paralepidocephalus translucens*, a New Species of Loach from a Cave in Eastern Yunnan, China (Teleostei: Cobitidae)." *Ichthyological Exploration of Freshwaters* 27:61–66.

Lyons, Kathleen Marie. 2019. "Exploring the Distribution of Groundwater Salamanders and Catfish with Environmental DNA." Master's thesis, University of Texas at Austin.

Ma, Li, Yahui Zhao, and Junxing Yang. 2019. "Cavefish of China." In *Encyclopedia of Caves*, 3rd ed., edited by William B. White, David C. Culver, and Tanja Pipan, 237–254. Amsterdam: Elsevier, Academic Press.

Ma, Ting, Siao Sun, Guangtao Fu, Jim W. Hall, Yong Ni, Lihan He, et al. 2020. "Pollution Exacerbates China's Water Scarcity and Its Regional Inequality." *Nature Communications* 11:1–9.

Mammola, Stefano, Pedro Cardoso, David C. Culver, Louis Deharveng, Rodrigo L. Ferreira, Cene Fišer, et al. 2019. "Scientists' Warning on the Conservation of Subterranean Ecosystems." *BioScience* 69:641–650.

Mirimin, L., D. O'Keeffe, A. Ruggiero, M. Bolton-Warberg, S. Vartia, and R. Fitzgerald. 2011. "A Quick, Least-Invasive, Inexpensive and Reliable Method for Sampling *Gadus morhua* Postlarvae for Genetic Analysis." *Journal of Fish Biology* 79:801–805.

Mousavi-Sabet, Hamed, and Soheil Eagderi. 2016. "*Garra lorstanensis*, A New Cave Fish from the Tigris River Drainage with Remarks on the Subterranean Fishes in Iran (Teleostei: Cyprinidae)." *FishTaxa* 1:45–54.

Mouser, Joshua. 2019. "Examining Occurrence, Life History, and Ecology of Cavefishes and Cave Crayfishes Using Both Traditional and Novel Approaches." Master's thesis, Oklahoma State University.

Nelson, Joseph S., Terry C. Grande, and Mark V. H. Wilson. 2016. *Fishes of the World*, 5th ed. Hoboken: John Wiley & Sons.

Niemiller, Matthew L., and Thomas L. Poulson. 2010. "Subterranean Fishes of North America: Amblyopsidae." In *Biology of Subterranean Fishes*, edited by Eleonora Trajano, Maria Elina Bichuette, and B. G. Kapoor, 169–280. Enfield, NH: Science Publishers.

Niemiller, Matthew L., and Daphne Soares. 2015. "Cave Environments." In *Extremophile Fishes: Ecology, Evolution, and Physiology of Teleosts in Extreme Environments*, edited by Rüdiger Riesch, Michael Tobler, and Martin Plath, 161–191. Switzerland: Springer International Publishing.

Niemiller, Matthew L., and Steven J. Taylor. 2019. "Protecting Cave Life." In *Encyclopedia of Caves*, 3rd ed., edited by William B. White, David C. Culver, and Tanja Pipan, 822–829. Amsterdam: Elsevier, Academic Press.

Niemiller, Matthew L., Maria E. Bichuette, Prosanta Chakrabarty, Danté B. Fenolio, Andrew G. Gluesenkamp, Daphne Soares, et al. 2019. "Cavefishes." In *Encyclopedia of Caves*, 3rd ed., edited by William B. White, David C. Culver, and Tanja Pipan, 227–236. Amsterdam: Elsevier, Academic Press.

[NSII] National Specimen Information Infrastructure. 2017. Accessed June 10, 2020. http://nsii.org.cn/2017/home-en.php.

Pan, Xiaofu, Xiaoyong Chen, and Junxing Yang. 2009. "Threatened Fishes of the World: *Sinocyclocheilus grahami* (Regan) 1904 (Cyprinidae)." *Environmental Biology of Fishes* 85:77–78.

Pei, Jiangguo, Maozhen Liang, and Zhen Chen. 2008. "Classification of Karst Groundwater System and Statistics of the Main Characteristic Values in Southwest China Karst Mountain." *Carsologica Sinica* 27:6–10.

Pellegrini, Thais Giovannini, and Rodrigo Lopes Ferreira. 2012. "Management in a Neotropical Show Cave: Planning for Invertebrates Conservation." *International Journal of Speleology* 41:359–366.

Peña-Herrejón, Guillermo A., Julieta Sanchez-Velazquez, Andrez Cruz-Hernández, Humberto Aguirre-Becerra, and Fernando García-Trejo. 2017. "Breeding System for *Astyanax mexicanus*." Paper presented at the XIII International Engineering Congress (CONIIN), Santiago de Queretaro, Mexico, May 15–19.

Poulson, Thomas L. 1976. "Management of Biological Resources in Caves." In *Proceedings of the National Cave and Karst Management Symposium*, 46–52. Albuquerque, New Mexico: National Cave and Karst Management Symposium.

Riddle, Misty, Brian Martineau, Megan Peavey, and Clifford Tabin. 2018. "Raising the Mexican Tetra *Astyanax mexicanus* for Analysis of Post-larval Phenotypes and Whole-Mount Immunohistochemistry." *Journal of Visualized Experiments* 142:1–7.

Romero, Aldemaro, and Steven M. Green. 2005. "The End of Regressive Evolution: Examining and Interpreting the Evidence from Cave Fishes." *Journal of Fish Biology* 67:3–32.

Romero, Aldemaro, and Kelly M. Paulson. 2001. "It's a Wonderful Hypogean Life: A Guide to the Troglomorphic Fishes of the World." *Environmental Biology of Fishes* 62:13–41.

Romero, Aldemaro, Yahui Zhao, and Xiaoyong Chen. 2009. "The Hypogean Fishes of China." *Environmental Biology of Fishes* 86:211–278.

Russell, Mick J., and Victoria L. MacLean. 2008. "Management Issues in a Tasmanian Tourist Cave: Potential Microclimatic Impacts of Cave Modifications." *Journal of Environmental Management* 87:474–483.

Sadoglu, Perihan. 1979. "A Breeding Method for Blind *Astyanax mexicanus* Based on Annual Spawning Patterns." *Copeia* 1979:369–371.

Secutti, Sandro, and Eleonora Trajano. 2009. "Reproductive Behavior, Development and Eye Regression in the Cave Armored Catfish, *Ancistrus cryptophthalmus* Reis, 1987 (Siluriformes: Loricariidae), Breed in Laboratory." *Neotropical Ichthyology* 7:479–490.

Secutti, Sandro, Roberto E. Reis, and Eleonora Trajano. 2011. "Differentiating Cave *Aspidoras* Catfish from a Karst Area of Central Brazil, Upper Rio Tocantins Basin (Siluriformes: Callichthyidae)." *Neotropical Ichthyology* 9:689–695.

Shao, Jing'an, Yangbing Li, Chaofu Wei, and Deti Xie. 2006. "The Characteristics of Landscape Patterns in Karst Area of Chongqing, China." *Progress in Geography* 25:31–40.

Shu, Shu-Sen, Wansheng Jiang, Tony Whitten, Junxing Yang, and Xiaoyong Chen. 2013. "Drought and China's Cave Species." *Science* 340:272.

Sket, Boris. 2005. "Diversity in Dinaric Karst." In *Encyclopedia of Caves*, 3rd ed., edited by William B. White, David C. Culver, and Tanja Pipan, 158–165. Amsterdam: Elsevier, Academic Press.

Smith, S. C. F., S. J. Meiners, R. P. Hastings, T. Thomas, and R. E. Colombo. 2017. "Low-Head Dam Impacts on Habitat and the Functional Composition of Fish Communities." *River Research and Applications* 33:680–689.

Soares, Daphne, and Matthew L. Niemiller. 2013. "Sensory Adaptations of Fishes to Subterranean Environments." *Bioscience* 63:274–283.

Soares, Daphne, Michelle Pluviose, and Yahui Zhao. 2019. "Ontogenetic Development of the Horn and Hump of the Chinese Cavefish *Sinocyclocheilus furcodorsalis* (Cypriniformes: Cyprinidae)." *Environmental Biology of Fishes* 102:741–746.

Song, Linhua, Wei Xiaoning, and Liang Fuyuan. 2000. "The Influence of Cave Tourism on CO_2 and Temperature in Baiyun Cave, Hebei, China." *International Journal of Speleology* 29:77–87.

Statistics Bureau of Guizhou Province. 2020. "Guizhou Macroeconomic Database." Guiyang: Statistics Bureau of Guizhou Province. Accessed August 10, 2020. http://hgk.guizhou.gov.cn/index.vhtml#/.

Stone, Fred D., and Francis G. Howarth. 2007. "Hawaiian Cave Biology: Status of Conservation and Management." In *National Cave and Karst Management Symposium*, October 31–November 4, 1–6. Huntsville, Alabama: National Cave and Karst Management Symposium.

Strecker, Ulrike, Bernhard Hausdorf, and Horst Wilkens. 2012. "Parallel Speciation in *Astyanax* Cave Fish (Teleostei) in Northern Mexico." *Molecular Phylogenetics and Evolution* 62:62–70.

Sweeting, Marjorie M. 1995. *Karst in China: Its Geomorphology and Environment.* Berlin: Springer.

Trajano, Eleonora. 2001. "Ecology of Subterranean Fishes: An Overview." *Environmental Biology of Fishes* 62:133–160.

Trajano, Eleonora, and Maria Elina Bichuette. 2010. "Subterranean Fishes of Brazil." In *Biology of Subterranean Fishes*, edited by Eleonora Trajano, Maria Elina Bichuette, and B. G. Kapoor, 331–355. Enfield: Science Publisher.

UNESCO. 2018. "World Heritage Fund." Accessed April 12, 2020. http://whc.unesco.org/en/world-heritage-fund/.

Vörös, Judit, Orsolya Márton, Benedikt R. Schmidt, Júlia Tünde Gál, and Dušan Jelić. 2017. "Surveying Europe's Only Cave-Dwelling Chordate Species (*Proteus anguinus*) Using Environmental DNA." *PLOS ONE* 12:e0170945.

Waltham, Tony. 2008. "Fengcong, Fenglin, Cone Karst and Tower Karst." *Cave and Karst Science* 35:77–88.

Wang, Junneng, Qianli Ma, Xuemin Zhao, Songxiong Zhong, and Zhencheng Xu. 2019. "Influence of Emergent Cadmium Pollution on Fish Species and Health Risk Assessment in Longjiang River in Guangxi Autonomous Region." *Ecology and Environmental Sciences* 28:974–982.

Wang, Yu, Shiyu Yang, and Daoxian Yuan. 2005. "The Status Quo of Karst Rocky Desertification and the Key Points for Harnessing Rocky Desertification in Yunnan Province." *Carsologica Sinica* 24:206–211.

Wei, Yuelong, Chengzhan Li, Weihai Chen, Qukan Luo, Dehao Zhu, and Tianwang Pan. 2018. "Characteristics and Formation and Evolution Analysis of the Karst Landscape of Guangxi." *Guangxi Sciences* 25:465–504.

White, Nicole E., Michelle T. Guzik, Andrew D. Austin, Glenn I. Moore, William F. Humphreys, Jason Alexander, et al. 2020. "Detection of Rare Australian Endemic Blind Cave Eel (*Ophisternon candidum*) with Environmental DNA: Implications for Threatened Species Management in Subterranean Environments." *Hydrobiologica* 847:3201–3211.

Whitten, Tony. 2009. "Applying Ecology for Cave Management in China and Neighbouring Countries." *Journal of Applied Ecology* 46:520–523.

Wu, Pan, and Guiquan Yan. 2006. "Exploitation and Utilization and Forming Conditions of Underground Reservoirs in Karst Area of Guizhou Province." *Guizhou Science* 24:26–27.

Wu, Tiejun, Jian Yang, and Lihui Xiu. 2015. "A New Species of *Bibarba* (Teleostei: Cypriniformes: Cobitidae) from Guangxi, China." *Zootaxa* 3905:138–144.

Wu, Yiping, and Ji Chen. 2013. "Investigating the Effects of Point Source and Nonpoint Source Pollution on the Water Quality of the East River (Dongjiang) in South China." *Ecological Indicators* 32:294–304.

Xing, Yingchun, Chunguang Zhang, Enyuan Fan, and Yahui Zhao. 2016. "Freshwater Fishes of China: Species Richness, Endemism, Threatened Species and Conservation." *Diversity and Distributions* 22:358–370.

Xiong, Kangning, and Yongkuan Chi. 2015. "The Problems in Southern China Karst Ecosystem in Southern of China and its Countermeasures." *Ecological Economy* 31:23–30.

Xiu, Lihui, Jian Yang, and Huifang. Zheng. 2014. "An Extraordinary New Blind Catfish, *Xiurenbagrus dorsalis* (Teleostei: Siluriformes: Amblycipitidae), from Guangxi, China." *Zootaxa* 3835:376–80.

Yang, Junxing, Yinrui Chen, and Jiahu Lan. 1994. "*Protocobitis typhlops*, A New Genus and Species of Cave Loach from China (Cypriniformes: Cobitidae)." *Ichthyological Exploration of Freshwaters* 5:91–96.

Yang, Junxing, Xiaofu Pan, Xiaoai Wang, Kunfeng Yang, and Wansheng Jiang. 2018. "A New Variety of *Sinocyclocheilus grahami,* Bayou No. 1." *China Fisheries* 2018:71–75.

Yang, Xiaoxia, Xu Xiang, Daoxian Yuan, and Jianbin Li. 2007. "Summary on Karst Cave Tourism Research." *Carsologica Sinica* 26:369–377.

Yuan, Daoxian, Yongjun Jiang, Licheng Shen, Junbing Pu, and Qiong Xiao. 2016. *Modern Karstology*. Beijing: Science Press.

Yuan, Daoxian, Dehao Zhu, and Xuewen Zhu. 1991. *Karst of China*. Beijing: Geological Publishing House.

Yue, Fujun, Siliang Li, Jun Zhong, and Jing Liu. 2018. "Evaluation of Factors Driving Seasonal Nitrate Variations in Surface and Underground Systems of a Karst Catchment." *Vadose Zone* 17:170071.

Zhang, Chunguang. 1986. "On the Ecological Adaptation and Geographical Distribution of the Barbine Fish *Varicorhinus (Scaphesthes) macrolepis* (Bleeker)." *Acta Zoologica Sinica* 32:266–272.

Zhang, Chunguang, and Yahui Zhao. 2016. *Species Diversity and Distribution of Inland Fishes in China*. Beijing: Science Press.

Zhang, Fan. 1998. "Comparison of Cave Tourism between Slovenia and Yunnan, China." *Yunnan Geographic Environment Research* (Supplement) 10:41–46.

Zhang, Fan, Jianzhen Teng, Weiquan Zhao, Kehua Wu, Jun Yuan, and Bo Jiang. 2011. "The Major Environmental Problems Caused by Exploitation of Mineral Resources in Karst Areas of Guizhou Province." *Guizhou Science* 29:65–71.

Zhang, Lansheng, Peihong Yin, and Di Zhang. 2012. "Karst Development and Formation of Modern Environments in Yunnan, Guizhou and Guangxi." In *Palaeogeography of China*, edited by Lansheng Zhang and Xiuqi Fang, 248–271. Beijing: Science Press.

Zhang, Ming, Liming Chen, Xiaolin Zhang, and Shenglin Yang. 2016. "Yunnan Endemic Animal and Grain and Their Medical Specialty in an Ancient Book Materia Medica of South Yunnan." *Yunnan Journal of Traditional Chinese Medicine and Materia Medica* 37:11–12.

Zhang, Qiang, Yubi Yao, Ying Wang, Suping Wang, Jinsong Wang, Jinhu Yang, et al. 2019. "Characteristics of Drought in Southern China under Climatic Warming, the Risk, and Countermeasures for Prevention and Control." *Theoretical and Applied Climatology* 136:1157–1173.

Zhang, Zhigan. 1980. "Karst Types in China." *GeoJournal* 4:541–570.

Zhao, Junfang, Houquan Lu, and Yijun Li. 2019. "Agricultural Adaptation to Drought for Different Cropping Systems in Southern China under Climate Change." *Journal of the American Water Resources Association* 55:1235–1247.

Zhao, Yahui, and Chunguang Zhang. 2006. "Cavefishes: Concept, Diversity and Research Progress." *Biodiversity Science* 14:451–460.

Zhao, Yahui, and Chunguang Zhang. 2009a. *Endemic Fishes of Sinocyclocheilus (Cypriniformes: Cyprinidae) in China—Species Diversity, Cave Adaptation, Systematics and Zoogeography*. Beijing: Science Press.

Zhao, Yahui, and Chunguang Zhang. 2009b. "Threatened Fishes of the World: *Sinocyclocheilus anophthalmus* (Chen and Chu, 1988) (Cyprinidae)." *Environmental Biology of Fishes* 86:163.

Zhao, Yahui, and Chunguang Zhang. 2013. "Validation and Re-description of *Sinocyclocheilus aluensis* Li et Xiao, 2005 (Cypriniforms: Cyprinidae)." *Zoological Research* 34:374–378.

Zhao, Yahui, R. E. Gozlan, and Chunguang Zhang. 2011. "Out of Sight Out of Mind: Current Knowledge of Chinese Cave Fishes." *Journal of Fish Biology* 79:154–162.

Zhao, Yahui, Chunguang Zhang, and Graham Proudlove. 2021. *Fishes of the genus Sinocyclocheilus (Cypriniformes: Cyprinidae) in China*. Peterborough, UK: Upfront Publishing.

Zhou, Zejian. 2012. "The Reason and Countermeasures for the Emergent Cadmium Contamination Event in the Longjiang River of Guangxi." *Overview of Disaster Prevention* 2012:58–61.

Zhu, Yu, Yejian Lü, Junxing Yang, and Sheng Zhang. 2008. "A New Blind Underground Species of the Genus *Protocobitis* (Cobitidae) from Guangxi, China." *Zoological Research* 29:452–454.

Index